T0092998

Intelligent Image and Video Analytics

Advances in image and video analytics has an endless array of successful applications with the recent progression in deep learning and rich computational platforms. Video has rich information including meta-data, visual, audio, spatial, and temporal data which can be analysed to extract a variety of low- and high-level features to build predictive computational models using machine-learning algorithms to discover interesting patterns, concepts, relations, and associations. This book includes a review of essential topics and a discussion of emerging methods and potential applications of image and video data mining and analytics. It integrates areas like intelligent systems, data mining and knowledge discovery, big data analytics, machine learning, neural network, and deep learning with more focus on multimodality image and video analytics and recent advances in research/applications.

Features:

- Provides up-to-date coverage of the state-of-the-art techniques in intelligent image and video analytics.

- Explores important applications that require techniques from both artificial intelligence and computer vision.

- Describes multimodality video analytics for different applications.

- Examines issues related to multimodality data fusion and highlights research challenges.

- Integrates various techniques from image and video processing, data mining, and machine learning which have many emerging indoor and outdoor applications of smart cameras in smart environments, smart homes, and smart cities.

This book aims at researchers, professionals, and graduate students in image processing, video analytics, computer science and engineering, signal processing, machine learning, and electrical engineering.

Intelligent Image and Video Analytics

Clustering and Classification Applications

Edited by
El-Sayed M. El-Alfy
George Bebis
MengChu Zhou

CRC Press
Taylor & Francis Group
Boca Raton London New York

CRC Press is an imprint of the
Taylor & Francis Group, an **informa** business

First edition published 2023
by CRC Press
6000 Broken Sound Parkway NW, Suite 300, Boca Raton, FL 33487-2742

and by CRC Press
4 Park Square, Milton Park, Abingdon, Oxon, OX14 4RN

CRC Press is an imprint of Taylor & Francis Group, LLC

ISBN: 978-0-367-51298-9 (hbk)
ISBN: 978-0-367-55301-2 (pbk)
ISBN: 978-1-003-05326-2 (ebk)

DOI: 10.1201/9781003053262

Typeset in CMR10
by KnowledgeWorks Global Ltd.

Publisher's note: This book has been prepared from camera-ready copy provided by the authors.

Contents

Preface vii

Editor's Biography xi

List of Figures xiii

List of Tables xix

Contributors xxi

1 Video Demographic Analytics of Social Media Users: Machine-Learning Approach 1
Sadam Al-Azani and El-Sayed M. El-Alfy

Bibliography 18

2 Toward Long-Term Person Re-identification with Deep Learning 23
Muna O. AlMasawa, Kawthar Moria, and Lamiaa A. Elrefaei

Bibliography 50

3 A Comprehensive Review of Crowd Behavior and Social Group Analysis Techniques in Smart Surveillance 57
Elizabeth B. Varghese and Sabu M. Thampi

Bibliography 76

4 Intelligent Traffic Video Analytics 85
Hadi Ghahremannezhad, Chengjun Liu, and Hang Shi

Bibliography 113

5 Live Cell Segmentation and Tracking Techniques 129
Yibing Wang, Onyekachi Williams, Nagasoujanya Annasamudram, and Sokratis Makrogiannis

Bibliography 162

6   Quantum Image Analysis – Status and Perspectives   **173**
    Konstantinos Tziridis, Theofanis Kalampokas, and
    George A. Papakostas
Bibliography   **210**

7   Visual Analytics for Automated Behavior Understanding
    in Learning Environments: A Review of Opportunities,
    Emerging Methods, and Challenges   **221**
    Zafi Sherhan Syed, Faisal Karim Shaikh, Muhammad Shehram Shah Syed,
    and Abbas Syed

Bibliography   **242**

8   Noise-Estimation-Based Fuzzy C-Means Clustering for
    Image Segmentation   **251**
    Cong Wang, MengChu Zhou, Witold Pedrycz, and Zhiwu Li

Bibliography   **296**

9   Sample Problems in Person Re-Identification   **303**
    Hua Han and MengChu Zhou

Bibliography   **327**

Index   **331**

Preface

We live in a new era of digital technology where many breakthroughs and applications have been witnessed due to the fusion of machine intelligence and image processing with video technology. Video contains rich information including meta-data, visual, audio, spatial, and temporal data. These data can be analyzed to extract a variety of low-level and high-level features to build predictive computational models using machine-learning algorithms in order to discover interesting patterns, concepts, relations, associations, etc. There are numerous potential applications including human–machine interactions, smart surveillance cameras, smartphones, social media analysis, entertainment industries, video games and sports, medicine and healthcare, intelligent traffic systems, crowd management, biometrics, demographic analysis, intelligent manufacturing, and intelligent instructional systems. This book includes contributions to the state of the art and practice in intelligent image and video analytics addressing some of the applications and challenges in this field, as well as prototypes, systems, tools, and techniques. It also includes surveys presenting the state of the art of various applications in image and video analytics and discussing future directions of research and technology development.

An area that can benefit significantly from intelligent image and video analytics is that of demographic data analysis of social media users. Analyzing all sorts of social media data and especially videos can provide useful socioeconomic and physiological information about different groups of people in various parts of the world. Such information could be leveraged by decision and policy makers to identify communities requiring special attention and/or planning. The chapter by Sadam Al-Azani and El-Sayed M. El-Alfy (Video Demographic Analytics of Social Media Users: Machine-Learning Approach) provides a review of multimodal video demographic analytics and presents a multimodal demographic detection system based on gender, age group, and Arabic dialect using a machine-learning approach.

Person re-identification is a challenging problem in visual surveillance assuming a wide network of non-overlapping cameras. In the work by Muna O. AlMasawa, Kawthar Moria, and Lamiaa A. Elrefaei (Toward Long-Term Person Re-identification with Deep Learning), an approach based on Resnet-50 is presented for long-term person re-identification by combining instance with batch normalization to better handle appearance variations. The proposed approach has been evaluated on several challenging datasets illustrating superior performance compared to state-of-the-art approaches. In a related chapter by Hua Han and MengChu Zhou (Sample Problems in Person Re-Identification), the authors consider the problem of small sample size in the context of person re-identification and investigate both traditional and deep-learning-based data augmentation techniques to improve performance.

Detecting and tracking crowds in videos to understand their behavior have important practical applications. The chapter by Elizabeth B Varghese and Sabu M Thampi (A Comprehensive Review of Crowd Behavior and Social Group Analysis Techniques in Smart Surveillance) provides a comprehensive overview of computer vision methods for crowd detection and analysis with emphasis on behavior and anomaly detection. A discussion of open research issues and future research directions concludes the chapter.

Modern traffic management systems rely on successful algorithms to automatically process traffic video data. The chapter by Hadi Ghahremannezhad, Chengjun Liu, and Hang Shi (Intelligent Traffic Video Analytics) presents an overview of recent developments in intelligent traffic analysis with emphasis on some of its core components related to object detection, classification, and tracking as well as detection of interesting traffic incidents. Current challenges and future research directions are also discussed to further motivate progress in the field. Medical applications of intelligent visual analytics represent an important research direction. The chapter by Yibing Wang, Onyekachi Williams, Nagasoujanya Annasamudram, and Sokratis Makrogiannis (Live Cell Segmentation and Tracking Techniques) provides an overview of automated segmentation and tracking of live cells in time-lapse microscopy. The chapter reviews both traditional and advanced methods based on deep learning with an emphasis on their strengths and weaknesses.

Quantum computing represents a promising alternative to speeding up methods that manipulate images and videos. The chapter by Konstantinos Tziridis, Theofanis Kalampokas, and George A. Papakostas (Quantum Image Analysis – Status and Perspectives) provides an overview on the application of quantum computing in the fields of image analysis and computer vision. Current limitations of quantum computing and scientific advances required to make it practical are discussed in the context of traditional and advanced techniques such as deep learning.

Inferring human behavior from visual cues in images and videos has been a very active research area of intelligent visual analytics. The chapter by Zafi Sherhan Syed, Faisal Karim Shaikh, Muhammad Shehram Shah Syed, and Abbas Syed (Visual Analytics for Automated Behavior Understanding in Learning Environments: A Review of Opportunities, Emerging Methods, and Challenges) provides an overview of visual sensing methods in classroom-based learning environments with specific focus on data collection and annotation, hand-crafted versus automated feature extraction, and machine-learning algorithms. The chapter concludes by discussing the benefits of combining non-visual cues, such as voice and physiological signals, to improve human behavior understanding.

A critical step in image and video analytics is segmenting objects of interest in images and videos. Since both images and videos are typically affected by different types of noise, segmentation results might not be satisfactory unless a noise reduction scheme is applied prior to segmentation. In the chapter by Cong Wang, MengChu Zhou, Witold Pedrycz, and Zhiwu Li (Noise-Estimation Based Fuzzy C-Means Clustering for Image Segmentation), a residual-related regularization term, derived from the distribution model of different types of noise, is integrated into Fuzzy C-Means to better address noise issues. The authors demonstrate the effectiveness of their approach both on synthetic and real data.

Finally, the chapter on (Sample Problems in Person Re-identification) introduces background on person re-identification problem in intelligent video analysis and a range of potential applications. Moreover, it presents a conventional idea to solve the small-sample-size (S^3) problem and presents two virtual sample generation methods. As is evident from this collection of chapters, there is a great variety of work taking place in the field of intelligent visual analytics. We hope that this book will provide further insight into the nature of this fast developing field. We are grateful to the authors who submitted their work to this special project and to CRC Press for their strong support and assistance.

El-Sayed M. El-Alfy
George Bebis
MengChu Zhou

Editor's Biography

El-Sayed M. El-Alfy is currently a Professor at the Information and Computer Science Department, Fellow of SDAIA-KFUPM Joint Research Center for Artificial Intelligence, Affiliate Interdisciplinary Research Center on Intelligent Secure Systems, King Fahd University of Petroleum and Minerals (KFUPM), Saudi Arabia. He has over 25 years of experience in industry and academia, involving research, teaching, supervision, curriculum design, program assessment, and quality assurance in higher education. He is an approved ABET/CSAB Program Evaluator (PEV), a Reviewer and a Consultant for NCAAA and several universities and research agents in various countries. He is an active Researcher with interests in fields related to machine learning and nature-inspired computing and applications to data science and cybersecurity analytics, pattern recognition, multimedia forensics, and security systems. He has published numerously in peer-reviewed international journals and conferences, edited a number of books published by reputable international publishers, attended and contributed in the organization of many world-class international conferences, and supervised master and Ph.D. students. He is a senior member of IEEE and was also a member of ACM, the IEEE Computational Intelligence Society, the IEEE Computer Society, the IEEE Communication Society, and the IEEE Vehicular Technology Society. His work has been internationally recognized and received a number of awards. He has served as a Guest Editor for a number of special issues in international journals and has been on the editorial board of a number of premium international journals.

George Bebis received his B.S. degree in mathematics and M.S. degree in computer science from the University of Crete, Crete, Greece, in 1987 and 1991, respectively, and Ph.D. degree in electrical and computer engineering from the University of Central Florida, Orlando, FL, USA, in 1996. He is currently a Foundation Professor with the Department of Computer Science and Engineering (CSE), University of Nevada, Reno (UNR), Reno, NV, USA, and the Director of the Computer Vision Laboratory. From 2013 to 2018, he served as a Department Chair of CSE, UNR. His research has been funded by NSF, NASA, ONR, NIJ, and Ford Motor Company. His research interests include computer vision, image processing, pattern recognition, machine learning, and evolutionary computing. Dr. Bebis is an Associate Editor of the Machine Vision and Applications Journal and serves on the Editorial Board of the International Journal on Artificial Intelligence Tools, the Pattern Analysis and Applications, and the Computer Methods in Biomechanics and Biomedical Engineering: Imaging and Visualization. He has served on the program committees of various national and international conferences and is the Founder/main Organizer of the International Symposium on Visual Computing.

MengChu Zhou received his B.S. degree in Control Engineering from Nanjing University of Science and Technology, Nanjing, China in 1983, his M.S. degree in Automatic Control from Beijing Institute of Technology, Beijing, China in 1986, and Ph.D. degree in Computer and Systems Engineering from Rensselaer Polytechnic Institute, Troy, NY in 1990. He joined the New Jersey Institute of Technology (NJIT), Newark, NJ in 1990, and is now a Distinguished Professor of Electrical and Computer Engineering. He is also with the School of Information and Electronic Engineering, Zhejiang Gongshang University, Hangzhou 310018, China. His research interests are in Petri nets, intelligent automation, Internet of Things, big data, web services, and intelligent transportation. He has over 1000 publications including 14 books, 700+ journal papers (600+ in IEEE TRANSACTIONS), 31 patents, and 32 book chapters. He is the founding Editor of the IEEE Press Book Series on Systems Science and Engineering and was Editor-in-Chief of the IEEE/CAA Journal of Automatica Sinica from 2018 to 2022. He is a recipient of Excellence in Research Prize and Medal from NJIT, the Humboldt Research Award for US Senior Scientists from the Alexander von Humboldt Foundation, the Franklin V. Taylor Memorial Award and the Norbert Wiener Award from IEEE Systems, Man, and Cybernetics Society. He is founding Chair/Co-chair of Enterprise Information Systems Technical Committee (TC), Environmental Sensing, Networking, and Decision-making TC, AI-based Smart Manufacturing Systems TC, and Humanized Crowd Computing TC of IEEE *Systems, Man, and Cybernetics* Society. He also serves as the Co-Chair of TC on Semiconductor Manufacturing Automation and TC on Digital Manufacturing and Human-Centered Automation of IEEE Robotics and Automation Society. He has been among most highly cited scholars for years and ranked top one in the field of engineering worldwide in 2012 by Web of Science. He is a life member of the Chinese Association for Science and Technology-USA and served as its President in 1999. He is a Fellow of the International Federation of Automatic Control (IFAC), the American Association for the Advancement of Science (AAAS), the Chinese Association of Automation (CAA), and the National Academy of Inventors (NAI).

List of Figures

1.1 High-level overview of the proposed multimodal demographic detection system from videos on social media. 6
1.2 Confusion matrix for multimodal (A-T-V) demographic detection systems: (a) gender, (b) age-group, and (c) dialect. . . . 12
1.3 Pairwise t-test of demographics (p-values). 14

2.1 Person re-identification categories. 25
2.2 Person Re-ID: (a) Long-term vs (b) Short-term. 26
2.3 The general procedure of the proposed method. 30
2.4 The framework of the proposed Deep LongReID. 31
2.5 DeepLabV3 [45]. 32
2.6 Examples of person segmentation: (a) Original RGB image from CASIA-B dataset, (b) The available silhouette image in the dataset, (c) The person segmentation by DeepLavV3, and (d) The GEI. 32
2.7 Instance Normalization vs Batch Normalization. The subplot is a feature map, N = batch axis, C = channel axis, H, W = spatial axis. The pixels that normalized by the same mean and variance are with blue color [48]. 34
2.8 Samples from CASIA-B dataset. (a) same person with normal appearance vs wearing coat. (b) same person carrying bag vs wearing coat. (c) same person carrying bag vs normal. (d) same person wearing coat vs normal. 37
2.9 Samples of GEI from CASIA-B for id 85 at Camera 90°: (a) Person with coat, (b) Person carrying bag, and (c) Person with normal appearance. 37
2.10 Samples from the Market-1501 dataset. 38
2.11 Example of random erasing data augmentation [54]. 39
2.12 CMC curve for long-term scenario. 45
2.13 CMC curve for long-term plus. 46
2.14 CMC curve for comparison with other recent models. 46
2.15 Example of top-ranked list of the integrated system: (a) Sample frame from query video, (b) The output of CL-NM, (c) The output of CL-CL, and (d) The output of CL-BG. (random frame is presented from output video). 48

3.1 Common applications of smart surveillance. 59

3.2   Taxonomy of crowd behavior analysis research.   60
3.3   Classification of visual-based approaches.   61

4.1   General flowchart of an intelligent traffic video analytics frame-
      work.   88
4.2   The foreground masks extracted using the MOG method and
      the GFM method, respectively. The GFM method extracts
      both the moving vehicles (blue) and the stopped vehicles (red)
      clearly [45].   94
4.3   The foreground mask after removing shadows. (a) Original
      video frame. (b) The results of the Hang and Liu [78] shadow
      detection method.   97
4.4   Example testing results of the YOLO and the Faster R-CNN
      deep-learning methods using traffic videos that are not seen
      during training. The frames in (a) and (c) show the vehicle clas-
      sification results by using the YOLOv3 deep-learning method.
      The frames in (b) and (d) show the vehicle classification results
      by using the Faster R-CNN deep-learning method [90].   99
4.5   Examples of road extraction in traffic videos [125].   106

5.1   The number of publications per year on traditional cell segmen-
      tation techniques. *Source: PubMed*   131
5.2   The number of publications per year using key words "Feature-
      based cell microscopy image segmentation" on PubMed. . . .   134
5.3   Segmentation results on Fluo-N2DL-GOWT1-01 sequence from
      the CTC dataset: (a) input frame, (b) diffused frame, (c) wa-
      tershed segmentation result, (d) segmentation mask, and (e)
      ground truth.   137
5.4   Segmentation results on Fluo-C2DL-MSC-02 sequence from the
      CTC dataset: (a) input frame, (b) segmentation mask, and (c)
      ground truth.   138
5.5   The number of publications per year on cell segmentation using
      deep learning. *Source: PubMed*   139
5.6   Structure of an artificial neural network.   140
5.7   Segmentation of Fluo-N2DH-SIM+01 by the method in [62].
      (a) input image, (b) reference frame, and (c) segmented image.   140
5.8   The original U-net architecture.   142
5.9   Timeline of published articles on cell segmentation categorized
      by network type *Source: PubMed.*   143
5.10 Illustration of cell events in a time-lapse sequence. Cells are
      represented as "*a, b, c, d*". At time $t = 2$, cell '*a*' and '*b*' collide;
      cell '*d*' migrates and leaves the field of view (disappearance),
      at time $t = 4$. Also, at time $t = 4$ cell c undergoes division to
      form cells 'c1' and 'c2'.   148
5.11 The number of published articles per year on cell tracking by
      detection. *Source: PubMed.*   149

5.12 The number of published articles per year on cell tracking by model evolution. *Source: PubMed* 150

5.13 The number of published articles per year on cell tracking by probabilistic approaches. *Source: PubMed.* 151

5.14 The number of published articles per year on cell tracking by deep learning. *Source: PubMed* 152

5.15 Flowchart of the CSTQ-MIVIC cell tracking method highlighting the main stages. 154

5.16 First row: estimation of optical flow (left) between frames 9 (left) and 10 (middle) of Fluo-C2DL-Huh701 dataset. Second row: Middlebury color coding of computed optical flow (left), the difference between cell indicator functions of reference cell masks of frames 9 and 10 (middle), and the difference between cell masks of frame 10 and warped frame 9 using the computed optical flow (right). 155

5.17 Example of cell event identification, and formation of cell lineages. 156

5.18 Lineage tree construction of Fluo-N2DH-SIM+01 dataset by CSTQ, where the cells are labeled by numbers. The dashed lines show migrating cells over time, and the thick lines show cell divisions. The yellow circle nodes denote new cells, or cells entering the field of view. The orange square nodes denote cell death, or cells exiting the field of view. 160

6.1 Bloch-sphere 178

6.2 Pipeline of Quantum Image Analysis 181

6.3 QSVM quantum circuit 196

6.4 QNN quantum circuit 198

6.5 QCNN quantum circuit 199

6.6 Quantum autoencoder 201

7.1 Conception of classroom behavior understanding system that can recognize various student/teacher activities based on facial appearance features and keypoints from the face, arms, and torso. 225

8.1 A comparison between the frameworks of FCM and RFCM. (a) FCM and (b) RFCM. 254

8.2 Comparison of technical cores of DSFCM and RFCM. (a) DSFCM and (b) RFCM. 254

8.3 Noise-free image and two observed ones corrupted by Gaussian and mixed noise, respectively. The first row: (a) noise-free image; (b) observed image with Gaussian noise; and (c) observed image with mixed noise. The second row portrays noise included in three images. 263

8.4   Distributions of Gaussian and mixed noise in different domains. (a) linear domain and (b) logarithmic domain.   263

8.5   Distributions of residual r_{il} and weighted residual $w_{il}r_{il}$, as well as the fitting Gaussian function in the logarithmic domain. .   264

8.6   Illustration of spatial information of pixel i.   265

8.7   Noise estimation of WRFCM. (a) noise-free image; (b) observed image; (c) segmented image of WRFCM; (d) noise in the noise-free image; (e) noise in the observed image; (f) noise estimation of WRFCM.   270

8.8   Convergence and robustness of WRFCM. (a) θ; (b) J; and (c) τ versus t.   270

8.9   Noise estimation comparison between DSFCM and RFCM. The row 1 shows a noise-free image. In other rows, from left to right: observed images (from top to bottom: Gaussian noise ($\sigma = 30$), impulse noise ($\zeta = 30\%$), and a mixture of Poisson, Gaussian ($\sigma = 30$) and impulse noise ($\zeta = 20\%$)) and the results of DSFCM and RFCM.   277

8.10 SA values versus ϕ.   281

8.11 Visual results for segmenting synthetic images ($\phi_1 = 5.58$ and $\phi_2 = 7.45$). From (a) to (j): noisy images, ground truth, and results of FCM_S1, FCM_S2, FLICM, KWFLICM, FRFCM, WFCM, DSFCM_N, and WRFCM.   282

8.12 Visual results for segmenting medical images ($\phi = 5.35$). From (a) to (j): noisy images, ground truth, and results of FCM_S1, FCM_S2, FLICM, KWFLICM, FRFCM, WFCM, DSFCM_N, and WRFCM.   284

8.13 Segmentation results on five real-world images in BSDS and MSRC ($\phi_1 = 6.05$, $\phi_2 = 10.00$, $\phi_3 = 9.89$, $\phi_4 = 9.98$, and $\phi_5 = 9.50$). From top to bottom: observed images, ground truth, and results of FCM_S1, FCM_S2, FLICM, KWFLICM, FRFCM, WFCM, DSFCM_N, and WRFCM.   285

8.14 Segmentation results on the first real-world image in NEO ($\phi = 6.10$). From (a) to (i): observed image and results of FCM_S1, FCM_S2, FLICM, KWFLICM, FRFCM, WFCM, DSFCM_N, and WRFCM.   287

8.15 Segmentation results on the second real-world image in NEO ($\phi = 9.98$). From (a) to (i): observed image and results of FCM_S1, FCM_S2, FLICM, KWFLICM, FRFCM, WFCM, DSFCM_N, and WRFCM.   288

8.16 Segmentation results on four real-world images in VOC2012 ($\phi_1 = 5.65$, $\phi_2 = 6.75$, $\phi_3 = 5.50$, $\phi_4 = 8.35$). From top to bottom: observed images, ground truth, and results of AMR_SC, PFE, and WRFCM.   291

9.1   Comparison among three KISS-related algorithms.   310

9.2   Eigenvalues (top 20 smallest ones) comparison between
        KISSME and KISS+.   311
9.3   Performance of G-KISS with different generations. (a) CMC
        curve of G-KISS with different generations, and (b) is the par-
        tial enlargement of (a).   314
9.4   Eigenvalue variation of covariance matrices. (a) Eigenvalues
        ($*10^{-3}$) of inverse matrix of intra-covariance matrix, (b) Eigen-
        values of inverse matrix of inter-covariance matrix, and (c)
        Eigenvalues of $\hat{M}$. The X-axis represents the sequence number
        of eigenvalues, and the Y-axis represents the eigenvalues. . . .   315
9.5   Overview of a graph approach for pseudo pairwise relationship
        estimation among unlabeled data and labeled ones.   319
9.6   Performance comparison of the proposed approach with base-
        line on Market-1501 with different settings of the number of
        generate images. Rank-1 recognition rates are shown in the leg-
        ends.   321
9.7   Performance comparison of the proposed approach with base-
        line on Duke MTMC-ReID with different settings of the number
        of generate images. Rank-1 recognition rates are shown in the
        legends.   323
9.8   Re-ID matching rates (%) comparison with different k and
        generate samples on Market-1501 and Duke MTMC-ReID. (a)
        and (b) shown the rank-1 accuracy obtained on Market-1501
        and Duke MTMC-ReID with different situations, respectively.
        Three systems are shown, which have 3, 4, and 5 generated
        samples for Market-1501 and 5, 6, and 7 generated samples for
        Duke MTMC-ReID, respectively.   325

List of Tables

1.1 Age-group description in the SADAM database. 5
1.2 Dialects description in the SADAM database. 5
1.3 The size of feature vectors. 8
1.4 Confusion matrix definition. 9
1.5 Results for demographic recognition systems. 11

2.1 Existing datasets used in previous long-term person Re-ID
     studies. 37
2.2 The accuracy % results our proposed model in long-term sce-
     nario. 41
2.3 The accuracy % results of our proposed model and baseline
     resnet-50. 42
2.4 The accuracy % results of softmax only and the combined loss
     functions. 42
2.5 The percentage accuracy results of our proposed model with
     and without using random erase. 43
2.6 The comparison with other state-of-the-art studies. '-' is not
     reported. 44
2.7 The accuracy % results of our proposed model in short-term
     using CASIA-B dataset. 47
2.8 The accuracy % results of our proposed model in short-term
     using market-1501 dataset. "-" is not reported. 47
2.9 The results of consuming time for using resnet-50 with INBN
     comparing by baseline resnet-50. 48

3.1 Motion-pattern methods. 64
3.2 Low-level feature and physics-based methods. 65
3.3 Cognitive-based methods. 67

5.1 Summary of traditional cell segmentation methods. 136
5.2 Cell segmentation methods using deep learning techniques. . . 144
5.3 Summary of cell tracking techniques. 153
5.4 Tracking performance of the level-set-based tracking method
     in [105]. 158
5.5 Tracking performance of the IMM and compared with the
     Kalman filter tracking method in [85]. 158
5.6 Cell tracking results produced by deep learning methods [106]. 158

5.7	Ranking of the TRA measure of CSTQ [52].	159

6.1	Quantum computing tools along with their providers.	203
6.2	Quantum cross platform frameworks.	205
6.3	Quantum computers architectures.	207

7.1	Summary of study objectives, behavioral attributes, recording setup, dataset type, and annotation methods for a selection of systems for classroom behavior understanding.	228
7.2	Summary of machine-learning objectives and methods for feature engineering and classification/regression for a selection of systems for classroom behavior understanding.	232

8.1	Data regularization models.	261
8.2	Segmentation performance (%) on synthetic images.	283
8.3	Segmentation performance (%) on medical images in Brian-Web.	283
8.4	Segmentation performance (%) on real-world Images in BSDS and MSRC.	286
8.5	Segmentation performance (%) on real-world images in NEO.	288
8.6	Average performance improvements (%) of WRFCM over comparative algorithms.	289
8.7	Computational complexity of all algorithms.	290
8.8	Comparison of execution time (in seconds) of all algorithms. .	290
8.9	Segmentation performance (%) on real-world images in VOC2012.	291
8.10	Comparison of execution time (in seconds) between WRFCM and two non-FCM methods.	292

9.1	Matching rate at rank = 1, 5, 10, 20 and PUR scores of G-KISS with different generations and the number of accepted virtual samples.	316
9.2	Performance (CMC: rank = 1, 5, 10, 20 precision) of proposed method, and comparison with some state-of-the-art semi-supervised methods on Market 1501 and Duke MTMC-ReID.	322
9.3	Re-ID accuracy of SSL with different neighbors select when 5 generated samples added on Market-1501.	324
9.4	Re-ID accuracy of SSL with different neighbors select when 7 generated samples added on Duke MTMC-reID.	324

Contributors

Sadam Al-Azani
King Fahd University of Petroleum
  & Minerals (KFUPM)
Dhahran, Saudi Arabia

Muna O. AlMasawa
King Abdulaziz University
Jeddah, Saudi Arabia

Nagasoujanya Annasamudram
Delaware State University
Dover, Delaware

El-Sayed M. El-Alfy
King Fahd University of Petroleum
  and Minerals (KFUPM)
Dhahran, Saudi Arabia

Lamiaa A. Elrefaei
King Abdulaziz University
Jeddah, Saudi Arabia
Benha University
Cairo, Egypt

Hadi Ghahremannezhad
New Jersey Institute of Technology
Newark, New Jersey

Hua Han
Shanghai University of Engineering
  Science
Shanghai, China

Theofanis Kalampokas
International Hellenic University
Kavala, Greece

Zhiwu Li
Xidian University
Xi'an, China
Macau University of Science and
  Technology
Macau, China

Chengjun Liu
New Jersey Institute of Technology
Newark, New Jersey

Sokratis Makrogiannis
Delaware State University
Dover, Delaware

Kawthar Moria
King Abdulaziz University
Jeddah, Saudi Arabia

George A. Papakostas
International Hellenic University
Kavala, Greece

Witold Pedrycz
University of Alberta
Edmonton, Alberta, Canada
Xidian University
Xi'an, China
King Abdulaziz University
Jeddah, Saudi Arabia

Faisal Karim Shaikh
Mehran University of Engineering
  and Technology
Sindh, Pakistan

Hang Shi
Innovative AI Technologies
Newark, New Jersey

Abbas Syed
University of Louisville
Louisville, Kentucky

Muhammad Shehram Shah Syed
RMIT University
Melbourne, Australia

Zafi Sherhan Syed
Mehran University of Engineering
    and Technology
Sindh, Pakistan

Sabu M. Thampi
Kerala University of Digital Sciences,
    Innovation and Technology
    (KUDSIT)
Trivandrum, Kerala, India

Konstantinos Tziridis
International Hellenic University
Kavala, Greece

Elizabeth B. Varghese
Kerala University of Digital Sciences,
    Innovation and Technology
    (KUDSIT)
Trivandrum, Kerala, India
Mar Baselios College of Engineering
    and Technology
Trivandrum, Kerala, India

Cong Wang
School of Artificial Intelligence,
    Optics, and ElectroNics (iOPEN)
Northwestern Polytechnical
    University
Xi'an, China
Shenzhen Research Institute
    of Northwestern Polytechnical
    University
Shenzhen, China

Yibing Wang
Delaware State University
Dover, Delaware

Onyekachi Williams
Delaware State University
Dover, Delaware

MengChu Zhou
Macau University of Science and
    Technology
Macau, China
New Jersey Institute of Technology
Newark, New Jersey

1

Video Demographic Analytics of Social Media Users: Machine-Learning Approach

Sadam Al-Azani

SDAIA-KFUPM Joint Research Center for Artificial Intelligence (JRC-AI), King Fahd University of Petroleum & Minerals (KFUPM), Dhahran, Saudi Arabia

El-Sayed M. El-Alfy

Professor, Information and Computer Science Department, Fellow of SDAIA-KFUPM Joint Research Center for Artificial Intelligence, Affiliate Interdisciplinary Research Center on Intelligent Secure Systems, College of Computing and Mathematics, King Fahd University of Petroleum and Minerals (KFUPM), Dhahran, Saudi Arabia

CONTENTS

1.1	Introduction	2
1.2	Related Work	3
1.3	Materials and Methods	4
	1.3.1 Dataset description	5
	1.3.2 Feature extraction and fusion	5
	1.3.2.1 Acoustic features	7
	1.3.2.2 Transcribed textual features	7
	1.3.2.3 Visual features	7
	1.3.2.4 Fusion	8
1.4	Evaluations	8
	1.4.1 Experimental settings	8
	1.4.2 Evaluation metrics	9
1.5	Results and Discussion	11
	1.5.1 Gender detection	12
	1.5.2 Age-group detection	13
	1.5.3 Dialect detection	13
	1.5.4 General discussion	13
1.6	Conclusions	15

DOI: 10.1201/9781003053262-1

1

1.1 Introduction

Nowadays, several social media platforms, such as YouTube, TikTok, Facebook/Meta, MySpace, Twitter, Instagram, Vimeo, and Flickr, enable people to upload, share, or live stream videos of their daily activities and debates. Whether it is restricted or public groups, the volume of unsolicited online video content is increasing at a rapid pace [1,2]. For instance, YouTube was founded 17 years ago but has experienced dramatic growth, e.g. to date there are over 2 billion unique users of 80 languages, over 5 billion videos shared, over 1 billion hours of videos watched daily, and over 500 hours of videos uploaded per minute[1]. These videos are very compelling and can convey significant multimodal multilingual information for social sensing to help governmental and non-governmental organizations and businesses in many purposes [3–5].

Demographic analysis refers to studying the compositional and changes of socioeconomic and physiological characteristics of a group of people based on factors such as age, gender, race, and location. These characteristics may include language or dialect (accent), education, occupation, marital status, fecundity (fertility or ability to produce offspring), disability, household income, nationality, neighborhood, obesity, mental status, attitude or sentiment, belief and opinion, and many others. These measurements are sometimes called soft biometrics since they provide some information about persons but lack uniqueness and distinctiveness [6,7]. Demographic data can be very important in various applications with huge impacts on people's life. Decision and policy makers can gain deeper insights into community structure and influence, groups requiring special attentions, relationships between members, expansion or contraction of groups, and betweenness and closeness centrality measures. For example, many of these factors can be very crucial to studies related to medicine and drug usage, government subsidies, wealth distribution, market analysis and segmentation, political campaign and participation, and strategic planning of future services [2, 8–16].

As an application of data analytics, demographic analytics can be categorized into four types: descriptive, diagnostic, predictive, and prescriptive. While descriptive and diagnostic analytics help to explore the data and attempt to understand what happened and why it happened, predictive analytics helps to draw insights into what will happen. In contrast, prescriptive analytics helps to provide instant recommendations and make optimal decisions to change the future [17,18].

This chapter reviews work related to multimodal video demographic analytics. Moreover, it presents a multimodal demographic detection system from videos on social media with focus on gender, age-group, and dialect of Arabic speakers as a case study. Several features are investigated for each modality and fused using a machine-learning approach.

---

[1]https://www.globalmediainsight.com/blog/youtube-users-statistics/ (Accessed: 3 March 2022).

1.2 Related Work

Detecting users' genders, age groups, dialects, and/or nationalities from videos on social media is very important and has several interesting applications. It is an excellent opportunity for large companies to capitalize on, by extracting users' sentiments, suggestions, and complaints on their products from video reviews while considering the demographic characteristics. Consequently, they can improve and enhance their products/services to meet the needs of their customers. Demographic analysis can also be useful to overcome the real-world gender, age, and racial biases of current content analysis systems. Several research works have been conducted to tackle a variety of issues related to these challenging problems. Studies vary in the way they treat demographic factors as predictors (inputs) or targets (outputs) of the analytic system.

A number of approaches have used machine learning for inferring demographic characteristics using different modalities (text, audio, and visual). For example, López-Santamaría et al. [19] developed different machine-learning based models to characterize Pinterest users based on two demographic factors (age and gender) using a variety of features extracted from imbalanced datasets of short texts. In another study, Lwowski and Rios [20] developed three machine-learning models (support vector machine, convolutional neural network, and bidirectional long-short-term memory) to analyze the risk of racial bias while tracking influenza-related content on Twitter from textual modality. In [21], a number of machine-learning methods were reviewed for detecting age and gender from face images. In [22], a model is proposed to detect gender using audio classifiers. Cesare et al. [23] reviewed a number of methods and related issues. They found that predicting gender from social data is relatively easier while detecting other characteristics such as race or ethnicity is more challenging. Moreover, there are some concerns related to individuals' privacy that need to be protected. In another study [24], machine-learning techniques, especially decision trees, were applied to analyze the demographic sequences in Russia, based on data from 1930 to 1984 for 11 generations. The main focus was to extract interesting patterns and knowledge from substantial datasets of demographic data.

Combining audio and visual information has been considered in the literature to detect gender. For example, an audio-visual gender recognition approach was presented by El-Shafey et al. [25]. Total Variability (i-vectors) and Inter-Session Variability modeling techniques were used for both unimodal and bimodal gender recognition. The approach was evaluated using the Mobile Biometrics (MoBio) database. It was reported that a bimodal system of audio-visual gender recognition achieved a higher accuracy than unimodal systems. On the other hand, another approach based on combining texts and images modalities to detect gender was proposed by Rangel et al. [26]. The texts and images were collected from Twitter. Three languages are evaluated, which are Arabic, English and Spanish.

Abouelenien et al. [27] presented an approach to analyze the capability of fusing thermal and physiological modalities with vision, acoustics, and language. Experimental results suggest that physiological and thermal information can be useful to recognize gender at reasonable accuracy levels, which are comparable to the accuracy of current gender prediction systems.

The authors in [28] presented a multimodal age-group recognition system for social media videos. The different modalities were combined at score level using an ensemble of ANNs. It differs from this work in terms of the fusion method, fusion level, and classification techniques. In addition, it just considered age-group detection while this chapter analyzed different demographic characteristics, namely: gender, age-group, and dialect.

Another bimodal of audio and textual approach for Arabic dialect detection was presented in [29]. Four Arabic dialects were considered, namely: Egyptian, Levantine, Gulf Arabic, and Maghrebi. This approach also analyzed the effect of speaker on detecting dialects through presenting speaker-dependent and speaker-independent approaches. Combining audio and textual modalities resulted in improving the performances of both speaker-independent and speaker-dependent models. The best performance was obtained using the speaker-dependent model with an F_1 score of 63.61%. This work differs from the work presented in this chapter in terms of the adopted modalities in addition to the evaluation method and evaluation measurements.

In some cases, bimodal approaches may perform better than combining all modalities of audio, textual, and visual information. For example, the bimodal of audio and visual systems for sentiment and gender detection presented in [30] achieved higher results than the multimodal recognition system (that combines audio, textual, and visual) for the same purpose to detect sentiment and gender of videos speakers. An F_1 of 91.61% was reported using audio-visual modality and Support Vector Machine (SVM). However, an F_1 of 90.63% was reported using the combined audio-textual-visual modality. This might be attributed to the small size of the textual corpus as the lowest results were obtained using the textual modality only.

1.3   Materials and Methods

An objective of the work presented in this chapter is to find out, when using the same experimental settings (dataset, preprocessing steps, features, fusion methods, classifiers, and evaluation measures), which of the considered demographics (gender, age-group, and dialect) can be more accurately detected? Another objective, given the three modalities (audio, textual, and visual), what is the most discriminative modality to detect the considered demographics? Additionally, what are the effects of combining different modalities to detect different demographic characteristics?

Figure 1.1 depicts the high-level structure of the proposed multimodal demographic detection system.

TABLE 1.1

Age-group description in the SADAM database.

Age-group	Age range (years)	# of samples	Percentage (%)
Young adults (**AGA**)	15–29	128	24.43
Middle-aged I (**AGB**)	30–39	159	30.34
Middle-aged II (**AGC**)	40–49	142	27.10
Senior (**AGA**)	>49	95	18.13
Total		524	

TABLE 1.2

Dialects description in the SADAM database.

Class	Dialects	# of samples	Percentage (%)
Egyptian	Egypt and Sudan	249	47.52
Gulf	Lebanon, Syria, Palestine, and Jordan	167	31.87
Levantine	Bahrain, Kuwait, Oman, Qatar, Iraqi, Saudi Arabia, United Arab Emirates, and Yemeni	79	15.08
Maghrebi	Morocco, Algeria, Tunisia, Libyan, and Mauritania	29	5.53
Total		524	

1.3.1 Dataset description

To evaluate various models in this study, we adopted the Sentiment Analysis Dataset for Arabic Multimodal (SADAM) [31]. This database contains user-generated videos collected from YouTube and is annotated in terms of sentiment, gender, age-group, nationalities, and dialect. It is composed of 63 opinion videos expressed by 37 males and 26 females. The videos were manually segmented into 524 utterances. The ages of speakers range between 15 and 65, approximately when videos were recorded. The speakers are from different countries with six different Arabic dialectics. Topics covered in the videos belong to different domains including reviews of products, movies, cultural views, etc. The videos were recorded in uncontrolled environments including houses, studios, offices, cars, or outdoors.

Age-groups and Arabic dialects considered in this study with the number of samples per class are presented in Tables 1.1 and 1.2, respectively. In case of gender detection, the number of male samples is 220 while the number of female samples is 304.

1.3.2 Feature extraction and fusion

In this section, we describe the feature engineering process for audio, textual and visual modalities, respectively.

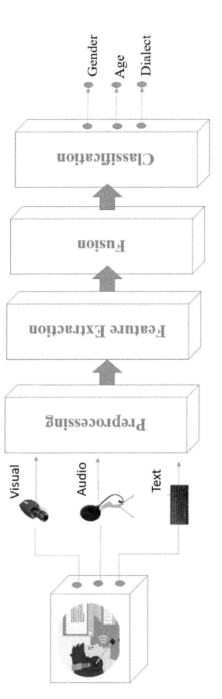

FIGURE 1.1
High-level overview of the proposed multimodal demographic detection system from videos on social media.

1.3.2.1 Acoustic features

Each input audio signal is segmented into frames of 50 millisecond in length and an overlap of 20 millisecond. Next, a set of 34 features are extracted from each segment, namely: (1) one value for Zero Crossing Rate (ZCR), (2) one value for Energy, (3) one value for Entropy of Energy, (4) one value for Spectral Centroid, (5) one value for Spectral Spread, (6) one value for Spectral Entropy, (7) one value for Spectral Flux, (8) one value for Spectral Rolloff, (9-21) the first 13 values of Mel Frequency Cepstral Coefficients (MFCCs), (22-33) 11 values for Chroma Vector, and (34) one value for Chroma Deviation. This is followed by calculating the arithmetic mean and standard deviation for all input audio's frames to be represented as one feature vector with a size of 68 (34×2) features.

1.3.2.2 Transcribed textual features

The skip-gram word embedding models [32] are adopted to extract the transcribed textual features. These models were learnt using a corpus of more than 77,600,000 Arabic tweets posted between 2008 and 2016. Tweets were written in modern standard Arabic and different Arabic dialects. A skip-gram model of 300 dimensionality is selected. Thus, each instance is represented by a feature vector of size 300 attributes. This is carried out by calculating the average of the embedding vectors of that sample.

1.3.2.3 Visual features

As our focus is to detect speakers demographics from their faces only while excluding other body parts, we need first to detect their faces. We applied the general frontal face and eye detectors [33] which are based on object detection using HAAR feature-based cascade classifiers [34]. In addition, the eye detector determines the eye positions which provide significant and useful values to crop and scale the frontal face to a size of the considered resolution of 240×320 pixels. This step is considered as a preprocessing step and run once on all data to reduce the computation time of feature extraction.

After detecting the whole face, it is possible to compute the optical flow to capture the evolution of complex motion patterns for the classification of facial expressions. Optical flow is considered to extract the visual features from the videos processed in the previous step. As a result, each point in the frame is represented by two values: magnitude and angle, which describe the vector representing the motion between two consecutive frames. This leads to a huge descriptor of size $NoF \times W \times H \times 2$ to represent each video, where NoF refers to the number of frames in the video and $W \times H$ is the resolution. This large size affects the performance and leads in curse of dimensionality issue. Thus, we need to summarize the generated descriptor as a reduced feature vector when constructing various machine-learning models. Several statistical methods can be used such as average, standard deviation, min, max, etc. In some earlier studies, Histogram of Optical Flow (HOF) has been

TABLE 1.3

The size of feature vectors.

Modality	# of features
Textual	300
Audio	68
Visual	800
Audio-textual	368
Audio-visual	868
Textual-visual	1100
Audio-textual-visual	1168

applied for detecting interaction-level of human activities from first-person videos [35] and for determining the movement direction of the object [36] and reported promising results. Similarly, a histogram is considered in this study to summarize the high dimensionality descriptor as a single feature vector. Each frame is divided into a grid of $s \times s$ bins which is smaller than the frame size. In addition to reducing the dimensionality size, this also accelerates computation.

1.3.2.4 Fusion

In this study, we considered feature-level fusion to combine all modalities. Four feature vectors are generated to combine the considered single modalities (Audio (A), Textual (T), and Visual (V)) with each other: Audio-Textual (A-T), Audio-Visual (A-V), Textual-Visual (T-V), and Audio-Textual-Visual (A-T-V).

1.4 Evaluations

This section describes the experimental settings, defines the used evaluation metrics, and presents the results.

1.4.1 Experimental settings

Since the number of samples per class is not large enough to divide the dataset into training and testing subsets, we considered 10-fold cross-validation as evaluation mode for all experiments. For each demographic characteristic considered in this study, three single models are generated for Audio, Textual, and Visual modalities. Then, every two modalities are combined to generate three models for Audio-Textual, Audio-Visual, and Textual-Visual modalities. Finally, one model is generated for the combination of the three modalities Audio-Textual-Visual. The number of features for each of the generated single and fused modalities are presented in Table 1.3. That means seven models are

TABLE 1.4
Confusion matrix definition.

	Positive prediction	Negative Prediction
Positive class	TP: true positive	FN: false negative
Negative class	FP: false positive	TN: true negative

generated using LibSVM with Linear Kernel classifier and evaluated for each demographic characteristic considered in this study.

The main components in the proposed system including preprocessing, feature extractors and fusion, classification, and evaluation modules are implemented using Python. We utilized Gensim package [37] for extracting textual features while PayAudioAnalysis [38] package was applied for extracting acoustic features. In addition, we used OpenCV [39] python package for extracting visual features. For classification and evaluation, we applied Scikit-learn package [40].

The imbalance class problem is also addressed by considering cost-sensitive classification. This is conducted through setting a penalty parameter of the error term C_0 of class i as $C_1 = class\_weight[i] * C_0$, where:

$$class\_weight[i] = n\_samples/(n\_classes \times n\_samples[i]). \quad (1.1)$$

1.4.2 Evaluation metrics

For the binary classification problem, the quality of each classifier is generally expressed in terms of the confusion matrix illustrated in Table 1.4 such that:

- True Positive (TP) indicates the number of positive examples that are classified correctly.

- False Positive (FP) indicates the number of negative examples that are classified incorrectly.

- True Negative (TN) indicates the number of negative examples that are classified correctly.

- False Negative (FN) indicates the number of positive examples that are classified incorrectly.

A variety of aggregate measures computed of these four quantities are very common in information retrieval, medical diagnosis, and machine-learning literature.

Accuracy is the most applied evaluation measure in the literature. However, it is not considered as a perfect measure, for imbalanced data-sets, since classifier is biased to ward the majority class to have higher accuracy [41, 42], which results in incorrect interpretations. For example, if we assume a dataset with imbalance ratio of 9.5 and all examples are classified incorrectly as negative class, then the accuracy will be 95%. Other measures include precision (Prc) and recall (Rec). These measures are computed as follows:

$$Accuracy = \frac{\text{number of instances classified correctly}}{\text{total number of instances}} = \frac{TP+TN}{TP+TN+FP+FN} \tag{1.2}$$

$$Prc = \frac{TP}{TP+FP} \tag{1.3}$$

$$Rec = \frac{TP}{TP+FN} \tag{1.4}$$

F_1 is another poplar measure used by the machine-learning community and is calculated from precision and recall as follows:

$$F_1 = 2 \times \frac{Precision \times Recall}{Precision + Recall} \tag{1.5}$$

Geometric mean (GM) is another good indicator for imbalanced dataset since it is independent of the distribution of examples between classes. This measure tries to maximize the accuracy on each of the two classes while keeping their accuracies balanced [43, 44]:

$$GM = \sqrt{TP_{rate}.TN_{rate}} = \sqrt{\frac{TP}{TP+FN}.\frac{TN}{FP+TN}} \tag{1.6}$$

Matthew's correlation coefficient (MCC) is recommended as another perfect metric for imbalanced dataset. It is computed as follows:

$$MCC = \frac{TP.TN - FP.FN}{\sqrt{(TP+FP)(TP+FN)(TN+FP)(TN+FN)}} \tag{1.7}$$

All aforementioned evaluation measures range from 0 to 1, except MCC which ranges from -1 to 1. Receiver Operating Characteristic (ROC) Curve is used to combine measures of positive class and negative class and visualize the trade-off between benefits (TP_{rate}) and costs (FP_{rate}). TP_{rate} represents the x-axes and FP_{rate} represents the y-axes. The best classifier will score in the upper left corner, coordinate (0,1), of the ROC space (FP_{rate}= 0, TP_{rate}=100%). However, the worst possible prediction method will score in the bottom right corner, coordinate (1,0), of the ROC space (FP_{rate}= 100%, TP_{rate}=0). For balanced datasets, a random guessing classifier would give a point somewhere along the diagonal line from the coordinate (0,0) to the coordinate (1,1) (FP_{rate}= TP_{rate}), since the model will throw up positive and negative examples at the same rate.

The area under curve (AUC) is computed by getting the area under the ROC plot to provide a single measure of a classifier's performance for

TABLE 1.5

Results for demographic recognition systems.

Demographic	Modality	Rec	Prc	F_1	GM	Acc	MCC	T_f	T_s	T_t
Gender	Audio	95.61	95.61	95.61	95.37	95.61	90.93	2.13	0.22	2.35
	Text	75.00	74.79	74.75	73.39	75.00	47.78	9.05	0.69	9.74
	Visual	90.65	90.63	90.64	90.25	90.65	80.67	20.08	1.32	21.40
	A-T	92.37	92.36	92.36	91.98	92.37	84.21	24.29	1.15	25.44
	T-V	91.22	91.21	91.21	90.80	91.22	81.84	48.00	4.38	52.38
	A-V	95.80	95.80	95.80	95.53	95.80	91.32	38.05	3.20	41.25
	A-T-V	95.42	95.44	95.43	95.41	95.42	90.59	43.61	3.96	47.57
Age	Audio	69.47	69.50	69.31	78.64	69.47	58.78	7.26	0.50	7.76
	Text	53.82	53.63	53.70	67.05	53.82	37.82	24.22	1.54	25.76
	Visual	74.81	75.08	74.84	82.50	74.81	66.10	69.28	3.82	73.10
	A-T	73.47	73.48	73.44	81.56	73.47	64.20	32.18	3.28	35.46
	T-V	75.76	75.78	75.77	83.25	75.76	67.32	55.30	4.78	60.08
	A-V	84.35	84.48	84.39	89.27	84.35	78.93	50.35	4.11	54.46
	A-T-V	86.64	86.66	86.65	90.87	86.64	81.99	61.53	5.18	66.71
Dialect	Audio	71.56	71.48	71.34	76.42	71.56	55.35	5.71	0.42	6.13
	Text	72.33	71.29	70.94	77.47	72.33	55.85	16.71	1.04	17.75
	Visual	83.97	83.97	83.79	86.71	83.97	74.98	62.41	3.23	65.64
	A-T	85.69	85.69	85.52	88.56	85.69	77.59	27.84	2.89	30.73
	T-V	87.98	88.17	87.88	90.15	87.98	81.28	50.47	5.47	55.94
	A-V	88.36	88.44	88.28	90.21	88.36	81.87	50.08	5.04	55.12
	A-T-V	90.65	90.79	90.56	92.19	90.65	85.42	58.95	5.93	64.88

evaluating which model is better on average [45]. An ideal classifier has an AUC of one:

$$AUC = \frac{1}{mn} \sum_{i=1}^{m} \sum_{j=1}^{n} V \tag{1.8}$$

such that i index counts the true positive samples (m) and j index counts the true negative samples (n). For each sample i and j, the predicted probabilities (scores) are pi and pj, respectively; while V is:

$$V = \begin{cases} 1 & if \ p_i > p_j \\ 0 & otherwise \end{cases}$$

Finally, the training (fitting) time (T_f), the testing (scoring) time (T_s), and the total time $(T_t = T_f + T_s)$ are also reported.

1.5 Results and Discussion

The performance of the evaluated modalities for all considered demographics is depicted in Table 1.5. The table is divided into three parts presenting the results of gender detection, age-group detection, and dialect detection, respectively. The corresponding confusion matrices are shown in Figure 1.2.

Predicted

Actual		Femal	Male	Total
	Male	294	14	308
	Female	10	206	216
	Total	304	220	524

(a)

Predicted

Actual		AGA	AGB	AGC	AGD	Total
	AGA	114	4	7	3	128
	AGB	4	141	9	5	159
	AGC	5	13	118	6	142
	AGD	3	2	9	81	95
	Total	126	160	143	95	524

(b)

Predicted

Actual		Egyptian	Gulf	LEV	Maghribi	Total
	Egyptian	239	9	1	0	249
	Gulf	16	148	3	0	167
	LEV	7	10	62	0	79
	Maghribi	2	1	0	26	29
	Total	264	168	66	26	524

(c)

FIGURE 1.2
Confusion matrix for multimodal (A-T-V) demographic detection systems: (a) gender, (b) age-group, and (c) dialect.

1.5.1   Gender detection

For single modality, audio achieved the best performance for detecting gender in terms of the evaluation measures and computational time. however, the textual modality had poor performance compared to audio and visual modalities. Combining audio and visual modalities leads to achieving the best results for gender detection. Additionally, combining visual and textual modalities enhanced the performance of both single modalities (i.e., audio modality and visual modality, individually). However, fusing audio and textual modalities dropped the performance of the audio modality significantly. Combining textual modality with audio and visual modalities caused insignificant drop in the results but increased the consumed time. This can be attributed to the poor performance of the textual modality.

1.5.2 Age-group detection

In case of single modalities, visual modality achieved the highest results for detecting the age-group of speakers compared to the other single modalities which is followed by the audio modality. However, the textual modality achieved the lowest results. Combining different modalities significantly improved the results of every single modality. Unlike gender detection, combining textual modality with the other modalities improved the results. The highest results were obtained when combining audio, textual, and visual modalities. As it can be observed in the confusion matrix in Figure 1.2(b), there is confusion between middle-aged I and middle-aged II. The misclassification between these two age-groups is reasonable because even expert annotators were confused as well.

1.5.3 Dialect detection

In case of the single modalities, visual modality reports the highest results for dialect detection which is followed by textual modality. Audio modality achieves the lowest results in this case. Combining each modality contributes significantly to enhancing the results of every single modality as shown in Table 1.5

The confusion matrix for dialect detection, depicted in Figure 1.2(c), shows that the most difficult challenges dialects are Egyptian and Gulf. There are many samples misclassified as either Egyptian or Gulf. Although Maghribi dialect has the lowest number of examples, it is classified well comparing to the other dialects considered in this study. Also, Maghribi dialect was the straightforward dialect to be labeled by expert annotators in the SADAM dataset [31].

1.5.4 General discussion

In order to explore whether the performance differences are significant or just achieved by chance, statistical tests are also conducted for all of the evaluated models and for each considered demographic characteristic. The experiments were re-run 10 times for each 10-fold classification mode. Multiple pairwise two-sided statistical t-tests with a 95% confidence interval were carried out. The null hypothesis is formulated as follows:

$H_{0(X,Y)}$: there is no significant difference between system X and system Y where X and Y can be any of the seven systems (A, T, V, A-T, T-V, A-V, and A-T-V). The p-values for pairwise t-test on F_1 scores are shown in Figure 1.3 where the cases in which no significant differences (i.e., p-value > 0.05) are represented in bold.

In general and based on results presented in Table 1.5 and Figure 1.3, it can be noticed that for multi-class problems (age-group and dialect), the models generated using the combined audio, textual and visual modalities (A-T-V) achieved the highest results, for example, the F_1 score reaches about 86.65%

Demographic Gender

A	T	V	AT	TV	AV	ATV	
94.68	74.62	90.45	92.51	91.53	96.45	96.34	AVG
2.75	6.22	3.67	3.17	3.41	2.37	2.40	STD
	1.85E-51	7.70E-14	1.79E-08	2.37E-11	4.20E-06	7.82E-06	A
		2.14E-38	5.48E-51	3.19E-44	4.67E-54	1.48E-58	T
			1.15E-04	2.79E-03	1.52E-30	9.43E-30	V
				3.52E-02	5.97E-17	1.76E-19	AT
					1.99E-22	1.35E-25	TV
						6.96E-01	AV

Age-group

A	T	V	AT	TV	AV	ATV	
69.22	54.41	76.08	71.45	75.94	84.98	86.24	AVG
5.25	7.21	6.12	6.02	5.49	4.41	4.64	STD
	1.46E-30	1.82E-14	1.72E-03	2.83E-15	2.66E-45	3.09E-47	A
		3.66E-42	5.45E-46	1.61E-48	7.08E-59	1.36E-64	T
			9.05E-08	7.60E-01	1.60E-33	9.61E-34	V
				4.54E-09	5.93E-36	1.25E-43	AT
					5.82E-34	5.84E-38	TV
						8.98E-04	AV

Dialect

A	T	V	AT	TV	AV	ATV	
71.71	72.06	82.94	84.39	84.59	88.41	90.63	AVG
5.90	5.20	5.40	4.21	4.92	4.24	3.63	STD
	6.64E-01	1.86E-27	1.96E-40	5.03E-32	1.16E-45	6.63E-50	A
		1.88E-28	5.07E-42	1.13E-33	1.33E-46	3.82E-57	T
			3.47E-02	1.08E-03	1.32E-20	9.44E-26	V
				7.56E-01	1.43E-10	1.47E-22	AT
					4.48E-12	8.04E-22	TV
						9.04E-06	AV

FIGURE 1.3
Pairwise t-test of demographics (p-values).

for age-group and 90.56% for dialect. However, compared to single modalities there has been an increase in the time complexity during training and testing but it remains around one minute for training and about five seconds for testing.

Audio modality takes the lowest time for either training or evaluation compared to models generated using visual and textual modalities. This can be attributed to that, audio models are generated with the smallest number of features. On the other hand, models generated using visual modalities took the largest computational time due to the large size of the visual feature vectors.

Considering single modalities as our baseline results for each demographic characteristic considered in this study, it can be observed that combining different modalities improved the results of the single modalities. Figure 1.3 shows that the reported improvements are statistically significant in nearly all cases. On the other hand, the negative effects when combining textual with both audio and visual modalities for gender detection leading to dropping the results are statistically insignificant.

The system detects the gender of video's speakers more accurately when compared to age-groups and dialects. This is because gender detection is a binary classification task while age-groups and dialects are multiple classification tasks. On the other hand, the system can detect dialects of speakers more accurately than their age groups with less fitting and evaluating time.

1.6   Conclusions

This chapter presents an empirical study to detect the demographic characteristics of videos on social media, namely: gender, age, and dialect of speakers. For textual modality, we evaluated neural word embedding features whereas a combination of prosodic and spectral features is extracted to represent audio modality. On the other hand, hybrid features of global and local descriptors are extracted to represent visual modality. These features were extracted from a video dataset for multimodal Arabic sentiment analysis called Sentiment Analysis Dataset for Arabic Multimodal (SADAM) and combined at feature + level fusion. The extracted features are used individually to train an SVM classifier as a baseline, then in combination to analyze the effects of fusing different modalities. The generated models were evaluated using several evaluation measures including: Recall, Precision, F1, Geometric mean, and Accuracy. Experimental results show that audio modality performs better than visual and textual modalities individually to detect the gender of speakers whereas visual modality achieves the highest results for detecting age-groups and dialects. In addition, the results confirm that combining different modalities can significantly improve the results. They also show that gender is relatively easier to be detected from videos whereas age-group is the most challenging task. The highest results for detecting age-group were obtained using combining the three modalities and achieved F_1 of 86.65%. Similarly, the highest results for dialect detection were obtained using the fusion of the three modalities with F_1 of 90.56%. In case of gender detection, the highest results were reported using audio-visual modalities with F_1 of 95.80%.

Acknowledgment

The authors would like to thank King Fahd University of Petroleum & Minerals (KFUPM), Saudi Arabia, for all support.

Authors' biographies

Sadam Al-Azani received B.Sc. degree (Hons.) in computer science from Thamar University, Dhamar, Yemen, in 2004, and the M.Sc. and Ph.D. degrees in computer science from King Fahd University of Petroleum & Minerals (KFUPM), Dhahran, Saudi Arabia, in 2014 and 2019, respectively. He is currently a Research Scientist III with SDAIA-KFUPM Joint Research Center for Artificial Intelligence (JRC-AI), KFUPM. He was a full-time Teaching Assistant with Thamar University, for five years. His research interests include artificial intelligence, natural language processing, video analytics, machine/deep learning, social network analysis, and pattern recognition. He has served as a Technical Program Committee Member and a Reviewer for a number of international conferences and journals.

El-Sayed M. El-Alfy is currently a Professor at the Information and Computer Science Department, Fellow of SDAIA-KFUPM Joint Research Center for Artificial Intelligence, Affiliate of Interdisciplinary Research Center on Intelligent Secure Systems, King Fahd University of Petroleum and Minerals (KFUPM), Saudi Arabia. He has over 25 years of experience in industry and academia, involving research, teaching, supervision, curriculum design, program assessment, and quality assurance in higher education. He is an approved ABET/CSAB Program Evaluator (PEV), and a Reviewer and a Consultant for NCAAA and several universities and research agents in various countries. He is an active Researcher with interests in fields related to machine learning and nature-inspired computing and applications to data science and cybersecurity analytics, pattern recognition, multimedia forensics, and security systems. He has published numerously in peer-reviewed international journals and conferences, edited a number of books published by reputable international publishers, attended and contributed to the organization of many world-class international conferences, and supervised's master's and Ph.D. students. He is a senior member of IEEE and was also a member of ACM, the IEEE Computational Intelligence Society, the IEEE Computer Society, the IEEE Communication Society, and the IEEE Vehicular Technology Society. His work has been internationally recognized and received a number of awards. He has served as a Guest Editor for a number of special issues in international journals and has been on the editorial board of a number of premium international journals, including *IEEE/CAA JOURNAL OF AUTOMATICA SINICA, IEEE TRANSACTIONS ON NEURAL NETWORKS AND LEARNING SYSTEMS, International Journal of Trust Management in Computing and Communications, and Journal of Emerging Technologies in Web Intelligence (JETWI).*

Acknowledgments

The authors would like to acknowledge the support received from Saudi Data and AI Authority (SDAIA) and King Fahd University of Petroleum and Minerals (KFUPM) under SDAIA-KFUPM Joint Research Center for Artificial Intelligence.

Bibliography

[1] Rana Al-Maroof, Kevin Ayoubi, Khadija Alhumaid, Ahmad Aburayya, Muhammad Alshurideh, Raghad Alfaisal, and Said Salloum. The acceptance of social media video for knowledge acquisition, sharing and application: A comparative study among youyube users and tiktok users' for medical purposes. *International Journal of Data and Network Science*, 5(3):197, 2021.

[2] Matthew A Gilbert. Strengthening Your Social Media Marketing with Live Streaming Video. In: Al-Masri, A., Curran, K. (eds) *Smart Technologies and Innovation for a Sustainable Future*. Advances in Science, Technology & Innovation. Springer, Cham, pages 357–365, 2019.

[3] Ganesh Chandrasekaran, Tu N Nguyen, and Jude Hemanth D. Multimodal sentimental analysis for social media applications: A comprehensive review. *Wiley Interdisciplinary Reviews: Data Mining and Knowledge Discovery*, 11(5):e1415, 2021.

[4] Hui Yuan, Yuanyuan Tang, Wei Xu, and Raymond Yiu Keung Lau. Exploring the influence of multimodal social media data on stock performance: An empirical perspective and analysis. *Internet Research*, 31(3):871–891, 2021.

[5] Jai Prakash Verma, Smita Agrawal, Bankim Patel, and Atul Patel. Big data analytics: Challenges and applications for text, audio, video, and social media data. *International Journal on Soft Computing, Artificial Intelligence and Applications (IJSCAI)*, 5(1):41–51, 2016.

[6] Antitza Dantcheva, Petros Elia, and Arun Ross. What else does your biometric data reveal? A survey on soft biometrics. *IEEE Transactions on Information Forensics and Security*, 11(3):441–467, 2015.

[7] Paula Lopez-Otero, Laura Docio-Fernandez, and Carmen Garcia-Mateo. Assessing speaker independence on a speech-based depression level estimation system. *Pattern Recognition Letters*, 68:343–350, 2015.

[8] Peggy T Cohen-Kettenis, Allison Owen, Vanessa G Kaijser, Susan J Bradley, and Kenneth J Zucker. Demographic characteristics, social competence, and behavior problems in children with gender identity disorder: A cross-national, cross-clinic comparative analysis. *Journal of Abnormal Child Psychology*, 31(1):41–53, 2003.

[9] Lei Shi and Alexandra I Cristea. Demographic indicators influencing Learning activities in moocs: Learning analytics of futurelearn courses. In *The 27th International Conference on Information Systems Development (ISD2018)*. Association for Information Systems, 2018. https://core.ac.uk/download/301376307.pdf

[10] Hamid Zaferani Arani, Giti Dehghan Manshadi, Hesam Adin Atashi, Aida Rezaei Nejad, Seyyed Mojtaba Ghorani, Soheila Abolghasemi, Maryam Bahrani, Homayoon Khaledian, Pantea Bozorg Savodji, Mohammad Hoseinian, et al. Understanding the clinical and demographic characteristics of second coronavirus spike in 192 patients in Tehran, Iran: A retrospective study. *Plos One*, 16(3):e0246314, 2021.

[11] Qing-Qing Wang, Yuan-Yuan Fang, Hao-Lian Huang, Wen-Jun Lv, Xiao-Xiao Wang, Tian-Ting Yang, Jing-Mei Yuan, Ying Gao, Rui-Lian Qian, and Yan-Hong Zhang. Anxiety, depression and cognitive emotion regulation strategies in Chinese nurses during the Covid-19 outbreak. *Journal of Nursing Management*, 29(5):1263–1274, 2021.

[12] Giuseppe Lippi, Camilla Mattiuzzi, Fabian Sanchis-Gomar, and Brandon M Henry. Clinical and demographic characteristics of patients dying from Covid-19 in Italy vs China. *Journal of Medical Virology*, 92(10):1759–1760, 2020.

[13] Shaul Kimhi, Hadas Marciano, Yohanan Eshel, and Bruria Adini. Resilience and demographic characteristics predicting distress during the Covid-19 crisis. *Social Science & Medicine*, 265:113389, 2020.

[14] Laura-Jayne Gardiner, Anna Paola Carrieri, Karen Bingham, Graeme Macluskie, David Bunton, Marian McNeil, and Edward O Pyzer-Knapp. Combining explainable machine learning, demographic and multi-omic data to inform precision medicine strategies for inflammatory bowel disease. *PloS One*, 17(2):e0263248, 2022.

[15] Dekel Taliaz, Amit Spinrad, Ran Barzilay, Zohar Barnett-Itzhaki, Dana Averbuch, Omri Teltsh, Roy Schurr, Sne Darki-Morag, and Bernard Lerer. Optimizing prediction of response to antidepressant medications using machine learning and integrated genetic, clinical, and demographic data. *Translational Psychiatry*, 11(1):1–9, 2021.

[16] Sudhanshu Kumar, Monika Gahalawat, Partha Pratim Roy, Debi Prosad Dogra, and Byung-Gyu Kim. Exploring impact of age and gender on sentiment analysis using machine learning. *Electronics*, 9(2):374, 2020.

[17] Norjihan Abdul Ghani, Suraya Hamid, Ibrahim Abaker Targio Hashem, and Ejaz Ahmed. Social media big data analytics: A survey. *Computers in Human Behavior*, 101:417–428, 2019.

[18] Ashish K Sharma, Durgesh M Sharma, Neha Purohit, Saroja Kumar Rout, and Sangita A Sharma. Analytics techniques: Descriptive analytics, predictive analytics, and prescriptive analytics. In *Decision Intelligence Analytics and the Implementation of Strategic Business Management*, pages 1–14. Springer, 2022.

[19] Luis-Miguel López-Santamaría, Juan Carlos Gomez, Dora-Luz Almanza-Ojeda, and Mario-Alberto Ibarra-Manzano. Age and gender identification in unbalanced social media. In *2019 International Conference on Electronics, Communications and Computers (CONIELECOMP)*, pages 74–80. IEEE, 2019.

[20] Brandon Lwowski and Anthony Rios. The risk of racial bias while tracking influenza-related content on social media using machine learning. *Journal of the American Medical Informatics Association*, 28(4):839–849, 2021.

[21] Jayaprada S Hiremath, Shantakumar B Patil, and Premjyoti S Patil. Human age and gender prediction using machine learning algorithm. In *Proceedings of the IEEE International Conference on Mobile Networks and Wireless Communications (ICMNWC)*, pages 1–5, 2021.

[22] Hadi Harb and Liming Chen. Gender identification using a general audio classifier. In *Proceedings of the International Conference on Multimedia and Expo. ICME'03.*, volume 2, pages II–733, 2003.

[23] Nina Cesare, Christan Grant, Quynh Nguyen, Hedwig Lee, and Elaine O Nsoesie. How well can machine learning predict demographics of social media users? *arXiv preprint arXiv:1702.01807*, 2017. https://doi.org/10.48550/arXiv.1702.01807

[24] Dmitry I Ignatov, Ekaterina Mitrofanova, Anna Muratova, and Danil Gizdatullin. Pattern mining and machine learning for demographic sequences. In *International Conference on Knowledge Engineering and the Semantic Web*, pages 225–239. Springer, 2015.

[25] Laurent El Shafey, Elie Khoury, and Sébastien Marcel. Audio-visual gender recognition in uncontrolled environment using variability modeling techniques. In *IEEE International Joint Conference on Biometrics (IJCB)*, pages 1–8, 2014.

[26] Francisco Rangel, Paolo Rosso, Manuel Montes-y Gómez, Martin Potthast, and Benno Stein. Overview of the 6th author profiling task at pan 2018: Multimodal gender identification in Twitter. *Working Notes Papers of the CLEF*, 2018.

[27] Mohamed Abouelenien, Verónica Pérez-Rosas, Rada Mihalcea, and Mihai Burzo. Multimodal gender detection. In *Proceedings of the 19th ACM International Conference on Multimodal Interaction*, pages 302–311, 2017.

[28] Sadam Al-Azani and El-Sayed M. El-Alfy. Multimodal age-group recognition for opinion video logs using ensemble of neural networks. *International Journal of Advanced Computer Science and Applications(IJACSA)*, 10(4), pages 371–378, 2019.

[29] Sadam Al-Azani and El-Sayed M El-Alfy. Audio-textual Arabic dialect identification for opinion mining videos. In *Proceedings of IEEE Symposium Series on Computational Intelligence (SSCI)*, pages 2470–2475, 2019.

[30] Sadam Al-Azani and El-Sayed M El-Alfy. Multimodal sentiment and gender classification for video logs. In *ICAART (2)*, pages 907–914, 2019.

[31] Sadam Al-Azani and El-Sayed M El-Alfy. Enhanced video analytics for sentiment analysis based on fusing textual, auditory and visual information. *IEEE Access*, 8:136843–136857, 2020.

[32] Abu Bakr Soliman, Kareem Eissa, and Samhaa R El-Beltagy. Aravec: A set of Arabic word embedding models for use in Arabic nlp. In *Proceedings of the 3rd International Conference on Arabic Computational Linguistics (ACLing)*, volume 117, pages 256–265, 2017.

[33] Paul Viola and Michael J Jones. Robust real-time face detection. *International Journal of Computer Vision*, 57(2):137–154, 2004.

[34] Paul Viola and Michael Jones. Rapid object detection using a boosted cascade of simple features. In *Proceedings of IEEE Conference on Computer Vision and Pattern Recognition (CVPR 2001)*, pages I-511–I-518, 2001.

[35] Michael S Ryoo and Larry Matthies. First-person activity recognition: What are they doing to me? In *Proceedings of IEEE Conference on Computer Vision and Pattern Recognition*, pages 2730–2737, 2013.

[36] Achmad Solichin, Agus Harjoko, and Agfianto Eko Putra. Movement direction estimation on video using optical flow analysis on multiple frames. *International Journal of Advanced Computer Science and Applications*, 9(6):174–181, 2018.

[37] Radim Řehůřek and Petr Sojka. Software framework for topic modelling with large corpora. In *Proceedings of the LREC 2010 Workshop on New Challenges for NLP Frameworks*, pages 45–50, 2010.

[38] Theodoros Giannakopoulos. Pyaudioanalysis: An open-source python library for audio signal analysis. *PloS One*, 10(12), pages 1–17, 2015.

[39] Gary Bradski and Adrian Kaehler. *Learning OpenCV: Computer Vision with the OpenCV Library*. O'Reilly Media, Inc., 2008.

[40] Fabian Pedregosa, Gaël Varoquaux, Alexandre Gramfort, Vincent Michel, Bertrand Thirion, Olivier Grisel, Mathieu Blondel, Peter Prettenhofer, Ron Weiss, Vincent Dubourg, et al. Scikit-learn: Machine learning in python. *Journal of Machine Learning Research*, 12:2825–2830, 2011.

[41] Mikel Galar, Alberto Fernandez, Edurne Barrenechea, Humberto Bustince, and Francisco Herrera. A review on ensembles for the class imbalance problem: Bagging-, boosting-, and hybrid-based approaches. *IEEE Transactions on Systems, Man, and Cybernetics, Part C (Applications and Reviews)*, 42(4):463–484, 2012.

[42] El-Sayed M El-Alfy and Sadam Al-Azani. Empirical study on imbalanced learning of arabic sentiment polarity with neural word embedding. *Journal of Intelligent & Fuzzy Systems*, 38(5):6211–6222, 2020.

[43] Kubat Miroslav, Stan Matwin, Addressing the curse of imbalanced training sets: one-sided selection. In *Proceedings of the Fourteenth International Conference on Machine Learning (ICML)*, 97(1), 1997.

[44] Victoria López, Alberto Fernández, Salvador García, Vasile Palade, and Francisco Herrera. An insight into classification with imbalanced data: Empirical results and current trends on using data intrinsic characteristics. *Information Sciences*, 250:113–141, 2013.

[45] Vaishali Ganganwar. An overview of classification algorithms for imbalanced datasets. *International Journal of Emerging Technology and Advanced Engineering*, 2(4):42–47, 2012.

2

Toward Long-Term Person Re-identification with Deep Learning

Muna O. AlMasawa
Computer Science Department, Faculty of Computing and Information Technology, King Abdulaziz University Jeddah, Saudi Arabia ORCID: 0000-0001-8309-9269

Kawthar Moria
Computer Science Department, Faculty of Computing and Information Technology, King Abdulaziz University Jeddah, Saudi Arabia ORCID: 0000-0001-6241-2658

Lamiaa A. Elrefaei
Computer Science Department, Faculty of Computing and Information Technology, King Abdulaziz University Jeddah, Saudi Arabia.
Electrical Engineering Department, Faculty of Engineering at Shoubra, Benha University, Cairo, Egypt ORCID: 0000-0001-5781-2251

CONTENTS

2.1	Introduction		24
2.2	Related Work		27
	2.2.1	Short-term person re-identification	27
	2.2.2	Long-term person re-identification	28
2.3	The Proposed Long-Term Person Re-Id Model (Deep LongReID)		30
	2.3.1	Preprocessing stage	30
		2.3.1.1 Person silhouette segmentation	30
		2.3.1.2 Person GEI generation	32
	2.3.2	Person re-identification stage	33
		2.3.2.1 Features extraction	33
		2.3.2.2 Combined loss and similarity computing	35
2.4	Experimental Evaluation		36
	2.4.1	Dataset	36
		2.4.1.1 CASIA-B dataset	36
		2.4.1.2 Market-1501 dataset	38
	2.4.2	Training setting	38
	2.4.3	Evaluation metric and experimental protocol	40

DOI: 10.1201/9781003053262-2

     2.4.4   Results and discussion 40

              2.4.4.1   Training and testing the re-identication stage   40

              2.4.4.2   Testing the integrated system 47

2.5   Conclusion 48

2.1   Introduction

The surveillance systems became everywhere in such a modern society such as in government buildings, airports, etc., for security and citizen safety especially, with the enormous increase in crimes and the rise of terrorist activities. With the growth of multi-camera surveillance systems, the huge data is recording daily that needing to process and monitor. So, there is an urgent need for intelligent surveillance systems that can help in automated the monitoring and decreasing the dependence on human effort and improve significantly the quality of surveillance and saving the time by speed up data processing [1]. Person re-identification (person Re-ID) is the major task within the intelligent surveillance systems. It has received considerable critical attention between the researchers in the computer vision field. It is defined as the process of matching a person that recorded by one camera with its correspondence that recorded by another non-overlapping camera. It is a challenging task because of the large variations in the appearance of persons in a different camera. Recently, instead of hand-crafted methods, deep-learning models add significant progress in several issues of computer vision. Several current person Re-ID systems proved that deep-learning models have improved the systems performance and they get promising results [2].

Person Re-ID systems, in general, can be divided into four categories based on different criteria as shown in Figure 2.1. First one is the system architecture where these systems can be traditional systems (hand-crafted systems) or deep-learning systems. The second one is the features type where the appearance or biometric features can be extracted from the input image to reidentify the person. The third one is the period of time that the systems support it if it is short-term or long-term. The last one is the type of probe and gallery if both are images or videos, or one is an image and the other is video. A traditional person Re-ID system architecture is getting the probe image of a person (query image) that has been recorded from one camera. Then, it extracts distinctive features by hand-crafted extraction methods such as color histograms, texture descriptor or deep-learning models such as ImageNet models and represented them as a feature vector. After that, person Re-ID try to find a probe image of person in the gallery whereas the gallery contains many images that were taken by other disjoint cameras. Finally, the top ranked list of gallery images is returned that have the lowest distance using a distance metric learning to minimize the distance between the similar image pairs and maximize the distance between the dissimilar pairs.

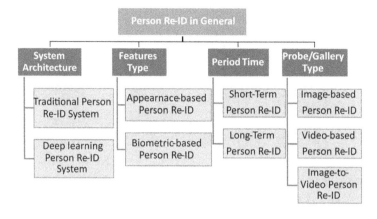

FIGURE 2.1
Person re-identification categories.

So, extracting the robust features and finding a strong distance metric is the main step toward built strong person Re-ID system [3–5]. Person Re-ID studies are classified into three types: image-based person Re-ID where the probe and gallery are a static image, video-based person Re-ID where the probe and gallery are the video frames, and image-to-video person Re-ID where the probe is a static image and the gallery is a video frames [6]. The appearance-based features like clothes color and texture, and the biometric-based features such as human gait and face are the two types of features that can be extracted from person image. The type of input image can be RGB images, RGB-depth images, and RGB-Infrared images, the last two types are used a special type of camera. The reidentification of person can be done in short-term or long-term as shown in Figure 2.2. Short-term reidentify a person that was appeared in camera A, and then after seconds or minutes reappeared in camera B; whereas the long-term reidentify a person that was appeared in camera A, and then after days he reappeared once more in camera A or B. Since the most studies used the appearance-based features, these studies applicable only for short-term person Re-ID.

Generally, appearance features have many limitations [5, 7]:

- In the real-world, persons sometimes wear similar clothes as dark clothes in winter. Also, same person may appear with variations in appearance. These variations happen because of environment changes such as illumination changes, pose changes, background changes, and so on.

- It can be used to reidentify a person in short time like when a person disappears from first camera, then reappear again after seconds or minutes in a second camera and in this situation, it does not change their appearance.

However, in long-term person Re-ID situation, the appearance features will drop and there is a need to more robust features over days. So, some studies start to adopt biometric features to support both the short-term and the long-term person Re-ID [8]. Human gait plays a crucial role in surveillance

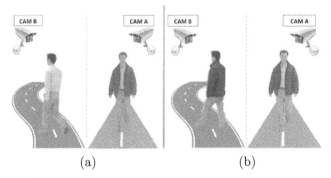

(a)   (b)

FIGURE 2.2
Person Re-ID: (a) Long-term vs (b) Short-term.

applications. It is among the most widely used in people recognition tasks, but it is a new trend and quite young is to adopt gait for person re-identification tasks. From the perspective of surveillance systems, the gait feature is more appropriate in re-identification task. Gait can be captured from a distance, in low resolution, without need to walker attention and it cannot be covered like other biometric features and it is inexpensive [7]. Gait contains both the physical shape of a person and the dynamic motions of the same person in a gait cycle [9]. There are some obstacles when considering gait as a feature for person identification include change in person's clothes, carrying situations and the speed of walking can impact on human gait by decreasing its discriminative as a biometric. Rather than hand-crafted methods, the deep-learning models added a significant improvement in such these problems. In this chapter, we introduced, to the best of our knowledge, the first video-based long-term person re-identification model based on deep learning called Deep LongReID. Our proposed model consists of two main stages: The preprocessing stage and the re-identification stage. The preprocessing stage consists of the person's silhouette segmentation and Gait Energy image (GEI) generation. The re-identification stage consists of features extraction and the combined loss functions and similarity learning. Our proposed model is one of the new and early works that adopted human gait for person Re-ID where the gait has good performances if persons change their appearance or not.

Also, this is the first implementation of Instance Normalization with Batch Normalization in Resnet-50 to extract discriminative gait features from silhouettes images. Our proposed model is video-based person Re-ID, but instead of other previous video-based models that built complex architecture with high computational resources to extract spatial-temporal features from all video frames for reidentification, the spatial-temporal features in our proposed model are embedded in one image, GEI (Gait Energy Image). So, our proposed model has straightforward steps and simple architectures with superior performance. The main contributions of this work are:

- We introduced, to the best of our knowledge, the first video-based long-term person Re-ID model based on deep learning, called Deep LongReID to address the challenge of appearance variations in long period of time.

- This is the first implementation of Instance Normalization with Batch Normalization in Resnet-50 to extract discriminative gait features from silhouettes images. This combination between instance and batch normalization is invariant to appearance changes and improves both the learning and generalization capabilities.

- The comprehensive evaluations show that the proposed system remarkably outperform the state-of-the-art models on the challenge benchmark CASIA-B dataset that has a different appearance for each identity, it achieved an accuracy from 59.7% to 88.1% in rank 1 and from 88.05% to 96.25% in rank 5.

The remaining of this chapter is structured as follows. In Section 2.2, we review the related works designed for person Re-ID based on the period of time. In Section 2.3, we proposed our model. In Section 2.4, we discuss the experiments and the results of the proposed model. Finally, we conclude this chapter in Section 2.5.

2.2   Related Work

This section will review the state-of-art studies and categorize them under short-term and long-term person Re-ID.

2.2.1   Short-term person re-identification

Since the short-term person Re-ID systems are trying to reidentify the same person over short time where the person did not change their appearance (hair color, clothes, etc.), the appearance-based features are the mostly used features for this task. Most of the current studies are categorized under this type and the current person Re-ID datasets were captured on short time. The appearance-based features can be extracted from the whole body of a person, or the body is divided into several parts, then extracting the features from these parts. Authors in [10] proposed deep-learning architecture for extracting the features from whole body of person and find the variances in the neighboring locations of the feature to compute the local relationship among the two images. In [11], authors introduced a parts-based deep-learning approach that get the human body and divided it into several parts, then the features are extracted from these parts. Since the features of full human body and parts body are complement each other, many studies introduced approaches combined the full and parts body features. In [12], the authors divided the body into seven body parts based on the 14 body joint locations. These parts are head with shoulder, upper and lower body, two arms and two legs. Then, one global features vector was extracted for the full human image and seven feature vectors for the seven body parts. Rather than developing new deep-learning models, a number of studies using an existing ImageNet models such as AlexNet, ResNet-50, and inception-v3, and also focus on the type of the loss function because it has a critical impact on the performance of person Re-ID systems. In [13], the authors used different types of triplet loss. This proved that triplet loss can enhance the performance of person Re-ID systems. To enhance the triplet loss, many loss functions were proposed such as lifted structured loss [14] and quadruplet loss [15]. All the aforementioned studies get the person image as static image. This type is called image-based person Re-ID. The most common datasets that used for evaluating this type of systems are VIPeR [16], CUHK01 [17], CUHK03 [18], and Market-1501 [19].

In real video surveillance system, the person recorded as video frames. From these frames we can get the temporal features beside the appearance features. This type called video-based person Re-ID [6]. Authors in [20] introduced architecture with a double stream, one stream for appearance features and in the other stream for temporal features. In [21], select only the useful frames using attention unit. Authors in [22] improved the local spatial representation to get more discriminative features by introducing the Siamese attention networks. Authors in [23] and [24] extended the area of interest for visual features in spatial and temporal dimensions by using 3D CNN rather than 2D CNN. The most common datasets that used for video-based systems evaluation are PRID2011 [25], iLIDSVID [26], and MARS [27]. All previous studies used RGB camera in their surveillance system and produced input image of person as RGB image. recently, RGB-Depth sensors like Kinect sensors become available and commonly used for many purposes. This brings a new type of features that can be obtained from the depth images like the anthropometric information and the shape of human body. These features are robust against many variations such as illumination changes. Many studies support multi-modalities were introduced. In [28] proposed an approach that extracted the appearance features and an anthropometric features. Since the infrared images can capture a clear image in low illumination such as at night, authors in [29] handled the RGB-Infrared cross-modality problem of person Re-ID system. The most common RGB-Depth dataset is Kinect-REID [30] and RGB-Infrared is SYSU-MM01 [29].

2.2.2 Long-term person re-identification

To expand the short-term person Re-ID systems, the long-term person Re-ID systems are introduced. They try to find and re-identify the same person over long period of time (days, month, etc.) where the persons change their appearance such as changing their clothes. Recent Studies toward long-term started by reducing dependence on appearance features by adopting biometric-based features with them. In [31], authors proposed the first study that was used biometric-based features by integrating a gait features with appearance features. Gait Energy image (GEI) is used to represent the gait information. Then, score-level and feature-level fusion are used to merge the appearance features and biometric features. In [32], authors proposed a graph learning framework to integrate the appearance features and soft biometric. To define the soft biometric, different attributes include gender, hair, backpack, and carrying are extracted. In [33], authors started to only depend on biometric features. In their approach, reidentifying a person based on spatial-temporal information of the gait in frontal video sequences using optical flow called Histogram of Flow Energy Image (HOFEI). In [34] and [35], authors adopted new type of features, the anthropometric features are extracted from RGB-D image that captured by using the Kinect sensor. These features described the size and shape of human body. In [36], authors proposed fine motion encoding model that used motion features. Also, they collected new long-term dataset called named Motion-ReID that consisted of video sequence. In [37], authors proposed gait-based person re-identification approach with handling of the covariate issues such as wearing coat or carrying a bag. This approach depended on a selection of the most significant parts dynamically rather than processed the full body for re-identification. Authors in [38] proposed an auto-correlation network that was invariant to start or end of the gait cycle. It was extracted human gait features from different viewpoints. The two public dataset for gait recognition, CASIA B [39] and OU-MVLP, that were recorded by multi-camera was used to evaluate the person Re-ID systems. In [40], authors introduced the first public long-term person Re-ID dataset, Celeb-reID, where the same person change their appearance. Also, they proposed a network that used vector neuron capsules as an alternative of the traditional scalar neurons which provide extra dimension. Since the success of soft-biometric, the newest works adopted them to make the model invariant against appearance variations, in [41] and [42] authors depended on extract the features of body shape. In [43] authors used architecture with two stream, image stream and gait stream which the features were extracted from single image with the support of gait features.

The reviewed related work shows that most of these studies are image-based, and few of them are video-based Re-ID, but video-based is closer to reality than image-based Re-ID systems. All these studies focused only on appearance features. There are some limitations: In the real-world, persons sometimes wear similar clothes as dark clothes in winter. Also, same person may appear with variation in his appearance. These variations happen because environment changes such as illumination changes, pose changes, background changes, and so on. So, it can be used to reidentify a person in short period of time such when a person disappears from first camera and reappear again after seconds or minutes in another camera and in this situation, person does not change their appearance Most of previous long-term studies built their system using hand-crafted methods, and few of them built their systems using deep-learning models the deep-learning achieved better accuracy results than hand-crafted. Most of these studies did not consider the real long-term Re-ID situation: multi-camera and multi-appearance for each person. Generally, the long-term person Re-ID problem is not widely explored.

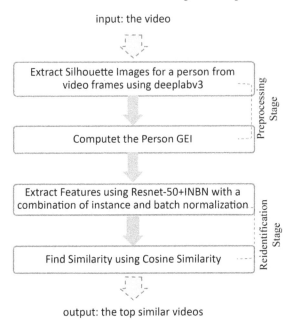

FIGURE 2.3
The general procedure of the proposed method.

2.3 The Proposed Long-Term Person Re-Id Model (Deep LongReID)

In this section, the proposed, to the best of our knowledge, first video-based long-term person re-identification model based on deep learning called Deep LongReID. The General procedures are presented in Figure 2.3. Deep LongReID addresses the challenge of the appearance variations in long period of time. Our proposed model consists of two main stages: The preprocessing stage and the re-identification stage. The details of Deep LongReID are described in Figure 2.4.

2.3.1 Preprocessing stage

This is the first stage in our proposed model. It consists of the person's silhouette segmentation step and person GEI generation step. The output of this stage will be an input to the re-identification stage.

2.3.1.1 Person silhouette segmentation

Gait analysis is very sensitive to person's silhouette segmentation. Mostly, to extract a person silhouettes, traditional computer vision, and machine-learning techniques are used. Image segmentation is one of the high-level tasks

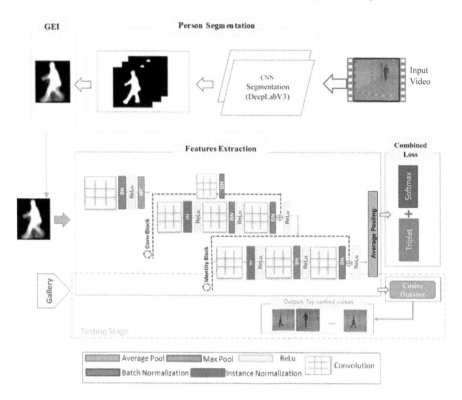

FIGURE 2.4
The framework of the proposed Deep LongReID.

that help in scene understanding. It is a process of assigning a label to each pixel in the image. Segmentation has made great progress in recent years using deep learning. Semantic segmentation is a type of image segmentation which each pixel is classified into one of the predefined classes, so all the objects of the same class will be labeled by this class. In surveillance system, not only pixel-level classification is done in the spatial domain but also across time [44]. Many approaches are built to handle the semantic segmentation and they achieved high accuracy results. One of these models is DeepLabV3 [45]. As illustrated in Figure 2.5, DeepLab used a ResNet that pretrained on ImageNet for extracting the features. It uses a special technique called atrous convolution for semantic image segmentation to capture multi-scale dense features by adopting multiple atrous rates. Atrous convolution is used on the input feature map v, at each location i on the output y with atrous rate r and filter wt and k is the kernel (Equation 2.1) where the standard convolution used rate $r = 1$ and atrous convolution used multiple rates values to expand the view:

$$y[i] = \sum_k \nu[i + r \cdot k]wt[k] \qquad (2.1)$$

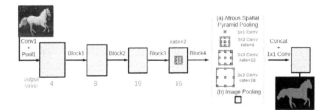

FIGURE 2.5
DeepLabV3 [45].

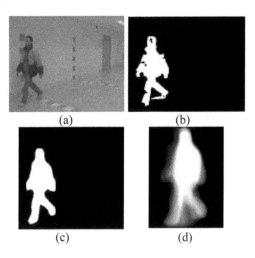

(a)   (b)

(c)   (d)

FIGURE 2.6
Examples of person segmentation: (a) Original RGB image from CASIA-B dataset, (b) The available silhouette image in the dataset, (c) The person segmentation by DeepLavV3, and (d) The GEI.

In our proposed model, each frame from the input video is segmented by Deeplabv3 to get person silhouette. Deeplabv3 with ResNet-101 backbone is adopted for the segmentation. This model has been trained on 20 different classes in COCO train 2017 dataset. One of these classes is a person, so we modify the model to detect only person class. DeepLabV3 achieved the performance result about 82.9% for person class, the examples of person segmentation by DeepLabV3 are shown in Figure 2.6. So, with the help of recently developed segmentation methods, we can obtain much more accurate person silhouettes whatever there is hard environment variations such as background clutter or illumination changes.

2.3.1.2   Person GEI generation

In this step, the spatial-temporal information of a gait cycle is encoded into one image, gait energy image (GEI). This provide rich features with simple processing needed. GEI represents both body shape and motion over a gait cycle. To get a GEI for a person, the silhouettes image is cropped to focus and center on the person silhouettes. Then, the average of the N generated silhouettes images $S(x, y)$ from previous step is computed to find the person GEI (Equation 2.2):

$$GEI = \frac{1}{N} \sum_{t=1}^{N} S(x, y, t) \tag{2.2}$$

where t is a frame number in the sequence. The high intensity pixels belong to the top parts of the person body that has the information of body shape. Low intensity pixels belong to the bottom parts of the person body that has the motion parts as shown in Figure 2.6(d).

2.3.2 Person re-identification stage

The re-identification stage is the second and main stage in our proposed model. It consists of features extraction step and the loss functions combination step, so the model can output the top-ranked list of similar videos.

2.3.2.1 Features extraction

a ResNet-50 is used as backbone network to extract visual features from person image, it is finetuning in our model and all layers are trained. The ResNet-50 is type of Residual Networks (ResNet). The ResNet-50 architecture consists of four stages, each stage has number of residual blocks, one called convolutional block and the others are identity blocks. Identity block is used when the input and output are with the same dimensions. Number of residual blocks are 3, 4, 6, and 3 in the stages 1–4, respectively. Each residual block consists of three layers, each layer has convolutional layers, batch normalization and ReLU activation functions. We replace the batch normalization with a combination of Instance and Batch normalization in our proposed model. At the end of the network, there are average pooling and fully-connected layer.

We replace the average pooling by adaptive average pooling which it can select the stride and kernel-size automatically whatever the input size. Finally, we replace fully-connected layer with the proposed Triplet-Softmax loss functions.

Batch normalization (BN) is used as the core in Resnet-50 model. Inspired by the great success of adopted instance normalization in [46], instead of batch normalization after first convolutional in residual block, Instance Normalization (IN) is used. So each residual block consisting of combination between batch normalization and instance normalization. The differences between instance and batch normalization in the computation method as shown in Figure 2.7. Batch normalization uses global statistics of a mini-batch to normalize each feature channels as Equation 2.3. It significantly accelerates training and preserves features discrimination, but it is weak against appearance variations:

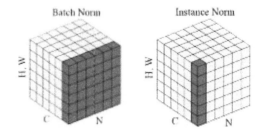

FIGURE 2.7

Instance Normalization vs Batch Normalization. The subplot is a feature map, N = batch axis, C = channel axis, H, W = spatial axis. The pixels that normalized by the same mean and variance are with blue color [56].

$$BNx_{nchw} = \frac{v_{nchw} - \mu_c}{\sqrt{\sigma_c^2 + \varepsilon}} \qquad (2.3)$$

$$\mu_c = \frac{1}{NHW} \sum_{n=1}^{N} \sum_{h=1}^{H} \sum_{w=1}^{W} v_{nchw} \qquad (2.4)$$

$$\sigma_c^2 = \frac{1}{NHW} \sum_{n=1}^{N} \sum_{h=1}^{H} \sum_{w=1}^{W} (v_{nchw} - \mu_c) \qquad (2.5)$$

where v is the feature map from a minibatch of samples, N is the batch size, c is the channel number, ε is learnable parameter, H is the height of the feature map and W is the width. The μ_c is the mean and σ_c is the variance for each feature map.

While the instance normalization is using the statistics of an single sample as an alternative of mini-batch to normalize features as Equation 2.6 [46]. This provides robust appearance invariance:

$$INx_{nchw} = \frac{v_{nchw} - \mu_{nc}}{\sqrt{\sigma_{nc}^2}} \qquad (2.6)$$

$$\mu_{nc} = \frac{1}{HW} \sum_{h=1}^{H} \sum_{w=1}^{W} v_{nchw} \qquad (2.7)$$

$$\sigma_{nc}^2 = \frac{1}{HW} \sum_{h=1}^{H} \sum_{w=1}^{W} (v_{nchw} - \mu_c) \qquad (2.8)$$

Usually, IN layers added to shallow layers while in deep layers BN should be inserted to preserve important content information. In our model instance normalization is added in first layer in first three stages of Resnet-50. The combination between batch normalization and instance normalization enhances the performance of the model by improving the learning and generalization capabilities by keeping the important style while normalizing a confusing one, so appearance variations will be reduced without increasing computational cost.

2.3.2.2   Combined loss and similarity computing

The loss function is a key part in person Re-ID. It helps in matching the same person after extracting features by reducing the distance between the same identity so it will be closer than different identities. At the training stage, the loss functions are mainly used to compute the error of the model. consider two types of losses at the training stage: the softmax with cross-entropy loss ($L_{softmax}$) and the triplet loss ($L_{triplet}$). Cross-entropy loss is simple and efficient loss for classification task. If P is softmax applied probabilities, the cross entropy is defined as Equation 2.9:

$$L_{softmax} = -\sum_{i=1}^{N} y^i log(P^i) \tag{2.9}$$

where $y = 1$ in true class and $y = 0$ if not. Adopting the label smoothing regularization method to modify the cross-entropy loss. Label smoothing changes the training target y for the network to y^{LS} (Equation 2.10) [47]:

$$y^{LS} = (1 - \alpha) * y + \alpha/C \tag{2.10}$$

where C is the number of label classes, and α is a hyperparameter that determines the amount of smoothing. The triplet loss is used as a ranking method. Inspired by the success of triplet loss to solve the person Re-ID in [13], it is adopted in our proposed model with the softmax to leverage the re-identification task.

It keeps the relative relationships between positive and negative pairs [48]. Batch Hard is an effective triplets mining method. It is selecting one hardest positive image and one hardest negative image for every image (anchor) in a batch (Equation 2.11). It shows a state-of-the-art performance in the experiments [49]:

$$L_{triplet} = \sum_{i=1}^{N} [Dist_{max}(f(x)^{ianchor}, f(x)^{ipositive})$$
$$- Dist_{min}(f(x)^{ianchor}, f(x)^{inegative}) + \mu] \tag{2.11}$$

where $Dist_{max}$ is the maximum distance between anchor and positive features vectors, $Dist_{min}$ is the minimum distance between anchor and negative features vectors. μ is a margin value. So, the final loss in our proposed model is (Equation 2.12):

$$L_{CombinedLoss} = L_{softmax} + L_{triplet} \tag{2.12}$$

At the testing stage, the cosine similarity metric is used and return the top ranked list of videos with the smallest distance. The similarity between two images is computed by cosine distance metric. It is measured by the cosine of the angle between two features vectors a and b by maximizing the cosine value for similar pairs and minimizing the cosine value for the different pairs (Equation 2.13):

$$cos\theta(a,b) = \frac{a \cdot b}{\|a\| \cdot \|b\|} = \frac{\sum_{i=1}^{n} a_i b_i}{\sqrt{\sum_{i=1}^{n} a_i} \sqrt{\sum_{i=1}^{n} b_i}} \qquad (2.13)$$

2.4 Experimental Evaluation

2.4.1 Dataset

Two datasets have been used in the proposed model. The first dataset is CASIA-B, it is used for long-term and short-term person Re-ID experiment. The second one is Market-1501, it is used for short-term person Re-ID experiment. All datasets are described in following sub-sections. To evaluate our proposed long-term person Re-ID model we need a video-based dataset with these three conditions:

1. Multiple camera views.

2. Highly appearance changes of each person.

3. A sequence frames for each person.

From the review of aforementioned long-term studies, most of the existing datasets that were used in previous long-term person Re-ID systems did not meet these conditions. As we present in Table 2.1, most of the datasets did not have a multi-appearance per identity, some of them captured from RGB-Depth camera such as Kinect sensor as BIWI RGBD-ID and RGB-D Person REID but this camera is not commonly used in surveillance systems, some of them is for image-based Re-ID or did not have a proper frames sequence to extract GEI such as Celeb-reID, some of them such as Motion-ReID is not for public used. CASIA-B [39, 50] dataset meets these conditions and it is available for used.

2.4.1.1 CASIA-B dataset

It is one of the largest and widespread cross-view datasets for gait analysis and related researches. It consisted of 124 identities at 11 different camera views (from 0 to 180) with 3 different appearances on 10 sequences for each person: carrying bag (BG), wearing coat (CL), and normal appearance without bag or coat (NM). So, for each identity, there are 11 images per sequence with different appearance, (BG, CL, or NM). As a result, this dataset has

TABLE 2.1
Existing datasets used in previous long-term person Re-ID studies.

Dataset	Year	Dataset Type	Camera Type	Identity Number	Multi Appearance	Multi Camera	Camera Number	Is public
CASIA-B [51]	2015	Video	RGB camera	124	√	√	11	√
BIWI RGBD-ID [65]	2014	Video	RGB-D camera	50	√	√	1	√
RGB-D Person REID [66]	2012	Video	RGB-D camera	79	√	√	1	√
Motion-ReID [36]	2018	Video	RGB camera	30	√	√	2	×
Celeb-reID [40]	2019	Image	RGB camera	1,052	√	√	-	√

FIGURE 2.8
Samples from CASIA-B dataset. (a) same person with normal appearance vs wearing coat. (b) same person carrying bag vs wearing coat. (c) same person carrying bag vs normal. (d) same person wearing coat vs normal.

13460 videos, each video has its GEI that it is used in the experiments of re-identification stage in the proposed model. Figure 2.8 shows some samples frames of CASIA-B dataset and the examples of CASIA-B GEI are shown in Figure 2.9.

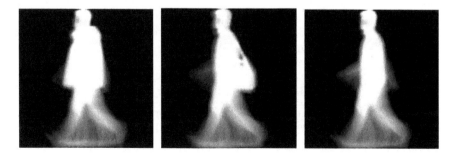

FIGURE 2.9
Samples of GEI from CASIA-B for id 85 at Camera 90°: (a) Person with coat, (b) Person carrying bag, and (c) Person with normal appearance.

CAM-1 CAM-3 CAM-6   CAM-1 CAM-3 CAM-6

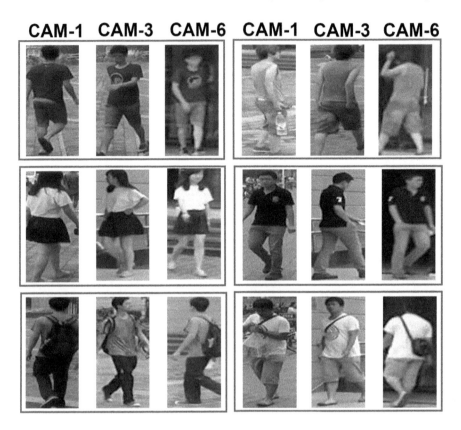

FIGURE 2.10
Samples from the Market-1501 dataset.

2.4.1.2 Market-1501 dataset

The Market1501 dataset [19] is one of the largest and most realistic image-based person Re-ID benchmark, it is used for short-term person Re-ID since each person did not change their appearance such as clothes in all camera views. It is containing 32,643 images of 1,501 identities captured by 6 cameras. Figure 2.10 shows some samples of Market1501.

2.4.2 Training setting

The pytorch package is used to implement our proposed model. All training is done in Lenovo Y720 with 16 RAM and 6GB NVIDIA GeForce GTX 1060. Deep learning is a very iterative process. We have to try out various combinations of the hyperparameters to find which combination has greatest results [51, 52]. Data augmentation is considered as regularization method. It is extensively used in the training of deep-learning models. It is helping in

FIGURE 2.11
Example of random erasing data augmentation [54].

expanding the dataset using several translations, such as flipping, cropping, adding noises, etc. In our model training, all images are resized to 180×180. The random erasing [53] is applied as complementary method to flip images horizontally. Since person in the images are occluded by other objects, Random erasing deal with the occlusion problem as shown in Figure 2.11. Batch size is 64. Adam optimization function is used to optimize the model. The optimization process learns the optimal decision variables of a function or a problem by minimize or maximize its objective function [54]. In model training, the goal is to reach the global minimum error. The initial learning rate is 0.00035 and is decreased by 0.1 at the 20th epoch and 40th epoch. The total training epoch is 100.

2.4.3   Evaluation metric and experimental protocol

For evaluating our proposed Deep LongReID, the top ranks accuracy is reported by Cumulative Matching Characteristic (CMC) curve. The CMC describes the probability of finding the true match at the top r matches where $r \in 1,5,10,20$. In other words, it is recognition precision for each rank r. It is represented as a plot of identification accuracy against the rank rate. For each rank r, the result of query Q is 0 when it is negative and 1 when it is positive, CMC is computed as:

$$CMC(r) = \frac{1}{N} \sum_{i=1}^{N} Q_{ri} \qquad (2.14)$$

where N is the queries number. The second metric for evaluation is the Mean Average Precision (mAP) which it used when considering the person re-identification as a retrieval task. To calculate mAP, first, the Average Precision AP for each query until position N is computed as:

$$AP = \frac{1}{GT} \sum_{i=1}^{N} \frac{TP}{i} \qquad (2.15)$$

where GT is the positive ground truth in the gallery. Then, the mean of all Average Precision, AP, of all queries N is computed as:

$$mAP = \frac{1}{N} \sum_{i=1}^{N} AP_i \qquad (2.16)$$

2.4.4   Results and discussion

Several experiments are done in this section to validate the effectiveness of our proposed model and the selected techniques that we used to build our model.

2.4.4.1   Training and testing the re-identication stage

In this experiment, the existing CASIA-B GEI is used as the input for the re-identification stage. At the training stage, the training setting that defined in previous section 2.4.2 is applied for training our model. At the testing stage, the cosine similarity metric is used and return the top ranked list of videos with the smallest distance.

The effectiveness of our proposed model for long-term person Re-ID: One of the main aims of this study is supporting the long-term person Re-ID. A comprehensive comparison of different appearances with different views is conducted to evaluate the robustness of our model. All experiments in long-term person Re-ID are conducted using in CASIA-B dataset. For this experiment, we define two scenarios:

1. Long-term setting: It considers only one appearance for probe and different appearance in gallery for the same identity and takes all 11 different camera views.

TABLE 2.2

The accuracy % results our proposed model in long-term scenario.

Scenario	Appearance Probe >> Gallery	Top Rank Accuracy			
		Rank-1	*Rank-5*	*Rank-10*	*Rank-20*
Long-term (including the probe camera in gallery)	NM>> CL	69.9	84.5	89.5	94.5
	NM >> BG	92.7	98.3	99.2	99.7
	BG >> CL	64.8	84.7	89.6	94.2
	BG >> NM	93.7	97.7	98.6	99.2
	CL >> NM	71.8	84.8	88.8	92.3
	CL >> BG	66.9	84.2	90.8	95
	NM >> BG, CL	90.2	96.6	98.3	99.4
	BG >> CL, NM	92.7	97.4	98.3	99.2
	CL >> BG, NM	73.0	87.1	90.5	94.6
Long-term Plus (excluding the probe camera from gallery)	NM>> CL	63.5	80.9	86.9	93
	NM >> BG	88.1	97.1	98.6	99.3
	BG >> CL	58.4	80.4	86.8	92.8
	BG >> NM	88.1	95.4	97.7	98.6
	CL >> NM	66.5	80.8	86.1	90.9
	CL >> BG	61	79.7	88.1	93.5
	NM >> BG, CL	85.4	95.1	97.3	99
	BG >> CL, NM	86.9	95	97.1	98.8
	CL >> BG, NM	65.5	83.3	87.5	92.9

2. Long-term setting plus: In addition to different appearance between probe and gallery for the same person, we exclude the camera view that appears in probe from gallery to add more reality to our long period experiment.

On long-term setting: Table 2.2 shows our proposed model achieved from 64.8% to 94.2% of accuracy at rank-1 to rank-20 when person in probe is carrying bag and gallery is the same person with coat. When person in probe is wearing coat and the gallery is the same person carrying bag, it achieved from 66.9% to 95% at rank-1 to rank-20. When a person in probe is carrying bag and gallery is the same person with normal appearance, it achieved from 93.7% to 99.2% at rank-1 to rank-20. When person in probe is wearing coat and gallery is the same person with a normal appearance, it achieved from 71.8% to 92.2% at rank-1 to rank-20. When person in probe is with normal appearance and the same person in the gallery is wearing coat, it achieved from 69.9% to 94.5% at rank-1 to 20. when person in probe is with normal appearance and the same person in the gallery is carrying bag, it achieved from 92.7% to 99.7% at rank-1 to 20.

On Long-term setting Plus: The proposed model achieved from 58.4% to 92.8% at rank-1 to rank-20 when person in probe is carrying bag and the gallery is the same person with coat. When a person in probe is wearing a coat and the gallery is the same person carrying bag, it achieved from 61% to 93.5% at rank-1 to rank-20. When person in probe is carrying bag and gallery

TABLE 2.3
The accuracy % results of our proposed model and baseline resnet-50.

Probe	Model	Gallery					
		NM		CL		BG	
		R1	R5	R1	R5	R1	R5
NM	Baseline Resnet-50	98.1	99.5	61.2	80.5	85.3	95.5
	Resnet-50+ INBN	98.9	99.8	63.5	80.9	88.1	97.1
CL	Baseline Resnet-50	65.6	81.1	96.7	99.8	55.8	78.1
	Resnet-50+ INBN	66.5	80.8	98.7	99.8	61	79.7
BG	Baseline Resnet-50	88.0	94.9	54.3	80.5	96.4	99.4
	Resnet-50+ INBN	88.1	95.4	58.4	80.4	97.3	99.5

TABLE 2.4
The accuracy % results of softmax only and the combined loss functions.

Probe	Loss Function	Gallery					
		NM		CL		BG	
		R1	R5	R1	R5	R1	R5
NM	Softmax	98.1	99.8	46.4	65.6	84.8	95.5
	Combined Loss	98.9	99.8	63.5	80.9	88.1	97.1
CL	Softmax	49.5	66.7	98.6	99.6	43.4	64.7
	Combined Loss	66.5	80.8	98.7	99.8	61	79.7
BG	Softmax	86.7	94.5	44.0	63.3	97.2	99.5
	Combined Loss	88.1	95.4	58.4	80.4	97.3	99.5

is the same person with normal appearance, it achieved from 88.1% to 98.6% at rank-1 to rank-20. When person in probe is wearing a coat and gallery is the same person with a normal appearance, it achieved from 66.5% to 90.9% at rank-1 to rank-20. When person in probe is with normal appearance and the gallery is the same person is wearing a coat, it achieved from 63.5% to 93% at rank-1 to rank-20. When a person in probe is with normal appearance and the same person in the gallery is carrying a bag, it achieved from 88.1% to 99.3% at rank-1 to rank-20.

To evaluate our model, we are now measuring the effectiveness of our setting selection:

The effectiveness of our proposed model comparing by the baseline Resnet-50: For this experiment, we follow the same our training strategy, but we adopted Resnet-50 without instance-batch normalization (INBN) in residual blocks. The results are presented in Table 2.3, we can find our proposed model improve the rank-1 and rank-5 more than Resnet-50 especially in the challenge appearance.

The effectiveness of our proposed model with using combined Triplet-Softmax loss functions: In the most existing available identification and gait recognition models, the softmax loss function is adopted for training deep

TABLE 2.5

The percentage accuracy results of our proposed model with and without using random erase.

Probe	Model	Gallery					
		NM		CL		BG	
		R1	R5	R1	R5	R1	R5
NM	without random erase	98	99.7	47.8	67.5	83.3	95.2
	with random erase	98.9	99.8	63.5	80.9	88.1	97.1
CL	without random erase	48.3	66.7	96.6	99.5	44.9	67.7
	with random erase	66.5	80.8	98.7	99.8	61	79.7
BG	without random erase	84.5	92.2	41.9	62.4	98.4	99.8
	with random erase	88.1	95.4	58.4	80.4	97.3	99.5

model. In this experiment, we follow the same our training strategy by using only softmax loss function and compare its results by the results of the combined Triplet-Softmax loss functions. In Table 2.4, we can show a significant improvement in rank-1 and rank-5 when adopted a triplet loss with softmax.

The effectiveness of our proposed model with and without using random erasing data augmentation: The person in the real world are maybe change their appearance and occluded by other person or objects. To address these problems, the random erasing is introduced as new data augmentation [53]. So, the random erasing is adopted as complementary to Random flipping in our model. For this experiment, we follow the same our training strategy with and without random erasing. Table 2.5 shows the adopting random erasing in our model achieving better results in most cases.

Comparisons with the state-of-the-art systems on CASIA-B dataset: To validate the performance of our approach, we compare with the following state-of-the-arts systems with the same dataset. System in [31] adopted gait features to overcome the limitations of person Re-ID when using only appearance-based features. Gait features and appearance features are integrated using both score-level fusion and feature-level fusion. The results of features-level fusion strategy are presented since is achieved better results than score level fusion in all cases. Systems in [37] selected only relevant body parts for person Re-ID and remove parts that are affected by semantic attributes. Then, P-LBP (partial LBP) was used for features extraction. Their experiments were done by two setting: simple setting with one view 90, and challenging setting with multi-view and multi-appearance. The results of challenging setting are presented since the same view mostly not consider in re-identification problem. System in [55] used new gait template called Patch Gait Feature (PGF) for spatial-temporal features extraction. Then, 1-NN classifier is applied for gait recognition. This work for gait recognition, so we consider only the Rank-1 results in case the person has different appearances. All these three systems are used hand-crafted methods. In Table 2.6, we show the accuracy results at Rank 1, 5, 10, and 20, respectively on CASIA-B for multi person appearance.

TABLE 2.6

The comparison with other state-of-the-art studies. '-' is not reported.

Methods	Re-ID Method	NM >>CL - CL>>NM				NM >>BG - BG >> NM				BG >> CL - CL>>BG				Nm/Bg/Cl - Nm/Bg/Cl			
		R1	R5	R10	R20	R1	R5	R10	R20	R1	R5	R10	R20	R1	R5	R10	R20
Z. Liu et al. 2015 [31]	Hand Crafted Method	20.28	42.64	56.87	75.02	31.81	53.59	64.14	77.03	33.05	54.21	64.66	77.69	25.49	45.098	62.74	78.43
Fendri, Chtourou, and Hammami 2019 [37]	Hand Crafted Method	-	-	-	-	-	-	-	-	-	-	-	-	52.35	78.24	85.49	93.53
Ghaeminia, Shokouhi, and Badiezadeh 2019 [56]	Hand Crafted Method	51.0	-	-	-	69.33	-	-	-	34.5	-	-	-	73.41	-	-	-
Ours	Deep Learning Method	65	80.85	86.5	91.95	88.1	96.25	98.15	98.95	59.7	80.05	87.45	93.15	94.5	98.6	99.2	99.6

Probe >> Gallery

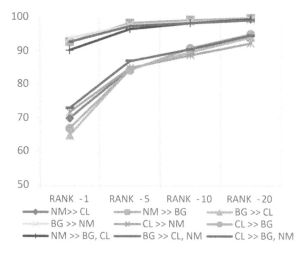

FIGURE 2.12
CMC curve for long-term scenario.

The results of appearance case and its opposite appearance are averaged. Our proposed model achieves the highest performance for rank-1 to rank-20 when there is changes in person appearance on different camera views, that is 59.7% at rank-1, 80.05% at rank-5, 87.45% at rank-10 and 93.15% at rank-20 in case of person wearing coat in probe and carrying bag in gallery and its opposite. It achieved 65% at rank-1, 80.85% at rank-5, 86.5% at rank-10 and 91.95% at rank-20 in case of person wearing coat in probe and with nothing in gallery and its opposite. It achieved 88.1% at rank-1, 96.25% at rank-5, 98.15% at rank-10 and 98.95% at rank-20 in case of person carrying bag in probe and with nothing in gallery and its opposite.

Our model is outperforming most of other state-of-the art methods by achieving about 21% improvement in rank-1 accuracy comparable by the best state-of-art method [55]. Figures 2.12, 2.13 and 2.14 represent the CMC curves for long-term, long-term plus experiments and for the comparison with other recent models.

The effectiveness of our proposed model for short-term person Re-ID: The short-term person Re-ID is a common scenario that was used in common previous works. It used the same appearance for person in probe and gallery but with different view. To validate the robustness of our model, we evaluate it in CASIA-B dataset with using the same appearance and in Market-1501 which it is the largescale short-term person Re-ID. In CASIA-B dataset, as we seen in Table 2.7, our proposed model is implemented when the person in both probe and gallery has same appearance. Rather than most of short-term Re-ID models that depended on extract appearance features such as color and textures, our model depends on gait features that extracted from silhouettes image. In Market-1501 dataset, as shown in Table 2.8, our model is compared to other short-term state-of-the-art models that depend on appearance

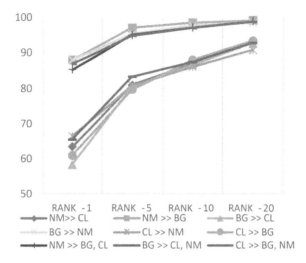

FIGURE 2.13
CMC curve for long-term plus.

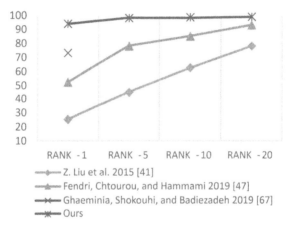

FIGURE 2.14
CMC curve for comparison with other recent models.

features. These systems are image-based person Re-ID and our model also here is image-based and it depends on appearance features.

The effectiveness of our proposed model based on time: In this experiment we do single query on 2152 images in gallery and compute the average of time. Table 2.9 shows the overall time of Resnet-50 with INBN that used in our model comparing by baseline Resnet-50. The baseline Resnet-50 improves the time better than our model by about 0.5 second but our model achieves a significant improvement in accuracy.

TABLE 2.7
The accuracy % results of our proposed model in short-term using CASIA-B dataset.

Scenario	Appearance	Top Rank Accuracy			
	Probe >> Gallery	*R-1*	*R-5*	*R-10*	*R-20*
Short-term (excluding	NM >> NM	98.9	99.8	100	100
the probe camera from	CL >> CL	98.7	99.8	99.9	99.9
gallery)	BG >> BG	97.3	99.6	100	100

TABLE 2.8
The accuracy % results of our proposed model in short-term using market-1501 dataset. "-" is not reported.

Method	Methods Top Rank (%) in Market-1501				
	R1	**R5**	**R10**	**R20**	**mAP**
Zhao et al., 2017 [12]	76.9	91.5	94.6	96.7	-
Li et al., 2017 [57]	80.31	-	-	-	57.53
Hermans, Beyer and Leibe, 2017 [13]	84.92	94.21	-	-	69.14
Wu et al., 2018 [61]	68.3	87.23	94.59	96.71	40.24
He et al., 2019 [14]	84.53	92.66	95.58	-	-
Z. Zheng, Zheng, and Yang 2017 [59]	79.51	-	-	-	59.87
L. Wu, Wang, Li, et al. 2018 [60]	67.15	-	95.67	97.08	-
L. Wu, Yang Wang, Li and Gao, 2019 [61]	64.23	-	-	-	41.36
L. Wu, Hong, Wang, et al. 2019 [62]	89.1	96.8	-	99.7	76.2
M. B. Jamal, J. Zhengang, and F. Ming. 2020* [63]	90.74	-	-	-	76.92
Zhu et al., 2019 * [64]	92.0	-	-	-	78.2
Ours (180×180)	**88.9**	**95.9**	**97.4**	**98.3**	**74.9**
Ours (256×128)	**90.1**	**96.7**	**98**	**96.6**	**78.3**

2.4.4.2   Testing the integrated system

In this experiment, we tested the integrated system using different videos from CASIA-B dataset. First, the video is entered to the preprocessing stage in which the first 20 frames of gait are entered to DeeplabV3 to detect the person in each frame and extract the silhouette image. Then, the GEI is calculated. The GEI is used as input to re-identification stage. The average of identification accuracy is ranging from 50% in rank-1 to 100% in rank-20. Figure 2.15 presents the example of the output of the integrated system.

TABLE 2.9

The results of consuming time for using resnet-50 with INBN comparing by baseline resnet-50.

Process	Our Resnet-50+ INBN	Baseline Resnet-50
Features Extraction from Query ≈	9.231	8.999
Features Extraction from Gallery ≈	18.402	18.08
Computing the distance ≈	0.038	0.038
Overall time ≈	27.671	27.117

(a)

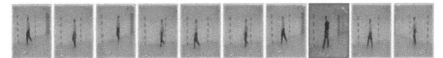

(b)

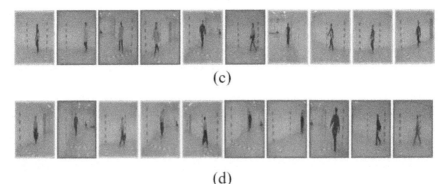

(c)

(d)

FIGURE 2.15

Example of top-ranked list of the integrated system: (a) Sample frame from query video, (b) The output of CL-NM, (c) The output of CL-CL, and (d) The output of CL-BG. (random frame is presented from output video).

2.5 Conclusion

Person re-identification is one of the important and critical tasks within intelligent video surveillance systems and it is still a challenging task. Until to date, most of the existing studies on person re-identification support only the short-term situation when the persons keep the same appearance in the different cameras over a short time, while the long-term situation when the persons change their appearance over a long time remains needed to a deep study. We proposed a model based on deep learning to solve the problem of long-term person re-identification by extracting discriminative features from human gait. Our proposed model used Resnet-50 with carefully adopting instance normalization with batch normalization to handle the appearance variations for the same person. A combination between softmax and triplet loss functions is used for training the model. CASIA-B dataset is used to evaluate and validate our proposed model in the long-term and short-term person Re-ID experiments and also the Market-1501 dataset is used for the short-term person Re-ID experiments. Comprehensive experiments that consider different appearance for each identity show the superior performance is achieved in both short-term and long-term person Re-ID as compared to other state-of-the-art methods. As future works for our research, since there is still no long-term video dataset, one of our future works is building a long-term person Re-ID dataset with multi-appearnces and multi-cameras that have a proper sequence of frames for each identity and then evaluate the segmentation stage and the re-identification stage in our proposed model. Also, trying another deep-learning segmentation models such as Mask R-CNN to extract person silhouette images. Finally, adopting the proposed instance-batch normalization into another deep-learning models and study its effect.

Bibliography

[1] Imed Bouchrika. On using gait biometrics for re-identification in automated visual surveillance. In *Developing Next-Generation Countermeasures for Homeland Security Threat Prevention*, pages 140–163. IGI Global, 2017.

[2] Di Wu, Si-Jia Zheng, Xiao-Ping Zhang, Chang-An Yuan, Fei Cheng, Yang Zhao, Yong-Jun Lin, Zhong-Qiu Zhao, Yong-Li Jiang, and De-Shuang Huang. Deep learning-based methods for person re-identification: A comprehensive review. *Neurocomputing*, 337:354–371, 2019.

[3] S Karanam, M Gou, Z Wu, A Rates-Borras, O Camps, and RJ Radke. A systematic evaluation and benchmark for person re-identification: features, metrics, and datasets. *IEEE Transactions on Pattern Analysis and Machine Intelligence*.

[4] Liang Zheng, Yi Yang, and Alexander G Hauptmann. Person re-identification: Past, present and future. *arXiv preprint arXiv:1610.02984*, 2016.

[5] Shaogang Gong, Marco Cristani, Chen Change Loy, and Timothy M Hospedales. The re-identification challenge. In *Person re-identification*, pages 1–20. Springer, 2014.

[6] Muna O Almasawa, Lamiaa A Elrefaei, and Kawthar Moria. A survey on deep learning-based person re-identification systems. *IEEE Access*, 7:175228–175247, 2019.

[7] Athira Nambiar, Alexandre Bernardino, and Jacinto C Nascimento. Gait-based person re-identification: A survey. *ACM Computing Surveys (CSUR)*, 52(2):1–34, 2019.

[8] Qingming Leng, Mang Ye, and Qi Tian. A survey of open-world person re-identification. *IEEE Transactions on Circuits and Systems for Video Technology*, 30(4):1092–1108, 2019.

[9] Changsheng Wan, Li Wang, and Vir V Phoha. A survey on gait recognition. *ACM Computing Surveys (CSUR)*, 51(5):1–35, 2018.

[10] Ejaz Ahmed, Michael Jones, and Tim K Marks. An improved deep learning architecture for person re-identification. In *Proceedings of the IEEE Conference on Computer Vision and Pattern Recognition*, pages 3908–3916, 2015.

[11] Yan Huang, Hao Sheng, Yanwei Zheng, and Zhang Xiong. Deepdiff: Learning deep difference features on human body parts for person re-identification. *Neurocomputing*, 241:191–203, 2017.

[12] Haiyu Zhao, Maoqing Tian, Shuyang Sun, Jing Shao, Junjie Yan, Shuai Yi, Xiaogang Wang, and Xiaoou Tang. Spindle net: Person re-identification with human body region guided feature decomposition and fusion. In *Proceedings of the IEEE Conference on Computer Vision and Pattern Recognition*, pages 1077–1085, 2017.

[13] Alexander Hermans, Lucas Beyer, and Bastian Leibe. In defense of the triplet loss for person re-identification. *arXiv preprint arXiv:1703.07737*, 2017.

[14] Zhangping He, Cheolkon Jung, Qingtao Fu, and Zhendong Zhang. Deep feature embedding learning for person re-identification based on lifted structured loss. *Multimedia Tools and Applications*, 78(5):5863–5880, 2019.

[15] Weihua Chen, Xiaotang Chen, Jianguo Zhang, and Kaiqi Huang. Beyond triplet loss: a deep quadruplet network for person re-identification. In *Proceedings of the IEEE Conference on Computer Vision and Pattern Recognition*, pages 403–412, 2017.

[16] Douglas Gray, Shane Brennan, and Hai Tao. Evaluating appearance models for recognition, reacquisition, and tracking. In *Proceedings of the IEEE International Workshop on Performance Evaluation for Tracking and Surveillance (PETS)*, volume 3, pages 1–7. Citeseer, 2007.

[17] Wei Li, Rui Zhao, and Xiaogang Wang. Human reidentification with transferred metric learning. In *Asian Conference on Computer Vision*, pages 31–44. Springer, 2012.

[18] Wei Li, Rui Zhao, Tong Xiao, and Xiaogang Wang. Deepreid: Deep filter pairing neural network for person re-identification. In *Proceedings of the IEEE Conference on Computer Vision and Pattern Recognition*, pages 152–159, 2014.

[19] Liang Zheng, Liyue Shen, Lu Tian, Shengjin Wang, Jingdong Wang, and Qi Tian. Scalable person re-identification: A benchmark. In *Proceedings of the IEEE International Conference on Computer Vision*, pages 1116–1124, 2015.

[20] Hao Liu, Zequn Jie, Karlekar Jayashree, Meibin Qi, Jianguo Jiang, Shuicheng Yan, and Jiashi Feng. Video-based person re-identification with accumulative motion context. *IEEE Transactions on Circuits and Systems for Video Technology*, 28(10):2788–2802, 2018.

[21] Zhen Zhou, Yan Huang, Wei Wang, Liang Wang, and Tieniu Tan. See the forest for the trees: Joint spatial and temporal recurrent neural networks for video-based person re-identification. In *Proceedings of the IEEE Conference on Computer Vision and Pattern Recognition*, pages 4747–4756, 2017.

[22] Lin Wu, Yang Wang, Junbin Gao, and Xue Li. Where-and-when to look: Deep siamese attention networks for video-based person re-identification. *IEEE Transactions on Multimedia*, 21(6):1412–1424, 2018.

[23] Jiawei Liu, Zheng-Jun Zha, Xuejin Chen, Zilei Wang, and Yongdong Zhang. Dense 3d-convolutional neural network for person re-identification in videos. *ACM Transactions on Multimedia Computing, Communications, and Applications (TOMM)*, 15(1s):1–19, 2019.

[24] Lin Wu, Yang Wang, Ling Shao, and Meng Wang. 3-d personvlad: Learning deep global representations for video-based person reidentification. *IEEE Transactions on Neural Networks and Learning Systems*, 30(11):3347–3359, 2019.

[25] Martin Hirzer, Csaba Beleznai, Peter M Roth, and Horst Bischof. Person re-identification by descriptive and discriminative classification. In *Scandinavian Conference on Image Analysis*, pages 91–102. Springer, 2011.

[26] Taiqing Wang, Shaogang Gong, Xiatian Zhu, and Shengjin Wang. Person re-identification by video ranking. In *European Conference on Computer Vision*, pages 688–703. Springer, 2014.

[27] Liang Zheng, Zhi Bie, Yifan Sun, Jingdong Wang, Chi Su, Shengjin Wang, and Qi Tian. Mars: A video benchmark for large-scale person re-identification. In *European Conference on Computer Vision*, pages 868–884. Springer, 2016.

[28] Liangliang Ren, Jiwen Lu, Jianjiang Feng, and Jie Zhou. Multi-modal uniform deep learning for rgb-d person re-identification. *Pattern Recognition*, 72:446–457, 2017.

[29] Ancong Wu, Wei-Shi Zheng, Hong-Xing Yu, Shaogang Gong, and Jianhuang Lai. Rgb-infrared cross-modality person re-identification. In *Proceedings of the IEEE International Conference on Computer Vision*, pages 5380–5389, 2017.

[30] Federico Pala, Riccardo Satta, Giorgio Fumera, and Fabio Roli. Multimodal person reidentification using rgb-d cameras. *IEEE Transactions on Circuits and Systems for Video Technology*, 26(4):788–799, 2015.

[31] Zheng Liu, Zhaoxiang Zhang, Qiang Wu, and Yunhong Wang. Enhancing person re-identification by integrating gait biometric. *Neurocomputing*, 168:1144–1156, 2015.

[32] Le An, Xiaojing Chen, Shuang Liu, Yinjie Lei, and Songfan Yang. Integrating appearance features and soft biometrics for person re-identification. *Multimedia Tools and Applications*, 76(9):12117–12131, 2017.

[33] Athira Nambiar, Jacinto C Nascimento, Alexandre Bernardino, and José Santos-Victor. Person re-identification in frontal gait sequences via histogram of optic flow energy image. In *International Conference on Advanced Concepts for Intelligent Vision Systems*, pages 250–262. Springer, 2016.

[34] Mohamed Hasan and Noborou Babaguchi. Long-term people reidentification using anthropometric signature. In *2016 IEEE 8th International Conference on Biometrics Theory, Applications and Systems (BTAS)*, pages 1–6. IEEE, 2016.

[35] Athira Nambiar, Alexandre Bernardino, Jacinto C Nascimento, and Ana Fred. Context-aware person re-identification in the wild via fusion of gait and anthropometric features. In *2017 12th IEEE International Conference on Automatic Face & Gesture Recognition (FG 2017)*, pages 973–980. IEEE, 2017.

[36] Peng Zhang, Qiang Wu, Jingsong Xu, and Jian Zhang. Long-term person re-identification using true motion from videos. In *2018 IEEE Winter Conference on Applications of Computer Vision (WACV)*, pages 494–502. IEEE, 2018.

[37] Emna Fendri, Imen Chtourou, and Mohamed Hammami. Gait-based person re-identification under covariate factors. *Pattern Analysis and Applications*, 22(4):1629–1642, 2019.

[38] Cassandra Carley, Ergys Ristani, and Carlo Tomasi. Person re-identification from gait using an autocorrelation network. In *Proceedings of the IEEE/CVF Conference on Computer Vision and Pattern Recognition Workshops*, pages 0–0, 2019.

[39] Shuai Zheng, Junge Zhang, Kaiqi Huang, Ran He, and Tieniu Tan. Robust view transformation model for gait recognition. In *2011 18th IEEE International Conference on Image Processing*, pages 2073–2076. IEEE, 2011.

[40] Yan Huang, Jingsong Xu, Qiang Wu, Yi Zhong, Peng Zhang, and Zhaoxiang Zhang. Beyond scalar neuron: Adopting vector-neuron capsules for long-term person re-identification. *IEEE Transactions on Circuits and Systems for Video Technology*, 30(10):3459–3471, 2019.

[41] Xuelin Qian, Wenxuan Wang, Li Zhang, Fangrui Zhu, Yanwei Fu, Tao Xiang, Yu-Gang Jiang, and Xiangyang Xue. Long-term cloth-changing person re-identification. In *Proceedings of the Asian Conference on Computer Vision*, 2020.

[42] Yu-Jhe Li, Xinshuo Weng, and Kris M Kitani. Learning shape representations for person re-identification under clothing change. In *Proceedings of the IEEE/CVF Winter Conference on Applications of Computer Vision*, pages 2432–2441, 2021.

[43] Xin Jin, Tianyu He, Kecheng Zheng, Zhiheng Yin, Xu Shen, Zhen Huang, Ruoyu Feng, Jianqiang Huang, Xian-Sheng Hua, and Zhibo Chen. Cloth-changing person re-identification from a single image with gait prediction and regularization. *arXiv preprint arXiv:2103.15537*, 2021.

[44] Swarnendu Ghosh, Nibaran Das, Ishita Das, and Ujjwal Maulik. Understanding deep learning techniques for image segmentation. *ACM Computing Surveys (CSUR)*, 52(4):1–35, 2019.

[45] Liang-Chieh Chen, George Papandreou, Florian Schroff, and Hartwig Adam. Rethinking atrous convolution for semantic image segmentation. *arXiv preprint arXiv:1706.05587*, 2017.

[46] Hyeonseob Nam and Hyo-Eun Kim. Batch-instance normalization for adaptively style-invariant neural networks. *Advances in Neural Information Processing Systems*, 31, 2018.

[47] Rafael Müller, Simon Kornblith, and Geoffrey E Hinton. When does label smoothing help? *Advances in Neural Information Processing Systems*, 32, 2019.

[48] Yuxin Wu and Kaiming He. Group normalization. In *Proceedings of the European conference on computer vision (ECCV)*, pages 3–19, 2018.

[49] Swathi Jamjala Narayanan, Boominathan Perumal, Sangeetha Saman, and Aditya Pratap Singh. Deep learning for person re-identification in surveillance videos. In *Deep Learning: Algorithms and Applications*, pages 263–297. Springer, 2020.

[50] Caihong Yuan, Jingjuan Guo, Ping Feng, Zhiqiang Zhao, Yihao Luo, Chunyan Xu, Tianjiang Wang, and Kui Duan. Learning deep embedding with mini-cluster loss for person re-identification. *Multimedia Tools and Applications*, 78(15):21145–21166, 2019.

[51] Shiqi Yu, Daoliang Tan, and Tieniu Tan. A framework for evaluating the effect of view angle, clothing and carrying condition on gait recognition. In *18th International Conference on Pattern Recognition (ICPR'06)*, volume 4, pages 441–444. IEEE, 2006.

[52] Hao Luo, Wei Jiang, Youzhi Gu, Fuxu Liu, Xingyu Liao, Shenqi Lai, and Jianyang Gu. A strong baseline and batch normalization neck for deep person re-identification. *IEEE Transactions on Multimedia*, 22(10):2597–2609, 2019.

[53] Kaiyang Zhou, Yongxin Yang, Andrea Cavallaro, and Tao Xiang. Learning generalisable omni-scale representations for person re-identification. *IEEE Transactions on Pattern Analysis and Machine Intelligence*, 2021.

[54] Zhun Zhong, Liang Zheng, Guoliang Kang, Shaozi Li, and Yi Yang. Random erasing data augmentation. In *Proceedings of the AAAI Conference on Artificial Intelligence*, volume 34, pages 13001–13008, 2020.

[55] Laith Abualigah, Ali Diabat, Seyedali Mirjalili, Mohamed Abd Elaziz, and Amir H Gandomi. The arithmetic optimization algorithm. *Computer Methods in Applied Mechanics and Engineering*, 376:113609, 2021.

[56] Mohammad H Ghaeminia, Shahriar B Shokouhi, and Ali Badiezadeh. A new spatio-temporal patch-based feature template for effective gait recognition. *Multimedia Tools and Applications*, 79(1):713–736, 2020.

[57] Dangwei Li, Xiaotang Chen, Zhang Zhang, and Kaiqi Huang. Learning deep context-aware features over body and latent parts for person re-identification. In *Proceedings of the IEEE conference on computer vision and pattern recognition*, pages 384–393, 2017.

[58] Lin Wu, Yang Wang, Junbin Gao, and Xue Li. Deep adaptive feature embedding with local sample distributions for person re-identification. *Pattern Recognition*, 73:275–288, 2018.

[59] Zhedong Zheng, Liang Zheng, and Yi Yang. A discriminatively learned cnn embedding for person reidentification. *ACM transactions on multimedia computing, communications, and applications (TOMM)*, 14(1):1–20, 2017.

[60] Lin Wu, Yang Wang, Xue Li, and Junbin Gao. What-and-where to match: Deep spatially multiplicative integration networks for person re-identification. *Pattern Recognition*, 76:727–738, 2018.

[61] Lin Wu, Yang Wang, Xue Li, and Junbin Gao. Deep attention-based spatially recursive networks for fine-grained visual recognition. *IEEE transactions on cybernetics*, 49(5):1791–1802, 2018.

[62] Lin Wu, Richang Hong, Yang Wang, and Meng Wang. Cross-entropy adversarial view adaptation for person re-identification. *IEEE Transactions on Circuits and Systems for Video Technology*, 30(7):2081–2092, 2019.

[63] Miftah Bedru Jamal, Jiang Zhengang, and Fang Ming. An improved deep mutual-attention learning model for person re-identification. *Symmetry*, 12(3):358, 2020.

[64] Xierong Zhu, Jiawei Liu, Hongtao Xie, and Zheng-Jun Zha. Adaptive alignment network for person re-identification. In *International Conference on Multimedia Modeling*, pages 16–27. Springer, 2019.

[65] Matteo Munaro, Andrea Fossati, Alberto Basso, Emanuele Menegatti, and Luc Van Gool. One-shot person re-identification with a consumer depth camera. In *Person Re-Identification*, pages 161–181. Springer, 2014.

[66] Igor Barros Barbosa, Marco Cristani, Alessio Del Bue, Loris Bazzani, and Vittorio Murino. Re-identification with rgb-d sensors. In *European Conference on Computer Vision*, pages 433–442. Springer, 2012.

3

A Comprehensive Review of Crowd Behavior and Social Group Analysis Techniques in Smart Surveillance

Elizabeth B. Varghese

Center for Research and Innovation in Cyber Threat Resilience (CRICTR), Kerala University of Digital Sciences, Innovation and Technology (KUDSIT), Trivandrum, Kerala, India,
Department of Computer Science & Engineering, Mar Baselios College of Engineering and Technology, Trivandrum, Kerala, India

Sabu M. Thampi

Center for Research and Innovation in Cyber Threat Resilience (CRICTR), Kerala University of Digital Sciences, Innovation and Technology (KUDSIT), Trivandrum, Kerala, India

CONTENTS

3.1 Introduction 58
3.2 Visual-Based Approaches 61
    3.2.1 Motion patterns 61
        3.2.1.1 Probabilistic models 61
        3.2.1.2 Optical flow 62
        3.2.1.3 Spatiotemporal descriptors 63
    3.2.2 Low-level patterns 63
        3.2.2.1 Histogram of gradient (HOG) 63
        3.2.2.2 Histogram of tracklets (HOT) 64
        3.2.2.3 Discrete cosine transform 64
        3.2.2.4 Spatiotemporal texture, graph 65
    3.2.3 Physics based 65
        3.2.3.1 Social force and particle dynamics 65
        3.2.3.2 Particle force and particle advection 66
        3.2.3.3 Fluid dynamics, energy models 66
    3.2.4 Cognitive based 66
        3.2.4.1 Machine learning 66
        3.2.4.2 Deep learning 68
        3.2.4.3 Fuzzy logic 69
        3.2.4.4 Psychological models 69

DOI: 10.1201/9781003053262-3

              3.2.4.5   Visual attention and saliency   69
3.3   Multimodal Approaches   70
       3.3.1   Video and sensor data   71
       3.3.2   Wearable sensors   71
3.4   Social Group Analysis   71
       3.4.1   Clustering   72
       3.4.2   Decision tree   72
       3.4.3   Friends-formation (F-formation)   72
3.5   Open Research Issues   73
3.6   Conclusion and Future Directions   74

3.1 Introduction

For the past two decades, the world has been experiencing massive urban growth, leading to an increase in the urban population. Such urbanization and population growth include the need for an effective and efficient monitoring system that monitors the activities and behaviors of objects and individuals in public places and provides security and protection for individuals and their property. The use of video cameras for surveillance and monitoring in government agencies and private institutions around the world is a conventional practice. Most existing monitoring systems require a human operator for continuous monitoring. Their effectiveness and response are mainly determined by the vigilance of the person observing the camera system [1]. In addition, the number of cameras and the surveillance area is limited due to the scarcity of people available. To overcome these limitations, traditional surveillance methods are being phased out using smart surveillance systems. In smart surveillance systems, the system becomes smart by analyzing the video intelligently and adding additional cognitive knowledge to the connected cameras so that they can make their decisions [1]. In short, smart surveillance depends on a better understanding of the captured data. Hence, current smart surveillance systems can be broadly classified into two based on the data analyzed as shown below:

- Visual Based: Video data from distributed cameras and unmanned aerial vehicles (UAV) are intelligently analyzed using computer vision algorithms to make real-time decisions. For example, IBM's Intelligent traffic monitoring system to detect accidents and produce emergency alerts, provide traffic management through optimization of routes, automatic license plate recognition, and face recognition [2].

- Multimodal: With the advent of the Industry 4.0 technologies such as Internet of Things (IoT) and Information and Communication Technologies (ICT), data from a wide range of sensors such as motion sensors, accelerometers, and audio sensors are also utilized along with video data for real-time information capturing. For example, Intel along with Hitachi

FIGURE 3.1
Common applications of smart surveillance.

for the detection of crime scenarios such as detection of gunshots and radioactive isotopes, and recognition of license plates [3]

Thus, the wide range of smart surveillance applications includes license plate recognition, person and object recognition, face recognition, etc., and some of them are shown in Figure 3.1. Even in the current pandemic situation, smart surveillance helps a lot to monitor people and their activities. Among the wide range of smart surveillance applications, one of the most complex and significant areas is to monitor the crowd and their activities. For instance, in events like concerts or public places such as pilgrimage spots or airports, where there are chances for unexpected catastrophes and crime-related incidents, pedestrian crowd monitoring, and crowd disaster prevention systems are necessary [1]. The increased rate of crime, terrorism, and disasters in crowded places has also entailed the need for robust algorithms for effective crowd management.

In a surveillance system, it is inevitable to analyze crowd behavior as it helps understand the crowd dynamics, develops crowd management systems, and reduces crowd-related disasters. The behavior of the crowd usually depends upon the circumstances in which the crowd is involved [4]. For example, consider the scenario of the main street that consists of a variety of shops. The people walk side by side resulting in an upward stream and downstream and

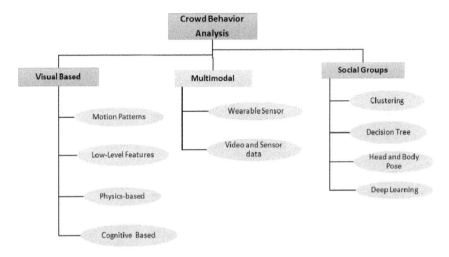

FIGURE 3.2
Taxonomy of crowd behavior analysis research.

enter the shops based on their preference. Most of them walk with relatives or friends, and some may walk alone. From a distant view, the street appears to be crowded by exhibiting a calm behavior. On the other hand, consider the context of a cricket match stadium. Fans and supporters gather in the stadium and gesture to show their support for their favorite players. When the game starts, the gaze of the crowd is in the direction of the pitch where the match happens. The reactions during their favorite team bag boundary are clapping, singing, and chanting. So this scenario marks the crowd behavior as either cheerful or excited. In this way, the crowd exhibits a variety of behaviors, including those in emergency conditions (panic behavior), riot situations (aggressive behavior), and so on. Hence monitoring crowd in smart surveillance needs the analysis of these types of diverse crowd behaviors.

Moreover, analysis and detection of social groups are indispensable in smart surveillance as it is a dominant component of a crowd. The socio-psychological aspects of humans having a social relationship with each other to form groups in a crowd may change the behavior of the crowd. Social gatherings such as free-standing conversation groups (FCG) also contribute to crowd behavior. Hence this chapter intends to uncover the various visual-based and multimodal techniques for analyzing crowd behavior as well as social groups. The taxonomy shown in Figure 3.2 portrays the areas of crowd behavior analysis research considered in this review.

The remaining sections are structured as Section 3.2 deals with computer vision-based approaches for analyzing crowd behavior. Multimodal techniques for crowd behavior and social group analysis are discussed in Sections 3.3 and 3.4. A detailed outlook of the open research issues is presented in Section 3.5. Finally, Section 3.6 portrayed the conclusion and future directions.

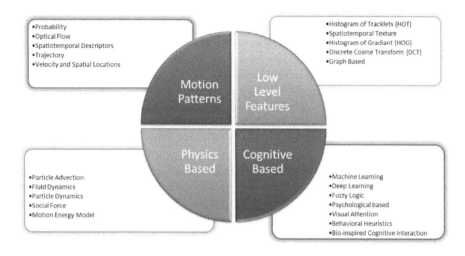

FIGURE 3.3
Classification of visual-based approaches.

3.2 Visual-Based Approaches

Video data analysis for crowd behavior detection is essential in smart surveillance because video cameras and visual sensors are integral components of a surveillance system. Several works have emerged in this area, and as shown in Figure 3.3, we have categorized them into four main groups based on motion patterns, low-level features, physics-based, and cognitive-based.

3.2.1 Motion patterns

Motion patterns are a rich source of information for performing visual tasks such as shape acquisition, recognition of objects, and scene understanding. To capture the information for crowd behavior analysis from video sequences, most of the works in the literature extracted the motion patterns from consecutive video frames. This section deals with different approaches based on the analysis of crowd motion patterns using probability, optical flow, spatiotemporal descriptors, trajectory, and movement velocity.

3.2.1.1 Probabilistic models

Early crowd behavior analysis researchers modeled crowd movement patterns using probabilistic approaches such as the Bayesian and Hidden Markov Model (HMM). Bayesian is based on the theorem, namely Bayes', which calculates the probability of an occurrence based on the conditional probability of the prior knowledge of that occurrence. On the other hand, HMM employed a

Markov property with hidden states and a transition from one state to another. They were used to analyze the interaction between synthetic agents [5] and detect crowd scenarios such as open-space fights and robbery [6]. The Bayesian framework proposed by Wu et al. [7] modeled crowd motions to detect escape events. Here, the destinations and divergent centers in escape and normal situations were detected using conditional probability from optical flow patterns. Also, Wang et al. [8] detected the activities and interactions among individuals in a crowd using the Bayesian model variants such as Hierarchical Dirichlet Processes (HDP), Dual HDP, and Latent Dirichlet Allocation (LDA).

The motion characteristics were extracted using a hierarchical Bayesian model by Rodriguez et al. [9] for crowd anomaly detection. Recently, Zitouni et al. [10] proposed a Gaussian Mixture Model (GMM) to detect individual crowd behaviors from video data. GMM is a probabilistic model for representing data as a mixture of Gaussian. In [10], the salient crowd motion patterns within a sequence of the video were modeled using GMM. Most of the above-discussed probabilistic approaches were aimed at identifying synthetic agent behaviors that fail in complex and uncertain real-time environments. Furthermore, all the methods were based on the movement patterns of pixel positions to find the interaction probability among individuals in a crowd which is feasible only in low-density crowd scenarios.

3.2.1.2   Optical flow

Most of the studies in the literature exploit motion patterns in the form of optical flow for crowd behavior analysis. Optical flow aims to find the movement patterns of objects in a site as a relative motion between a viewer and the viewed scene. [1]. A flow descriptor was generated from optical flow velocities by Hassner et al. [11] to differentiate violent and non-violent crowd scenes. The descriptor was a histogram that represents the change in the magnitude of flow velocities. Histogram of Optical Flow (HOF) was also employed in [12] to classify anomalous events from video data. In contrast, Rao et al. [13] categorized crowded events such as running and walking using optical flow vector lengths as optical flow manifolds. Pennisi et al. [14] utilized Kanade-Lucas-Tomasi (KLT), the most common optical flow method to segment moving blobs for crowd behavior detection. KLT extracts a displacement vector h from visual signals and minimizes their difference between consecutive frames within a region R. The region-wise optical flow patterns were also used by Gnouma et al. [15] to create the DMOF (Distribution of Magnitude of Optical Flow) descriptor for detecting anomalies in crowd events. Since all of these methods calculate optical flow per pixel in a frame, they are considered to be efficient for extracting crowd movement patterns. However, they are very sensitive to brightness change, which makes it difficult to find a threshold to distinguish between foreground and background pixels.

3.2.1.3 Spatiotemporal descriptors

A few methods utilized the motion information of video data as spatiotemporal descriptors for crowd behavior analysis. In [16], local motion features were extracted in the form of spatial and temporal gradients. Initially, the Sobel edge detector was utilized to calculate spatial edge, and then gradients were computed by considering the difference in time between the moving average of the edge image and the current spatial edge image. This method works well in low-resolution scenarios but is not as good if there is noise. Another method developed by Balasundaram et al. [17] employed the space and time volume from consecutive video frames as the spatiotemporal descriptor. Absolute temporal derivatives of all pixels with their depth values created a spatiotemporal volume from one frame. Moreover, velocity and direction variations of pixels were considered for creating local and mid-level visual descriptors in [18] and for trajectory generation in [19] and [20]. As the motion analysis of velocity and trajectory extraction methods focuses on each subject individually, they are applicable for situations with few moving objects.

In a nutshell, motion patterns from video data are largely explored for crowd behavior detection in a wide variety of ways. One common shortcoming among all these approaches is that these methods only consider low-density crowd scenarios that may fail when crowd density increases. Besides, the fact that some approaches ignore spatial properties and consider only temporal information has led to misinterpretations of crowd behavior. Most of the methods used background subtraction methods, and as a result, each set of input videos needs to create a new background model each time, which takes time to distinguish the foreground in varying illumination and noise. Table 3.1 shows the summary of motion pattern methods from the literature.

3.2.2 Low-level patterns

Low-level features are specific characteristics that help to classify events from videos. The features specific to crowd videos are pixel gradient descriptors, edge pixels, trajectory, blobs, location of key points that can identify humans, spatiotemporal texture patterns, tracklets, etc. This section discusses techniques in the literature that employ low-level features for crowd anomaly detection and are summarized in Table 3.2.

3.2.2.1 Histogram of gradient (HOG)

HOG is usually used as a low-level feature extraction method [20, 23]. In HOG, the video frames are divided into spatiotemporal patches. Each patch is further divided into cells for the calculation of histogram descriptors. The histograms are normalized and put together to get the final descriptor and the edge orientations as histogram bins. HOG features are normally used to detect humans in surveillance systems. But in highly occluded scenarios, HOG tends to fail as it detects non-humans as humans.

TABLE 3.1

Motion-pattern methods.

Type	Authors	Description	Remarks
Probabilistic models	Oliver et al. [5], 2000	Interaction between Synthetic Agents	Identify synthetic agent behaviors. Fail in complex and uncertain real-time environments.
	Wang et al. [8], 2008	Model & detect interactions and activities among individuals	Based on the movement pattern of pixel positions.
	Drews et al. [6], 2010	Detect robbery and open space fights	Find interaction probability among individuals.
	Rodriguez et al. [9], 2011	Detect crowd anomalies	Feasible only in low-density crowd scenarios.
	Wu et al. [7], 2013	Crowd motion modeling for escape event detection	
	Zitouni et al. [10], 2020	Estimate salient crowd motion patterns	
Optical flow	Hassner et al. [11], 2012	Differentiate violent and non-violent crowd scenes	Calculate optical flow per pixel. Computationally complex.
	Rao et al. [13], 2015	Classify walking and running using optical flow manifolds	Sensitive to brightness change. Difficult to find threshold to distinguish between foreground and background pixels.
	Colque et al. [12], 2016	Classify anomalous crowd events using HOF	Fail when crowd density increases. Consider only temporal information.
	Pennisi et al. [14], 2016	Use KLT to segment moving blobs for crowd behavior detection	
	Gnouma et al. [15], 2018	DMOF descriptor for crowd anomaly detection	
Spatiotemporal descriptors	O'Gorman et al. [16], 2014	Spatial and temporal gradients	Works well in low resolution scenarios.
	Fradi et al. [18], 2016	Velocity and direction variation of pixels	Not good when noise is present. Focus on each object individually.
	Almeida et al. [21], 2016	Trajectory generated from velocity direction variation of pixels	Applicable to situation with a few moving objects.
	Rabiee at al. [22], 2016		
	Balasundaram et al. [17], 2018	Space and time volume from consecutive video frames	

3.2.2.2 Histogram of tracklets (HOT)

Another common low-level feature vector suggested in [24] used the Oriented Tracklets Histogram (HOT) to analyze the movement of rigid objects in a scene. Tracklets are the trajectory representation or sequence of points extracted from moving objects. Tracklets were computed by dividing the input video frames into cuboids, and for each temporal window magnitude and orientation were calculated. The histogram depicts the frequency of the magnitude-orientation pair. Since HOT encodes the magnitude-orientation pairs, it efficiently represents the crowd motions But, in the presence of occlusion, it suffers tracking drift resulting in the incorrect classification of abnormal motion.

3.2.2.3 Discrete cosine transform

The method proposed by Yuan et al. [25] employed a 3D-Discrete Cosine Transform (3D-DCT) as the feature extraction mechanism for crowd anomaly detection. A set of consecutive video frames were taken as 3D volume to calculate DCT coefficients. The frames were reconstructed considering only low-frequency DCT components, ignoring the high-frequency values. Multiple objects were identified by examining the likelihood and minimizing the reconstruction error. Although multiple humans can be identified from a single scene, this method may fail in noisy and high-density situations.

TABLE 3.2

Low-level feature and physics-based methods.

Type	Authors	Description	Remarks
HOG	Rabiee et al. [22], 2016 Amraee et al. [23], 2018	Appearance and motion extracted using HOG for crowd behavior classification	Used to detect humans. Detect non-humans as humans in high-density occluded scenarios.
HOT	Mousavi et al. [24], 2015	Analyze movement of rigid objects	Efficiently represent crowd motions. Tracking drift in occlusion.
DCT	Yuan et al. [25], 2014	Multiple Object Detection	Fail in noisy and high-density scenarios.
Texture	Wang et al. [26], 2016	Texture patterns from spatiotemporal volume for anomaly detection.	Insufficient to define all high-level crowd behaviors.
Graph	Chaker et al. [27], 2017	Form social network graph from tracklets of obje [26]cts.	
Social force	Mehran et al. [28], 2009	Each object treated as particle and interaction between them are simulated as forces	Require prior knowledge of the scenario. Only consider the temporal characteristics of data.
	Yang et al. [29], 2012	Histogram of Oriented Pressure calculated from social force model	
Particle force	Singh et al. [30], 2009	Repulsive forces between particles in the form of a Discrete Element Model.	Applicable for situations with few moving objects.
	Yuan et al. [25], 2014 Chen et al. [31], 2018	Structural context descriptor created from inter-force particles	
Fluid dynamics	Su et al. [32], 2013	Crowd as flowing fluids	When density of crowd increases, particle identification, and tracking become a challenge.
	Xue et al. [33], 2017	Lattice Boltzmann Model for crowd particle collision. Identify velocity by particle direction.	

3.2.2.4 Spatiotemporal texture, graph

Certain studies extract the spatiotemporal texture patterns [26] and social network graphs [27] for abnormal crowd classification. The former was based on a spatiotemporal volume to extract common spatiotemporal textures. Any deviation from the normal texture pattern was an anomaly. But the latter employed tracklets of objects to form a social network graph. Graph similarity measures were applied to find crowd abnormalities.

In short, almost all methods in the literature depend on low-level descriptors to classify crowd characteristics. But these descriptors are inadequate to classify all the diverse behaviors because the behaviors of the crowd are contextual and very ambiguous on a semantic level. Hence, there exists a semantic gap between crowd behaviors and low-level descriptors.

3.2.3 Physics based

Analyzing moving crowds can be done by adopting physics concepts such as particle force and fluid dynamics. Most of the methods in the literature used the social force mechanism to identify the crowd dynamics and patterns. The summary of the review is shown in Table 3.2.

3.2.3.1 Social force and particle dynamics

The social force mechanism used the interaction between individuals to diagnose crowd behavior. It is an application of particle dynamics where each object was treated as a particle, and the interaction between them was simulated as forces. Newton's equation was then applied to compute the acceleration, velocity, and position from the summed-up particle forces [28]. In addition to the computation of displacement velocity and acceleration, the Histogram of Oriented pressure was calculated from the Social Force model by combining the interaction force between objects [29]. These models required prior knowledge of the scenario and only considered the temporal characteristics of the data.

3.2.3.2 Particle force and particle advection

The force between particles or objects can also be consolidated as an energy function to analyze the crowd behavior [25,31]. A structural context descriptor (SCD) was created by adopting the potential energy function of the inter-force between particles to connect the visual context. By investigating the SCD variation of the crowd, the relationship between individuals was identified to localize the abnormality. The repulsive force between particles was also used to classify crowd behavior [30]. Each person in the crowd was represented by three overlapping circles, namely the head, the upper body, and the lower body. The psychological and physical forces of these parts were grouped in the form of a Discrete Element Model (DEM) for further analysis.

3.2.3.3 Fluid dynamics, energy models

The fluid dynamics concept was applied in crowd behavior analysis [32] as well. Individuals were considered particles in the fluid, and crowd movement was modeled based on Lattice Boltzmann Model [33]. Crowd particle collision and particle direction were analyzed to calculate the velocity of the crowd. The abnormal events were identified through particle streaming and collisions.

All of the above approaches focused on each subject in the crowd individually, applicable in situations with few moving objects. As the crowd density increases, individual object identification and tracking become a challenge. Also, in a high-density crowded scenario, sole human nature is vanished and exhibited as collaborative behavior [34].

3.2.4 Cognitive based

Researchers and scientists have adopted the cognitive capability of humans to develop techniques and algorithms for surveillance systems that can extract large amounts of data and process them intelligently through memory processes. This section discusses approaches based on fuzzy logic, psychology, visual attention, behavioral heuristics, deep learning, and machine learning that simulate the procedures and judgment of the human brain for crowd behavior analysis and are summarized in Table 3.3.

3.2.4.1 Machine learning

The emergence of automated surveillance entails cognitive-based algorithms for crowd behavior classification/detection. Besides, the overabundance of surveillance data envisaged the need for sophisticated algorithms in machine

TABLE 3.3

Cognitive-based methods.

Type	Algorithm	Authors	Description	Remarks
Machine learning	EM	Baig et al. [35], 2014	Crowd emotions from topographical maps	The reliability depends on the correct extraction of the features.
		Zhou et al. [36], 2015	Collective crowd behavior	
		Kumar et al. [37], 2018	Track crowd motion patterns	
	SVM	Fradi et al. [18], 2016	Classify crowd behaviors such as splitting, walking, evacuation, running, formation, and dispersion.	Feature engineering is tedious in the case of voluminous video data.
		Amrace et al. [23], 2018	Classify normal and abnormal crowd events using one-class SVM.	Fails to satisfy real-time requirements.
	Linear Classifier	Tudor et al. [38], 2017	Detect normal and abnormal crowded event with 3D gradients and CNN as feature extractors.	
	AdaBoost	Zhang et al. [39], 2016	Identify crowd mood from valence and arousal features.	
	Clustering	Yang et al. [40], 2018	Crowd emotion and behavior detection.	
Deep learning	CNN	Conigliaro et al. [41], 2015	Analyze crowds present in a stadium.	Only partial modeling of crowd behavior. Behaviors that are ambiguous on a semantic level cannot be classified. Difficult to track and segment objects in high-density crowd scenarios. A context-level analysis is not performed
		Zhou et al. [42], 2016	Detect events that are abnormal in a crowded scenario	
		Majhi et al. [43], 2019	Two parallel CNN frameworks for abnormal crowd detection.	
		Belhadi et al. [44], 2021	Collective abnormal crowd behaviors.	
	3DCNN	Dupont et al. [45], 2017	Analyze crowd behaviors based on motion	
	3DCNN + SAE	Sabokrou et al. [46], 2017	Detect abnormal crowd events.	
	SAE	Wang et al. [47], 2018	Detect anomaly from crowd video.	
	3DGMM	Feng et al. [48], 2017	Abnormal and panic behavior detection in the crowd.	
	GAN	Ravanbakhsh et al. [49], 2017		
	ConvLSTM	Lazarides et al. [50], 2018		
Fuzzy logic	Fuzzy rule-base	Banarjee et al. [51], 2005	Modeling emotions of crowd in panic situations.	Good at modeling uncertain human behavior. Used only for agent simulation approaches and datasets. Fail in real-world situations. Not explored real videos for analyzing crowd behavior
		Saifi et al. [52], 2016	Modeling emotions of crowd for analyzing critical emotions in unusual situations.	
		Xue et al. [33], 2017	Analyze pedestrian behavior using personality traits.	
		Zakaria et al. [53], 2019	Detect human behaviors from individual emotions.	
Psychological models	Three-factor Eysenck personality model	Guy et al. [54], 2011 Bera et al. [55], 2017	Derive various crowd agent behaviors.	Utilized simulated environments in modeling agent behavior. Not applicable in real situations.
	Artificial Bee Colony Algorithm	Zhang et al. [56], 2018	Crowd evacuation emotional contagion detection as negative and positive	
	OCC theory + OCEAN model	Durupinar et al. [57], 2015	Identify crowd behaviors when audience changes to mob.	
		Saifi et al. [52], 2016	Simulated agent behavior detection.	
Visual attention	Saliency	Mancas et al. [58], 2010	Saliency index in the form of blobs. Low, medium, and high speed based on the bob size.	Only considers moving objects Behavior associated with static crowd areas cannot be identified.
		Li et al. [59], 2013	Discriminant saliency map to create spatial maps for crowd anomaly detection.	Pixel-level computations required Only suitable for low-density crowd patterns.

learning and deep learning. Although many machine-learning algorithms are available, the powerful and most common among them for crowd analysis is Support Vector Machine (SVM). In [23], normal and abnormal crowd events were classified using a one-class SVM. The main features for the classification of crowd anomaly were drawn using HOG-LBP and HOF. Normal and abnormal crowd events were also classified by Tudor et al. [38] using a linear classifier that utilized 3D gradients and CNN as feature extractors. The approach proposed by Fradi et al. [60] employed SVM for classifying crowd behaviors such as dispersion, walking, formation, running, splitting, and evacuation. Here, the direction, density, and speed of the crowd were employed as features for crowd behavior classification. Another prevalent supervised algorithm, called the AdaBoost, was utilized in [39] to classify crowd mood as valence and arousal.

Unsupervised machine-learning methods such as Expectation-Maximization (EM) [35–37], K Nearest Neighbor(KNN) [37], and Clustering [40] were used for emotion detection and behavior analysis in the crowd. EM is observed to be useful in situations where a large number of important features are present and, among them only a small subset is useful. Also, in EM, values of instances that are not observable are predicted based on the probability distribution of observable latent values. In [37], KNN clustered the available features extracted using Kalman filtering and Kullback–Leibler divergence, and EM tracked the crowd motion patterns. Crowd emotions were analyzed based on topographical maps using EM in [35], whereas in [36], EM was used to identify collective crowd behaviors. The former employed an instantaneous topological map while the latter extracted crowd behaviors using KLT keypoint tracker. The reliability of machine-learning algorithms depends on accurate feature extraction. However, feature engineering is tedious in the case of voluminous video data and therefore fails to satisfy real-time requirements.

3.2.4.2 Deep learning

On the contrary, deep-learning algorithms have recently been widely adopted for object detection and segmentation [61] which have the advantage of working on raw data rather than extracting features. The deep-learning framework is similar to that of a human system, initially extracting local feature descriptors and eventually integrating them into better abstract integrated features. Researchers have also exploited this property in analyzing crowd behavior. Convolutional Neural Network (CNN) and its variants are one of the most widely used deep frameworks for crowd behavior analysis. Conigliaro et al. [41] employed CNN for analyzing crowd in a stadium. A variant of CNN, namely spatial-temporal CNN proposed by Zhou et al. [42] effectively utilized the properties of spatial and temporal video data for detecting abnormal crowd events. On the contrary, Majhi et al. [43] employed two parallel CNN frameworks to extract the spatial and temporal properties. Furthermore, abnormal crowd events were classified using 3DCNN by extracting the temporal properties based on motion patterns [45, 46]. A recent work proposed by Belhadi et al. [44] utilized CNN for detecting collective abnormal crowd behaviors.

Studies based on an ensemble of deep frameworks are available for the analysis of spectators [41] and abnormal activity detection [46]. A CNN was employed in [41] for extracting low-level features, which were then classified as spectators using the Event Attention CatcH (EACH), a probabilistic algorithm. On the other hand, Sabokrou et al. [46] employed a 3DCNN to extract the significant features, and crowd anomalies were classified with the help of a Stacked Autoencoder (SAE). SAE works in an unsupervised manner that consists of an encoder and a decoder, where input data is mapped to the hidden layers via encoding, and outputs are classified from hidden layers via decoding [62]. In [41], SAE alone and a combination of SAE and EACH was utilized for classifying spectators. Wang et al. [47] employed a variant of SAE for crowd anomaly detection. Frameworks such as Generative Adversarial Network (GAN) [63], Convolutional LSTM [50], and 3D Gaussian Mixture Model (3DGMM) [48] are also available in the literature for panic behavior detection, abnormal event classification, and so on. While the foremost choice of crowd behavior analysis is to classify and predict all (most of) the divergent crowd behaviors, a universal shortcoming of existing deep-learning works is that not all behaviors that are ambiguous on a semantic level can be classified.

3.2.4.3 Fuzzy logic

Another cognitive-based classification is the theory of fuzzy logic. It is ideal for modeling ambiguous and uncertain crowd behaviors. Banarjee et al. [51] proposed a system based on fuzzy rules for modeling the emotions of a panic crowd. In addition, Xue et al. [64] utilized personality traits to model the behavior of heterogeneous strollers using fuzzy logic. Fuzzy logic was also employed by Saifi et al. [52] to analyze critical emotions of the crowd during abnormal events. Zakaria et al. [53] developed a fuzzy-based model to extract human behaviors and individual emotions during emergency evacuation situations such as confusion and panic. All the approaches discussed above were based on simulations of agents that could not be applied in real-world situations. While fuzzy logic is acceptable in modeling uncertain behaviors of humans, they have not been further investigated for smart crowd behavior analysis based on real videos.

3.2.4.4 Psychological models

Psychological factors such as emotions and personality traits were explored by early researchers in the simulation and graphics field. One of the prominent psychological theories, known as the three-factor Eysenck personality theorem, was wielded by Bera et al. [55] and Guy et al. [54] to extract crowd behaviors, namely, aggressive, tense, active, shy, impulsive, and assertive. The former figured out the behavior of each pedestrian based on Bayesian estimation, while the latter employed factor analysis to map simulation factors to behaviors. Another approach proposed by Zhang et al. [56] studied the negative and positive emotional contagion and employed an artificial bee colony algorithm to calculate optimal strategies and positions for safe crowd evacuation. Durupinar et al. [57] used the combination of two psychological models, the OCEAN (openness, conscientiousness, extraversion, agreeableness, and neuroticism) personality model and OCC (Ortony, Clore, and Collins) theory of emotions for identifying behaviors when a crowd transforms from audience to mob. Saifi et al. [52] also used this combination to detect agent behaviors such as anxious, leader, and quiet. Similar to the case of fuzzy-based approaches, these methods utilized simulated environments for modeling agent behaviors, which could not be applied in real-world situations.

3.2.4.5 Visual attention and saliency

Ultimately, the discussion reviews a few methods that used the cognitive visual attention property based on saliency. Mancas et al. [58] utilized a saliency index to identify appropriate dense crowd behavior using bottom-up and top-down attention mechanisms. Only motion speed in the form of blobs was employed as the saliency index, and the speed was divided into high, medium, and low based on the blob size. This method only considers moving objects, and hence the behavior associated with static crowd areas cannot be identified. Li et al. [59] proposed a discriminant saliency map to create spatial maps for crowd anomaly detection. The saliency map was obtained from small frame patches that discriminate from the background pixels. Since pixel-level computations were required, this method is only suitable for low-density crowd patterns.

In general, from the fruitful discussion on visual-based approaches, we perceived that most of the motion pattern approaches relied on temporal characteristics of video data by ignoring the spatial properties. Furthermore, methods based on low-level feature descriptors failed to define the high-level behaviors of all diverse crowds. Also, the methods based on low-level features failed to define all high-level diverse crowd behaviors. Thus, most approaches were partial modeling of crowd behavior, such as misdirection and panic. Also, feature engineering from voluminous data is tedious in cases of machine learning, whereas fuzzy and psychological methods were used only in a simulated environment for agent behavior modeling.

3.3   Multimodal Approaches

Studies and researches in the modern fusion of cross-modality data have been laid their foundation in the first half of the last century. Later, the development of techniques such as canonical correlation analysis, tensor decomposition, and parallel factor analysis helped to achieve a significant leap in the development of new algorithms in many fields, especially chemometrics and psychometrics [65]. Recently, technological advances in many domains, such as smart surveillance, social networks, Internet of Things (IoT), vehicular networks, etc., have been generating massive amounts of data, which tends to the increasing interest in information fusion from multiple modalities. There are several advantages to using multimodal data in smart surveillance. Data from multiple modals brings a global and unified mechanism for improving decision-making, and through exploratory research, it is possible to provide answers for specific questions, identify distinct and usual temporal components, and thus provide knowledge from the integration of multimodal data for paramount purposes [65]. Therefore, the integration of multimodal data helps extract and integrate outstanding information from a variety of modalities and to improve performance over a single modality. This section discusses a few works available in the literature that utilized data from multiple modalities for behavior analysis.

3.3.1   Video and sensor data

A few studies were done using multiple modality data to detect free-standing conversation groups (FCGs) in the context of social gatherings. Alameda et al. [66] calculated the body and head pose required to identify FCGs using data from distributed cameras and wearable sensors (audio, accelerometer, and infrared). Data fusion was a matrix completion problem for optimizing the body and head poses of people in a group in terms of the temporal stability of the data. Another method proposed by Cabrera et al. [67] also employed video and wearable acceleration to identify hand gestures in crowded mingle scenarios. This method detected 205 hand gestures of individuals in a social group during mingling using a decision level fusion of multiple modalities of data. Here, a bag of words was created from each unimodal data, and finally, they were combined in a probabilistic manner to take the decision level fusion.

3.3.2   Wearable sensors

Apart from the above approaches, only wearable sensors were used in several works to detect the social mingling of individuals. Body-worn accelerometer and audio data were utilized by Hung et al. [68] to identify social actions of individuals during conversations in low-density crowded scenarios. Wearable acceleration was used to obtain labeled activities such as speech, actions, and hand and head gestures. Similar to the above approach, Gedik et al. [69] proposed a method to extract pairwise features for recognizing the interaction between people during conversations employing the actions from wearable acceleration. The authors in [70] culled proxemics and dynamics from wearable devices and detect whether individuals create F-formation (Friends-Formation). From sensor data, pairwise representations of participants in a conversation group were extracted and given to an LSTM to uncover F-formation.

Although fusion of multiple modality data provides more degrees of freedom in detecting behaviors, their research exploration is less popular. This is due to the lack of annotated datasets and the impracticality of deploying sensors in public spaces. However, the prior approaches extract data from wearable sensors and distributed cameras. But, they were only used for the detection of FCGs regardless of the social behavior of individuals. Besides, these methods ignored the spatial properties and temporal compatibility of all the modalities for F-formation detection that resulted in wrong group formation.

3.4   Social Group Analysis

Social groups are human's natural tendency to form mini-groups that engage in harmony and share similar characteristics. This socio-psychological passion was significantly reconstructed in the field of computer vision by examining the dynamic movement patterns of individuals.

3.4.1   Clustering

Most researchers wielded motion patterns and trajectories of individuals in a crowd to identify the group formation among them. Trajectory clustering was performed in [71] and [72] to determine social groups. The former employed agglomerative hierarchical clustering to group the strollers, while the latter computes the distance between each individual in a cluster to decide the group membership. Also, Chang et al. [73] detect social groups in a crowd using the notion of agglomerative clustering. Here, the tracked individuals were brought together into cohesive groups, and then clustering was performed based on the distance between individuals.

3.4.2   Decision tree

In addition to trajectories, Chamveha et al. [74] extracted the speed, displacement, walking direction, and head pose to classify social groups from surveillance videos. Attention-based features were derived from the difference in head pose, relative position, and walking speed whereas position-based features were acquired from the displacement, velocity difference, and direction difference of trajectories. The social group behavior was modeled from the detected features using a decision tree. On the contrary, stationary groups in a moving crowd were analyzed by Yi et al. [75] by analyzing the trajectories. The stationary time was monitored from the trajectories of moving people using twelve crowd descriptors to detect four types of stationary crowd.

3.4.3   Friends-formation (F-formation)

According to socio-psychology principles, individuals who share a common space utilize F-formation concepts to maintain orientation and spatial relationships [76]. Studies were also performed in the literature to identify F-formations from surveillance data. In [77] and [78], a semi-supervised optimization algorithm was employed to estimate the head and body pose of individuals and their participation in F-formations. Features related to head and body pose were determined using the HOG feature map. Furthermore, multimodal data were exploited for the F-formation analysis and discussed in Section 3.3.

Briefly, the studies discussed in this section explored the formation of socially interactive groups in crowded scenarios. The methods that rely on trajectories are not reliable in high-density crowd areas because of the difficulty of segment, track, and detect individuals. In addition, all the methods were based on low-level feature patterns that require more time within flat intensity regions. Occlusion and overlapping of objects in crowded scenes also posses a hindrance in extracting the body and head poses, especially for the identification of F-formation.

3.5   Open Research Issues

In recent years, due to the population growth in urban areas and persistent crowd disasters, efficient and active monitoring of crowds has become more significant. Even though emerging behaviors in a crowd are often unpredictable, researchers from computer vision and other related fields are seeking to identify this emerging behavior. The extensive review provided in this chapter is deliberated to analyze the shared insights from existing crowd behavior analysis studies. The comprehensive discussion and insights from the review pave the way for the following open research issues.

- In the past few years, several works came into the light to analyze crowded scenarios, especially those that use crowd movement patterns. Most of them focused only on limited patterns of crowd behavior such as panic escape and misguided movement. Most of the methods ignore spatial information and rely only on temporal characteristics resulting in disregarding the crucial features of a stationary crowd.

- To a greater extent, almost all approaches in the literature use low-level descriptors for visual analysis. But these features are insufficient to describe all the varying high-level crowd behaviors because crowd behaviors are contextual and semantically ambiguous. This leads to the semantic gap between high-level crowd behaviors and low-level descriptors.

- Most methods in the literature only consider low and medium-density crowds that rely upon individual motion pattern extraction and tracking. But, tracking and recognizing distinct objects is impractical in a high-dense crowd. Besides, methods that combine information about appearance and motion spend a lot of time on feature representation and feature extraction that fail to meet real-time requirements.

- The critical challenges related to visual analysis, such as illumination variation and occlusion among subjects, were not considered in many approaches. In particular, low-light situations and occlusion were not considered in the social group detection, which led to unreliable and erroneous group inclusion.

- The detection of crowd anomaly approaches in the literature suffers from the impracticability of detecting anomalies from non-existing classes. That is, crowd anomaly is contextual, and the subjective definition of anomalies is ambiguous. Besides, learning contextual abnormal patterns and occlusion among individuals in a high-density crowd from voluminous data impose difficulty in describing all possible anomalies.

- The extensive proliferation of monitoring devices leads to an enormous amount of data production to be analyzed and transmitted in real-time. Therefore, the surveillance platform must be smart in providing low-latency communication, enabling long-term data processing, and handling large volumes of data. The major works for crowd behavior analysis do not cover the difficulties of high network bandwidth usage and high latency.

3.6 Conclusion and Future Directions

Smart crowd analysis has become a vital concern in a public surveillance system due to unpredictable and complex human behavior. In this chapter, we uncovers the wide variety of methods in the literature for determining the behavior of the crowd, social group, and conversation group detection. This thorough literature review shows that low-level features and motion patterns are the major factors used by researchers for crowd behavior analysis. The review in this chapter entails deliberation of open research issues and the rest of this section presents a brief description of some concrete issues that can be considered as future research directions.

Analysis of socially interacting groups among crowd: Social interaction depicts the interconnection between two or more individuals in a society. The analysis of the tendency of humans to form social groups helps to formulate rules for controlling crowds in public places.

Detection of social group behavior: The individuals in a crowd usually focus their attention on a common concern or cause. The ideas and objectives about the common cause spread rapidly among the participants. Moreover, there is no clear definition of the behavior of the participants, and no one knows what will happen next. The general tendency of individuals in a crowd is to form groups and interact with others, which influence their views and opinions. These may lead to the development of homogeneous groups within a crowd. Therefore more attention needs to be paid to finding the social group behavior that helps plan crowd management strategies in a surveillance environment.

Explainable AI and Mobile Crowdsensing for anomaly detection in crowd: In the field of computer vision, detecting anomalous crowd events is an arduous domain. In this context, the explainable AI will help better understand the contextual anomalies and assist in making proactive decisions. Besides, mobile crowdsensing uses spatiotemporal patterns to analyze data from a group of mobile devices. In the case of crowd behavioral analysis, such knowledge of the spatial distribution of individuals is indispensable to analyze their movements. Since the spatial patterns are disseminated over various time scales, temporal information is also accessible. This insight enables better coordination and behavior analysis of the crowd, depending on the present situation, as well as long-term planning to avoid crowd disasters and suspicious activities.

Real-time multimodal cognitive IoT for crowd behavior analysis: Today's intelligent world requires the live processing of an enormous amount of unstructured data, making decisions like the human brain, grasping contextual elements such as purpose, location, and time, for decision framing, and interacting with multiple devices and processors. The platform should also provide communication with low-latency, long-term data processing by using the services such as cloud, fog, and edge. Hence cognitive IoT-based smart platforms are essential for crowd management that help make context-conscious initiative decisions based on real-time multiple-modality data.

Bibliography

[1] Sabu M Thampi and Elizabeth B Varghese. IoT-based smart surveillance: Role of sensor data analytics and mobile crowd sensing in crowd behavior analysis. In *Crowd Assisted Networking and Computing*, pages 45–76. CRC Press, 2018.

[2] Ying-li Tian, Lisa Brown, Arun Hampapur, Max Lu, Andrew Senior, and Chiao-fe Shu. IBM smart surveillance system (S3): Event based video surveillance system with an open and extensible framework. *Machine Vision and Applications*, 19(5-6):315–327, 2008.

[3] Data integration helps smart cities fight crime. White Paper, Soluton Safe, Intel IOT, Government, 2015.

[4] N Wijermans. Understanding crowd behaviour. *Thesis, University of Groningen, Groningen*, 2011.

[5] Nuria M Oliver, Barbara Rosario, and Alex P Pentland. A bayesian computer vision system for modeling human interactions. *IEEE Transactions on Pattern Analysis and Machine Intelligence*, 22(8):831–843, 2000.

[6] P Drews, João Quintas, Jorge Dias, Maria Andersson, Jonas Nygårds, and Joakim Rydell. Crowd behavior analysis under cameras network fusion using probabilistic methods. In *2010 13th International Conference on Information Fusion*, pages 1–8. IEEE, 2010.

[7] Si Wu, Hau-San Wong, and Zhiwen Yu. A bayesian model for crowd escape behavior detection. *IEEE Transactions on Circuits and Systems for Video Technology*, 24(1):85–98, 2013.

[8] Xiaogang Wang, Xiaoxu Ma, and W Eric L Grimson. Unsupervised activity perception in crowded and complicated scenes using hierarchical bayesian models. *IEEE Transactions on Pattern Analysis and Machine Intelligence*, 31(3):539–555, 2008.

[9] Mikel Rodriguez, Josef Sivic, Ivan Laptev, and Jean-Yves Audibert. Data-driven crowd analysis in videos. In *2011 International Conference on Computer Vision*, pages 1235–1242. IEEE, 2011.

[10] M Sami Zitouni, Andrzej Sluzek, and Harish Bhaskar. Towards understanding socio-cognitive behaviors of crowds from visual surveillance data. *Multimedia Tools and Applications*, 79(3):1781–1799, 2020.

[11] Tal Hassner, Yossi Itcher, and Orit Kliper-Gross. Violent flows: Real-time detection of violent crowd behavior. In *2012 IEEE Computer Society Conference on Computer Vision and Pattern Recognition Workshops*, pages 1–6. IEEE, 2012.

[12] Rensso Victor Hugo Mora Colque, Carlos Caetano, Matheus Toledo Lustosa de Andrade, and William Robson Schwartz. Histograms of optical flow orientation and magnitude and entropy to detect anomalous events in videos. *IEEE Transactions on Circuits and Systems for Video Technology*, 27(3):673–682, 2016.

[13] Aravinda S Rao, Jayavardhana Gubbi, Slaven Marusic, and Marimuthu Palaniswami. Crowd event detection on optical flow manifolds. *IEEE Transactions on Cybernetics*, 46(7):1524–1537, 2015.

[14] Andrea Pennisi, Domenico D Bloisi, and Luca Iocchi. Online real-time crowd behavior detection in video sequences. *Computer Vision and Image Understanding*, 144:166–176, 2016.

[15] Mariem Gnouma, Ridha Ejbali, and Mourad Zaied. Abnormal events' detection in crowded scenes. *Multimedia Tools and Applications*, 77(19): 24843–24864, 2018.

[16] Lawrence O'Gorman, Yafeng Yin, and Tin Kam Ho. Motion feature filtering for event detection in crowded scenes. *Pattern Recognition Letters*, 44:80–87, 2014.

[17] A Balasundaram and C Chellappan. An intelligent video analytics model for abnormal event detection in online surveillance video. *Journal of Real-Time Image Processing*, pages 1–16, 2018.

[18] Hajer Fradi, Bertrand Luvison, and Quoc Cuong Pham. Crowd behavior analysis using local mid-level visual descriptors. *IEEE Transactions on Circuits and Systems for Video Technology*, 27(3):589–602, 2016.

[19] Igor R de Almeida, Vinicius J Cassol, Norman I Badler, Soraia Raupp Musse, and Cláudio Rosito Jung. Detection of global and local motion changes in human crowds. *IEEE Transactions on Circuits and Systems for Video Technology*, 27(3):603–612, 2016.

[20] Hamidreza Rabiee, Javad Haddadnia, Hossein Mousavi, Maziyar Kalantarzadeh, Moin Nabi, and Vittorio Murino. Novel dataset for fine-grained abnormal behavior understanding in crowd. In *2016 13th IEEE International Conference on Advanced Video and Signal Based Surveillance (AVSS)*, pages 95–101. IEEE, 2016.

[21] Xavier Alameda-Pineda, Jacopo Staiano, Ramanathan Subramanian, Ligia Batrinca, Elisa Ricci, Bruno Lepri, Oswald Lanz, and Nicu Sebe. Salsa: A novel dataset for multimodal group behavior analysis. *IEEE*

Transactions on Pattern Analysis and Machine Intelligence, 38(8):1707–1720, 2015.

[22] Hamidreza Rabiee, Javad Haddadnia, and Hossein Mousavi. Crowd behavior representation: an attribute-based approach. *SpringerPlus*, 5(1):1–17, 2016.

[23] Somaieh Amraee, Abbas Vafaei, Kamal Jamshidi, and Peyman Adibi. Abnormal event detection in crowded scenes using one-class svm. *Signal, Image and Video Processing*, 12(6):1115–1123, 2018.

[24] Hossein Mousavi, Sadegh Mohammadi, Alessandro Perina, Ryad Chellali, and Vittorio Murino. Analyzing tracklets for the detection of abnormal crowd behavior. In *2015 IEEE Winter Conference on Applications of Computer Vision*, pages 148–155. IEEE, 2015.

[25] Yuan Yuan, Jianwu Fang, and Qi Wang. Online anomaly detection in crowd scenes via structure analysis. *IEEE Transactions on Cybernetics*, 45(3):548–561, 2014.

[26] Jing Wang and Zhijie Xu. Spatio-temporal texture modelling for real-time crowd anomaly detection. *Computer Vision and Image Understanding*, 144:177–187, 2016.

[27] Rima Chaker, Zaher Al Aghbari, and Imran N Junejo. Social network model for crowd anomaly detection and localization. *Pattern Recognition*, 61:266–281, 2017.

[28] Ramin Mehran, Alexis Oyama, and Mubarak Shah. Abnormal crowd behavior detection using social force model. In *2009 IEEE Conference on Computer Vision and Pattern Recognition*, pages 935–942. IEEE, 2009.

[29] Hua Yang, Yihua Cao, Shuang Wu, Weiyao Lin, Shibao Zheng, and Zhenghua Yu. Abnormal crowd behavior detection based on local pressure model. In *Proceedings of The 2012 Asia Pacific Signal and Information Processing Association Annual Summit and Conference*, pages 1–4. IEEE, 2012.

[30] Harmeet Singh, Robyn Arter, Louise Dodd, Paul Langston, Edward Lester, and John Drury. Modelling subgroup behaviour in crowd dynamics dem simulation. *Applied Mathematical Modelling*, 33(12):4408–4423, 2009.

[31] Tianyu Chen, Chunping Hou, Zhipeng Wang, and Hua Chen. Anomaly detection in crowded scenes using motion energy model. *Multimedia Tools and Applications*, 77(11):14137–14152, 2018.

[32] Hang Su, Hua Yang, Shibao Zheng, Yawen Fan, and Sha Wei. The large-scale crowd behavior perception based on spatio-temporal viscous fluid field. *IEEE Transactions on Information Forensics and Security*, 8(10):1575–1589, 2013.

[33] Yiran Xue, Peng Liu, Ye Tao, and Xianglong Tang. Abnormal prediction of dense crowd videos by a purpose-driven lattice boltzmann model. *International Journal of Applied Mathematics and Computer Science*, 27(1), 2017.

[34] Ralph H Turner, Lewis M Killian, et al. *Collective Behavior*, volume 3. Prentice-Hall Englewood Cliffs, NJ, 1957.

[35] Mirza Waqar Baig, Emilia I Barakova, Lucio Marcenaro, Matthias Rauterberg, and Carlo S Regazzoni. Crowd emotion detection using dynamic probabilistic models. In *International Conference on Simulation of Adaptive Behavior*, pages 328–337. Springer, 2014.

[36] Bolei Zhou, Xiaoou Tang, and Xiaogang Wang. Learning collective crowd behaviors with dynamic pedestrian-agents. *International Journal of Computer Vision*, 111(1):50–68, 2015.

[37] Santosh Kumar, Deepanwita Datta, Sanjay Kumar Singh, and Arun Kumar Sangaiah. An intelligent decision computing paradigm for crowd monitoring in the smart city. *Journal of Parallel and Distributed Computing*, 118:344–358, 2018.

[38] Radu Tudor Ionescu, Sorina Smeureanu, Bogdan Alexe, and Marius Popescu. Unmasking the abnormal events in video. In *Proceedings of the IEEE International Conference on Computer Vision*, pages 2895–2903, 2017.

[39] Yanhao Zhang, Lei Qin, Rongrong Ji, Sicheng Zhao, Qingming Huang, and Jiebo Luo. Exploring coherent motion patterns via structured trajectory learning for crowd mood modeling. *IEEE Transactions on Circuits and Systems for Video Technology*, 27(3):635–648, 2016.

[40] Meng Yang, Lida Rashidi, Aravinda S Rao, Sutharshan Rajasegarar, Mohadeseh Ganji, Marimuthu Palaniswami, and Christopher Leckie. Cluster-based crowd movement behavior detection. In *2018 Digital Image Computing: Techniques and Applications (DICTA)*, pages 1–8. IEEE, 2018.

[41] Davide Conigliaro, Paolo Rota, Francesco Setti, Chiara Bassetti, Nicola Conci, Nicu Sebe, and Marco Cristani. The s-hock dataset: Analyzing crowds at the stadium. In *Proceedings of the IEEE Conference on Computer Vision and Pattern Recognition*, pages 2039–2047, 2015.

[42] Shifu Zhou, Wei Shen, Dan Zeng, Mei Fang, Yuanwang Wei, and Zhijiang Zhang. Spatial–temporal convolutional neural networks for anomaly detection and localization in crowded scenes. *Signal Processing: Image Communication*, 47:358–368, 2016.

[43] Snehashis Majhi, Ratnakar Dash, and Pankaj Kumar Sa. Two-stream cnn architecture for anomalous event detection in real world scenarios. In *International Conference on Computer Vision and Image Processing*, pages 343–353. Springer, 2019.

[44] Asma Belhadi, Youcef Djenouri, Gautam Srivastava, Djamel Djenouri, Jerry Chun-Wei Lin, and Giancarlo Fortino. Deep learning for pedestrian collective behavior analysis in smart cities: A model of group trajectory outlier detection. *Information Fusion*, 65:13–20, 2021.

[45] Camille Dupont, Luis Tobias, and Bertrand Luvison. Crowd-11: A dataset for fine grained crowd behaviour analysis. In *Proceedings of the IEEE Conference on Computer Vision and Pattern Recognition Workshops*, pages 9–16, 2017.

[46] Mohammad Sabokrou, Mohsen Fayyaz, Mahmood Fathy, and Reinhard Klette. Deep-cascade: Cascading 3d deep neural networks for fast anomaly detection and localization in crowded scenes. *IEEE Transactions on Image Processing*, 26(4):1992–2004, 2017.

[47] Tian Wang, Meina Qiao, Zhiwei Lin, Ce Li, Hichem Snoussi, Zhe Liu, and Chang Choi. Generative neural networks for anomaly detection in crowded scenes. *IEEE Transactions on Information Forensics and Security*, 14(5):1390–1399, 2018.

[48] Yachuang Feng, Yuan Yuan, and Xiaoqiang Lu. Learning deep event models for crowd anomaly detection. *Neurocomputing*, 219:548–556, 2017.

[49] Mahdyar Ravanbakhsh, Moin Nabi, Enver Sangineto, Lucio Marcenaro, Carlo Regazzoni, and Nicu Sebe. Abnormal event detection in videos using generative adversarial nets. In *2017 IEEE International Conference on Image Processing (ICIP)*, pages 1577–1581. IEEE, 2017.

[50] Lazaros Lazaridis, Anastasios Dimou, and Petros Daras. Abnormal behavior detection in crowded scenes using density heatmaps and optical flow. In *2018 26th European Signal Processing Conference (EUSIPCO)*, pages 2060–2064. IEEE, 2018.

[51] Soumya Banarjee, Crina Grosan, and Ajith Abraham. Emotional ant based modeling of crowd dynamics. In *Seventh International Symposium on Symbolic and Numeric Algorithms for Scientific Computing (SYNASC'05)*, pages 8–pp. IEEE, 2005.

[52] Lynda Saifi, Abdelhak Boubetra, and Farid Nouioua. An approach for emotions and behavior modeling in a crowd in the presence of rare events. *Adaptive Behavior*, 24(6):428–445, 2016.

[53] Wahida Zakaria, Umi Kalsom Yusof, and Syibrah Naim. Modelling and simulation of crowd evacuation with cognitive behaviour using fuzzy logic.

International Journal of Advance Soft Computing Application, 11(2), 2019.

[54] Stephen J Guy, Sujeong Kim, Ming C Lin, and Dinesh Manocha. Simulating heterogeneous crowd behaviors using personality trait theory. In *Proceedings of the 2011 ACM SIGGRAPH/Eurographics Symposium on Computer Animation*, pages 43–52, 2011.

[55] Aniket Bera, Tanmay Randhavane, and Dinesh Manocha. Aggressive, tense or shy? identifying personality traits from crowd videos. In *IJCAI*, pages 112–118, 2017.

[56] Guijuan Zhang, Dianjie Lu, and Hong Liu. Strategies to utilize the positive emotional contagion optimally in crowd evacuation. *IEEE Transactions on Affective Computing*, 11(4):708–721, 2018.

[57] Funda Durupınar, Uğur Güdükbay, Aytek Aman, and Norman I Badler. Psychological parameters for crowd simulation: From audiences to mobs. *IEEE Transactions on Visualization and Computer Graphics*, 22(9):2145–2159, 2015.

[58] Matei Mancas and Bernard Gosselin. Dense crowd analysis through bottom-up and top-down attention. In *Proceedings of the Brain Inspired Cognition System (bics)*, pages 1–12. Citeseer, 2010.

[59] Weixin Li, Vijay Mahadevan, and Nuno Vasconcelos. Anomaly detection and localization in crowded scenes. *IEEE Transactions on Pattern Analysis and Machine Intelligence*, 36(1):18–32, 2013.

[60] Hajer Fradi and Jean-Luc Dugelay. Spatial and temporal variations of feature tracks for crowd behavior analysis. *Journal on Multimodal User Interfaces*, 10(4):307–317, 2016.

[61] Elizabeth Varghese, Sabu M Thampi, and Stefano Berretti. A psychologically inspired fuzzy cognitive deep learning framework to predict crowd behavior. *IEEE Transactions on Affective Computing*, 2020.

[62] Elizabeth B Varghese and Sabu M Thampi. Towards the cognitive and psychological perspectives of crowd behaviour: A vision-based analysis. *Connection Science*, pages 1–26, 2020.

[63] Mahdyar Ravanbakhsh, Enver Sangineto, Moin Nabi, and Nicu Sebe. Training adversarial discriminators for cross-channel abnormal event detection in crowds. In *2019 IEEE Winter Conference on Applications of Computer Vision (WACV)*, pages 1896–1904. IEEE, 2019.

[64] Zhuxin Xue, Qing Dong, Xiangtao Fan, Qingwen Jin, Hongdeng Jian, and Jian Liu. Fuzzy logic-based model that incorporates personality traits for heterogeneous pedestrians. *Symmetry*, 9(10):239, 2017.

[65] Elizabeth B Varghese and Sabu M Thampi. A multimodal deep fusion graph framework to detect social distancing violations and fcgs in pandemic surveillance. *Engineering Applications of Artificial Intelligence*, 103:104305, 2021.

[66] Xavier Alameda-Pineda, Yan Yan, Elisa Ricci, Oswald Lanz, and Nicu Sebe. Analyzing free-standing conversational groups: A multimodal approach. In *Proceedings of the 23rd ACM International Conference on Multimedia*, pages 5–14, 2015.

[67] Laura Cabrera-Quiros, David MJ Tax, and Hayley Hung. Gestures in-the-wild: detecting conversational hand gestures in crowded scenes using a multimodal fusion of bags of video trajectories and body worn acceleration. *IEEE Transactions on Multimedia*, 22(1):138–147, 2019.

[68] Hayley Hung, Gwenn Englebienne, and Laura Cabrera Quiros. Detecting conversing groups with a single worn accelerometer. In *Proceedings of the 16th International Conference on Multimodal Interaction*, pages 84–91, 2014.

[69] Ekin Gedik and Hayley Hung. Detecting conversing groups using social dynamics from wearable acceleration: Group size awareness. *Proceedings of the ACM on Interactive, Mobile, Wearable and Ubiquitous Technologies*, 2(4):1–24, 2018.

[70] Alessio Rosatelli, Ekin Gedik, and Hayley Hung. Detecting f-formations & roles in crowded social scenes with wearables: Combining proxemics & dynamics using lstms. In *2019 8th International Conference on Affective Computing and Intelligent Interaction Workshops and Demos (ACIIW)*, pages 147–153. IEEE, 2019.

[71] Weina Ge, Robert T Collins, and Barry Ruback. Automatically detecting the small group structure of a crowd. In *2009 Workshop on Applications of Computer Vision (WACV)*, pages 1–8. IEEE, 2009.

[72] Francesco Solera, Simone Calderara, and Rita Cucchiara. Socially constrained structural learning for groups detection in crowd. *IEEE Transactions on Pattern Analysis and Machine Intelligence*, 38(5):995–1008, 2015.

[73] Ming-Ching Chang, Nils Krahnstoever, Sernam Lim, and Ting Yu. Group level activity recognition in crowded environments across multiple cameras. In *2010 7th IEEE International Conference on Advanced Video and Signal Based Surveillance*, pages 56–63. IEEE, 2010.

[74] Isarun Chamveha, Yusuke Sugano, Yoichi Sato, and Akihiro Sugimoto. Social group discovery from surveillance videos: A data-driven approach with attention-based cues. In *BMVC*. Citeseer, 2013.

[75] Shuai Yi, Xiaogang Wang, Cewu Lu, Jiaya Jia, and Hongsheng Li. *l*_0 regularized stationary-time estimation for crowd analysis. *IEEE Transactions on Pattern Analysis and Machine Intelligence*, 39(5):981–994, 2016.

[76] Adam Kendon. The f-formation system: The spatial organization of social encounters. *Man-Environment Systems*, 6(01):1976, 1976.

[77] Jagannadan Varadarajan, Ramanathan Subramanian, Samuel Rota Bulò, Narendra Ahuja, Oswald Lanz, and Elisa Ricci. Joint estimation of human pose and conversational groups from social scenes. *International Journal of Computer Vision*, 126(2):410–429, 2018.

[78] Elisa Ricci, Jagannadan Varadarajan, Ramanathan Subramanian, Samuel Rota Bulo, Narendra Ahuja, and Oswald Lanz. Uncovering interactions and interactors: Joint estimation of head, body orientation and f-formations from surveillance videos. In *Proceedings of the IEEE International Conference on Computer Vision*, pages 4660–4668, 2015.

4

Intelligent Traffic Video Analytics

Hadi Ghahremannezhad

New Jersey Institute of Technology, Newark, NJ, USA

Chengjun Liu

New Jersey Institute of Technology, Newark, NJ, USA

Hang Shi

Innovative AI Technologies, Newark, NJ, USA

CONTENTS

4.1	Introduction	86
4.2	Main Steps in Intelligent Traffic Video Analytics	87
	4.2.1 Vehicle detection	88
	4.2.2 Shadow removal	94
	4.2.3 Vehicle classification	97
	4.2.4 Vehicle tracking	100
	4.2.4.1 Region-based tracking	101
	4.2.4.2 Contour-based tracking	101
	4.2.4.3 Feature-based tracking	102
	4.2.4.4 Model-based tracking	102
	4.2.5 Region of interest determination	102
	4.2.6 Incident detection	105
	4.2.6.1 Stopped vehicle detection	107
	4.2.6.2 Wrong-way vehicle detection	107
	4.2.6.3 Traffic congestion detection	107
	4.2.6.4 Traffic accident detection	108
	4.2.6.5 Traffic anomaly detection	109
4.3	Main Challenges in Intelligent Traffic Video Analytics	109
	4.3.1 Performance and reliability	110
	4.3.2 Flexibility and versatility	110
	4.3.3 Efficiency and responsiveness	111
4.4	Conclusions	111

DOI: 10.1201/9781003053262-4

4.1   Introduction

Road transportation has been one of the most common modes of land transport in the world. Since the twentieth century, the developments in road transportation and the related infrastructures, vehicles, and operations have had a key impact on economic growth [1]. However, the rapid expansion in the number of motor vehicles and the increasing demand for road transportation has risen several concerns about the high fuel consumption, traffic jams, negative environmental effects, safety, and travel delays. Therefore, there is an inevitable need to monitor road traffic in order to improve the efficiency and safety of road transportation. During the 1980s, the technological advancements in information, control, communication, GPS, computers, microprocessors, and sensing, has led to the development of intelligent transportation systems (ITS) [2]. Intelligent transportation systems have broad applications in emergency vehicle notification, automatic road enforcement, setting various speed limits, collision avoidance, car navigation, traffic signal control, parking guidance, and automatic number plate recognition. The performance of these systems depends on the accuracy and efficiency of data collection and processing.

Various sensing devices have been used along with different forms of short-range and long-range wireless communication technologies in order to collect useful and reliable data. These sensing devices are generally categorized into two groups, namely in-vehicle and intra-road sensing platforms [3]. In-vehicle sensors and actuators, such as fuel, pressure, radar, ultrasonic, and A/C sensors, GPS, microphone, and video cameras are mounted in the vehicles by manufacturers in order to monitor the status of the vehicle and assist the drivers. Intra-road sensors are deployed along the road and within the transportation network in order to gather environmental data in real-time. Based on the location of intra-road sensors they are divided into two groups of intrusive (in-roadway) and non-intrusive (over-roadway) sensors [4, 5]. Intrusive intra-road sensors, such as pneumatic road tubes, Inductive Loop Detectors (ILDs), magnetometers, micro-loop probes, and piezoelectric cables are installed over the pavement surfaces. They have relatively high accuracy in counting and classifying the vehicles and measuring the vehicles' weight and speed. However, the installation and maintenance of these sensors are relatively expensive and cause traffic disruption. On the other hand, non-intrusive sensors, such as radar, infrared, ultrasonic, video cameras, microphones, and acoustic array sensors are installed at different locations along the road. These sensors are used for vehicle detection and classification, speed measurement, vehicle counting, and collecting information on weather conditions. Although non-intrusive sensors are less expensive and easier to install and maintain, they can be affected by weather conditions and be spotted by drivers.

One of the main applications of the ITS is in Advanced Traffic Management Systems (ATMS). The information is collected from the sensors and transmitted to Transportation Management Center (TMC) in real-time where it is processed in an attempt to improve safety and traffic flow. When it comes to making a cost-efficient choice among various sensor technologies, camera sensors are preferred in most cases due to simple installation steps, rich visual information, and a relatively vast area of coverage provided by each device. In general, there are two types of camera sensors used in traffic management applications, namely, traffic enforcement cameras and traffic surveillance cameras. The former category refers to the cameras that are mounted beside or over a road and capture high-resolution pictures to detect motoring offenses and enforce traffic rules, whereas the latter refers to the remotely controllable cameras that are typically mounted on high poles and street lights and capture live lower-resolution videos with the goal of observing the traffic and detecting incidents.

The data received from the traffic surveillance cameras are used for vehicle counting, object classification, speed estimation, and detecting events such as congestion, stopped vehicles, wrong-way vehicles, and accidents. Manually processing the huge amount of video data by human operators is a tedious and impractical task and the results are not always reliable. With the advances in fields of artificial intelligence, pattern recognition, and computer vision it is of no surprise that automated techniques in intelligent video analytics have gained a lot of attention in traffic management systems [6,9]. In this chapter, we discuss the core components of intelligent video analytics and their applications and challenges in processing traffic video data. The remainder of the chapter is structured as follows: Section 4.2 contains a brief overview of the main steps and the various approaches applied in vision-based traffic monitoring systems. Section 4.3 outlines the main challenges faced by the intelligent video analytics techniques. Finally, further discussions and conclusions are given in Section 4.4.

4.2   Main Steps in Intelligent Traffic Video Analytics

Traffic flow monitoring based on video analytics has been one of the main applications of computer vision. There are several steps taken towards developing a fast and reliable system in order to process traffic videos. Traffic surveillance cameras, commonly powered by electricity or solar planes, are installed along major roads in urban areas, such as highways, motorways, freeways, and intersections to provide consistent live imagery data. The live video is sent to a monitoring center and is processed by applying intelligent video analytic techniques in order to recognize traffic patterns, collect traffic flow parameters, and detect incidents and anomalies. In terms of intelligent video analytics, several steps are taken to process the raw video data and generate useful information in real-time. The major steps after camera calibration and pre-processing are composed of vehicle detection, classification,

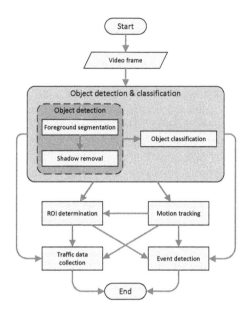

FIGURE 4.1
General flowchart of an intelligent traffic video analytics framework.

and tracking, region of interest determination, and incident detection [10,11]. Figure 4.1 illustrates the general steps in an intelligent traffic video analytics framework. This section contains an overview of the different approaches used for each step.

4.2.1 Vehicle detection

Object detection is a fundamental task in many computer vision applications [12]. As opposed to single images, videos captured by stationary cameras contain moving objects which have significant motion information compared to a static background. This information can be utilized for segmenting the image into foreground and background in order to use the segmentation results for the task of object detection. Traffic surveillance cameras overlook roads, highways, and urban traffic environments and the objects of interest are all road users including different types of moving vehicles, cyclists, and pedestrians. Among different object detection techniques, background modeling has been proven to be an efficient approach to segment the foreground and obtain the location of moving objects in traffic videos [6,13–15]. In order to subtract the stationary background from the moving foreground, each video frame is compared with a background model and the pixels with considerably different values from the current background are classified as foreground. Background subtraction methods are generally categorized into five groups, namely, basic, statistical, fuzzy, neural networks, and non-parametric methods [16]. Statistical methods have shown notable performance and efficiency which has made these methods most popular in real-time surveillance applications. In most used statistical approaches the background is estimated using frame averaging [17], single Gaussian [18–21], or a mixture of Gaussian distributions [22–27].

Statistical methods based on Gaussian distribution have been the most used pixel-based approaches for modeling the background in videos due to the compromise between their computational cost and performance. Initially, each pixel was modeled using a single Gaussian distribution [18]. Later, to remove the effect of noise, camera jitter, and background texture, Gaussian Mixture Model (GMM) was proposed and each pixel was modeled by a mixture of K Gaussian distributions [22,28]. The GMM method was further improved upon by efficient parameter updating in adaptive GMM (AGMM) [23,29,30] and other pattern recognition techniques [31–34]. Several non-parametric methods have also been introduced to deal with the problem of foreground segmentation. Some of the representative non-parametric background modeling methods are Vibe [35], PBAS [36], Codebook [37,38], SOBS [26], and SACON [39]. Here, we review the GMM method as the most commonly used approach in the applications of traffic surveillance videos.

In most real-world scenarios, videos captured by stationary cameras contain animated texture, such as the movements of trees and water waves. Object detection methods based on foreground segmentation tend to discard these types of motion and classify them as background. Stauffer et al. [22] used a mixture of K Gaussian distributions to model the background pixels by taking the animated texture into consideration. By using a mixture of K Gaussian distributions for each pixel, the GMM method can deal with slow changes in illumination, slow-moving objects, repetitive motions, and cluttered regions. In the GMM method, the values of each pixel are modeled separately by a mixture of Gaussian distributions, and some Gaussians are chosen to represent the background colors according to their persistence and variance. The probability of each input pixel value x at time t is computed as follows:

$$P(x_t) = \sum_{k=1}^{K} \omega_{k,t} \cdot \eta\left(x_t, \mu_{k,t}, \Sigma_{k,t}\right) \tag{4.1}$$

where K is the total number of Gaussian distributions, $\omega_{k,t}$ is an estimate of the weight of the kth distribution in the mixture at time t which indicates the portion of the data contained by the kth Gaussian, $\mu_{k,t}$ is the mean value of the kth distribution in the mixture at time t, $\sum_{k,t}$ is the covariance matrix of the kth distribution in the mixture at time t, and $\eta\left(x_t, \mu_{k,t}, \Sigma_{k,t}\right)$ is the kth Gaussian probability density function which is defined as follows:

$$\eta\left(x_t, \mu, \Sigma\right) = \frac{1}{(2\pi)^{\frac{n}{2}}|\Sigma|^{\frac{1}{2}}} e^{-\frac{1}{2}(x_t - \mu_t)^T \Sigma^{-1}(x_t - \mu_t)} \tag{4.2}$$

In practice, K is set to a number between 3 and 5 depending on how much jitter or small movements occur at the pixels. In the GMM method, the red, green, and blue channels of the RGB video frames are assumed to be independent and mostly uncorrelated. Therefore, the changes in the intensity levels are presumed to have uniform standard deviations and the covariance matrix $\Sigma_{k,t} = \sigma_{k,t}^2 I$ is assumed to be diagonal.

Each new pixel x_t is checked against the K Gaussian distributions currently modeled for that pixel in order to find a match for it. A match is defined to be a Gaussian distribution that the new pixel value is within 2.5 standard deviations from its mean. If no match is found the Gaussian distribution with the lowest weight (the least probable) is replaced with a distribution with the current pixel value x_t set as its mean value and with an initial high variance and low weight. If a pixel matches with one of the current K Gaussian distributions, the prior weights are updated as follows:

$$\omega_{k,t} = (1 - \alpha)\omega_{k,t-1} + \alpha \cdot M_{k,t} \tag{4.3}$$

where α is a learning rate and $M_{k,t}$ indicates weather the new observation matches the model k or not:

$$M_{k,t} = \begin{cases} 1 & \text{, if } x_t \text{ matches with } k^{th} \text{ distribution} \\ 0 & \text{, otherwise} \end{cases} \tag{4.4}$$

The parameters of unmatched distributions do not change unless they match with the new value. When a Gaussian distribution matches with the new observation, it's mean and standard deviation parameters are updated as follows:

$$\mu_{k,t} = (1 - \rho)\mu_{k,t-1} + \rho x_{k,t}$$
$$\sigma_{k,t}^2 = (1 - \rho)\sigma_{k,t-1}^2 + \rho (x_{k,t} - \mu_{k,t})^T (x_{k,t} - \mu_{k,t}) \tag{4.5}$$

where ρ is another learning rate that is defined as follows:

$$\rho = \alpha \eta (X_t | \mu_k, \sigma_k) \tag{4.6}$$

In order to classify the pixels as foreground and background, the K Gaussian distributions are ordered based on the value of ω/σ, and the first B distributions are considered to be of the background model. Every Gaussian distribution that has a weight larger than a designated threshold T is considered as part of the background model and the rest are considered as foreground:

$$B = argmin_b \left(\sum_{k=1}^{b} \omega_k > T \right) \tag{4.7}$$

where T is a predefined threshold. If a pixel is not classified as the background it will be assigned to the foreground class until there is a Gaussian distribution that includes its value with sufficient evidence supporting it.

As opposed to simple and fast background subtraction methods such as frame differencing and median filtering that is not sufficient for complex real-world scenarios, The adaptive Gaussian mixture model is capable of handling some challenging cases, such as long-term changes in the scene, repetitive small motions in the clutter, and multi-modal backgrounds. In the GMM method, each pixel has separate thresholds that are adapting by time and there is no need to find unified global thresholds for all pixels. Objects can become part of the background after they are stopped for a period of time. Also, using a mixture of Gaussian distributions makes it possible to subtract the background in videos with jitter and gradual illumination changes. Nevertheless, the GMM method has a number of disadvantages which many studies have attempted to address. One of the main disadvantages of the GMM method is the inability to deal with sudden illumination changes and slow recovery from failures. Also, the parameters, such as the number of Gaussian distributions(K), learning rate (α), and the classification threshold (T) are predefined and are supposed to be initialized and tuned manually. In addition, when a moving object is stopped its pixel values become part of the background model which results in the object being removed from the foreground. Background subtraction methods including GMM face several challenging situations, such as dynamic background, foreground aperture, camouflage, moving cast shadows, sudden illumination changes, stopped and slow-moving objects, camera jitter, and bootstrapping [40].

The shortcomings of the GMM method have been addressed in many studies in an attempt to improve the classification results [23, 30, 34, 41, 42]. For example, Zivkovic et al. [23] proposed a model selection criterion in order to automatically choose the right number of components for each pixel. In this method which is called efficient GMM, the background model is estimated by the B largest components as follows:

$$B = argmin_b \left(\sum_{k=1}^{b} \omega_k > (1 - c_f) \right) \tag{4.8}$$

where c_f is a measure that indicates the maximum portion of the data classified as foreground without effecting the background model. The weight update procedure is adapted to limit the influence of the older frames by an exponentially decaying envelope $\alpha = \frac{1}{T}$ as follows:

$$\omega_{k,t} = \omega_{k,t-1} + \alpha \left(M_{k,t} - \omega_{k,t-1} \right) - \alpha c_T \tag{4.9}$$

where c_T is a negative prior evidence weight that is introduced to suppress the components that are not supported by the data and to ensure that components with negative weights are discarded and the mixing weights are not negative [43]. The weights are normalized after each update so that the summation of all weights is one. In the beginning, there is only one component for each pixel centered on the first pixel's value. For each new value, the Mahalanobis distance between the new sample and the mean of each component is calculated and compared with a threshold. The ownership $M_{k,t}$ is set to one for the matching component and set to zero for the remaining ones. In case no component matches the new sample, a new component with $\omega_{k+1,t} = \alpha$, $\mu_{k+1,t} = x_t$, and $\sigma_{k+1,t} = \sigma_0$ is constructed where σ_0 is an initialized value. In case of reaching the maximum number of components, the Gaussian distribution with the smallest weight is discarded.

The efficient GMM method only uses a single learning rate which can reduce the accuracy. If the learning rate is set to a relatively high value, the update speed is increased and slow-moving or stopped objects are quickly absorbed in the background model. This problem will increase the number of false negatives as the slow-moving or stopped objects are incorrectly classified as background. On the other hand, a relatively low learning rate will result in the model not being able to adapt to sudden illumination changes which in turn increases the number of false positives as the pixels affected by sudden illumination changes are classified as foreground. These issues are common among most background subtraction methods and various studies have applied several techniques to deal with the challenging scenarios. The Global Foreground Modeling method (GFM) [34, 44] has increased the dimensionality of the feature vector in order to improve the discriminatory power of the feature vectors. In addition to the RGB values, the color values in the YIQ and YCbCr color-spaces, the horizontal and vertical Haar wavelet features, and the temporal difference features are integrated in order to enhance the discriminating power of the pattern vectors for better classification performance. Another improvement in the GFM method is the foreground being modeled globally in addition to the local background modeling. A Gaussian distribution is derived for each foreground pixel instead of every location. None of the components in the foreground model is eliminated, even if the weight is relatively low; as each component represents a foreground feature vector. As a result, the classification performance of the GFM method is improved compared to the other GMM-based methods. Additionally, the GFM method is able to detect the moving objects even after they have slowed down or stopped due to the use of two separate models for background and foreground. The Bayes classifier is applied to classify the pixels into the foreground and background categories.

In the GFM method, K Gaussian distributions are defined for all foreground pixels and there is a counter for each distribution. For a new feature vector, the GFM model is first searched to find a component with zero weight. If such a component exists the background model is checked to see if the new feature vector can be absorbed by the background GMM model. In case it can be absorbed by the background model, the GFM model remains unchanged. Otherwise, if the new feature vector is not absorbed by the background model or by any of the positive-weight components in the foreground model, a new Gaussian distribution is created with the input feature vector as the mean and the identity matrix multiplied by a predefined value as the covariance matrix and the counter of the new distribution is set to one. In case all the components in the foreground model have positive weights the Bayes decision rule for minimum error is applied in order to identify a Gaussian density for the input feature vector and the Gaussian density becomes the conditional probability density function for the foreground as follows:

$$p(\mathbf{x}|\omega_f)P(\omega_f) = \max_{i=1}^{K}\{p(\mathbf{x}|\omega_i)P(\omega_i)\} \qquad (4.10)$$

where $\mathbf{x}$ is the feature vector and $p(\mathbf{x}|\omega_f)$ is the foreground conditional probability density function in the next segmentation step. The Gaussian density function and the weights in the GFM model are updated as follows:

$$\mathbf{M}'_k = (n_k\mathbf{M}_k)/(n_k+1)$$
$$\sigma'_k = (n_k\Sigma_k + (\mathbf{x} - \mathbf{M}_k)(\mathbf{x} - \mathbf{M}_k)^t)/(n_k+1)$$
$$n'_k = n_k + 1 \qquad (4.11)$$
$$\alpha'_k = n'_k / \sum_{k=1}^{K} n'_k$$

where $\mathbf{M}_k$ is the mean vector for the k_{th} Gaussian distribution, Σ_k is the corresponding covariance matrix, n_k is the counter, and α_k is the weight. In order to model the background locally for each pixel, the most significant Gaussian distribution in the GMM model is chosen as the conditional probability density function for the background: $p(\mathbf{x}|\omega_b) = N(\mathbf{M}'_1, \Sigma'_1)$, where ω_b represents the background class. The prior probability for the background, $P(\omega_b)$, is estimated using the weight for the most significant Gaussian density $P(\omega_b) = \alpha_1$ and the prior probability for the foreground, $P(\omega_f)$, is estimated as follows: $P(\omega_f) = 1 - \alpha_1$. The feature vector is classified to either one of the foreground or background classes according to the following discriminant function:

$$h(\mathbf{x}) = p(\mathbf{x}|\omega_f)P(\omega_f) - p(\mathbf{x}|\omega_b)P(\omega_b) \qquad (4.12)$$

The feature vector $\mathbf{x}$ is classified to the foreground class if $h(\mathbf{x}) > 0$, and is classified to the background class otherwise.

One of the advantages of the GFM method is the increased classification performance due to the application of the Bayes decision rule for minimum error. Another advantage is the ability to continuously detect the foreground objects even after they are stopped. This is because a separate global model is trained for the foreground objects and the GFM model will maintain an accurate Gaussian distribution to model the foreground pixels. These advantages are useful in the applications of security cameras and traffic surveillance videos where the moving people or vehicles are detected even after they have slowed down or stopped. As an example, the slow or stopped vehicles on the side of highways pose danger to high-speed traffic and should be detected as abnormal driving behavior. Figure 4.2 shows and example of the foreground segmentation applied on a real traffic video data. Note that the GFM method extracts a more accurate foreground mask with both the moving vehicles (blue) and the stopped vehicles (red) clearly detected in the binary mask. In comparison, the MOG method fails to detect the stopped vehicles.

(a) Original frame   (b) MOG foreground   (c) GFM foreground

FIGURE 4.2
The foreground masks extracted using the MOG method and the GFM method, respectively. The GFM method extracts both the moving vehicles (blue) and the stopped vehicles (red) clearly [45].

4.2.2   Shadow removal

One of the main challenges in traffic surveillance videos is the cast shadows of vehicles which has negative effects on further video analytics tasks, such as vehicle tracking and classification. Most foreground detection methods classify the shadows caused by moving objects as foreground due to the similarities in the motion patterns among the shadows and the moving objects. This misclassification tends to deteriorate the performance of further steps of the video analytics. For example, we can consider the adverse effects of the moving cast shadows in vehicle tracking methods that are based on the foreground blobs as the shadowed pixels may connect the foreground blobs of vehicles and cause multiple vehicles to be tracked as one. Many studies have attempted to detect the moving cast shadows and remove them from the foreground class [46–49] using various approaches, such as applying color information [50–54], using statistical methods [55, 56], applying texture information [57–59], or other information such as shape, size, and direction [60–62]. Also, the applicability of supervised methods based on deep neural networks has been studied for removing shadows from single images and video sequences [63–65].

Most shadow detection methods utilize two reference points for the same surface, namely shadowed and lit samples. In the case of video data captured by a stationary camera the moving objects are detected by applying foreground segmentation techniques and the two reference points are taken from the same location at different times. The lit sample is taken from the background model and the shadowed sample is taken from the foreground model when there is attenuation in illumination. A large number of studies assume that the shadow region is darker and has lower brightness compared to the corresponding lit region with slight changes in chromaticity. Some methods tend to exploit the shadow properties at the pixel level and classify the foreground pixels into two groups, namely objects and shadows. Cucchiara et al. [66] analyze the background and shadow properties in the Hue-Saturation-Value (HSV) color-space for shadow removal in traffic control and surveillance applications based on the assumptions that the background is darkened in the luminance component, while the hue and saturation components are less affected within certain limits. Horprasert et al. [67] propose a model to compare the intensity component with the chromaticity component at each pixel. A combination of three thresholds, defined over a single Gaussian distribution, is used to classify each pixel as background, shaded, highlighted, or moving object. McKenna et al. [68] made similar assumptions about the significant attenuation in intensity with slight changes in chromaticity and modeled the pixel chromaticity and the first-order gradient using mean and variance values to classify shadowed pixels in the background class. The main drawback of pixel-based methods is the classification errors due to dark objects exhibiting similar behavior to the shadowed regions. When the attenuation in brightness and the changes in chromaticity of a pixel are similar to a shadowed pixel, the pixel-based methods are unable to classify that pixel as a moving object. The other problem with most shadow removal methods that only work at the pixel level is the requirement to set and tune explicit parameters and thresholds which can differ from one video to another.

Some studies have focused on statistical modeling of shadow properties using temporal information [69–72]. These methods benefit from the repetitions in shadowed values of the pixels throughout the video frames in order to use the statistical information for modeling the extracted features and classify the foreground into object and shadow categories. For instance, Huang et al. [72] refer to the physical properties of reflection models in order to extract discriminatory features for each pixel and model the extracted features by applying the Gaussian mixture model. Martel-Brisson et al. [69] present a non-parametric framework to model the extracted physical features in order to identify the direction in the RGB space representing the background pixel values under the shadow. The statistical methods require a longer training period and their performance depends on the frequency of moving objects. Also, most methods that use physics-based heuristics in order to extract shadow features are dependent on three color channels, especially in the case of chromatic shadows where the ambient illumination is not negligible.

A number of studies have attempted to detect and remove cast shadows at region or block-level [73–76]. Stander et al. [73] propose a luminance model based on physics-based properties to describe illumination changes. The covered background regions are integrated temporally while the uncovered background regions are subtracted. The results of detecting changes, static edges, penumbra, and shading are combined in order to detect cast shadows. Toth et al. [74] apply color and shading information to propose a shadow detection method. The mean-shift algorithm is applied in order to segment the input image into several regions and color ratio constancy is assumed over a small shadowed region. Similarly, Amato et al. [76] propose a general model to detect chromatic and achromatic shadows even in situations when the foreground objects are very similar in color to the shadowed regions. Local color ratio constancy is exploited by dividing the values of the subtracted background values by the foreground values in RGB color-space in order to identify segmented regions with low gradient constancy and distinguish shadows from moving objects. Segmentation-based and block-based methods are computationally more expensive than other shadow detection methods. However, in comparison, they provide better results in different situations in terms of performance measures [77].

In terms of moving cast shadow detection in traffic surveillance videos, there are a few specific considerations worth noting. The entire process of the intelligent video analytics system is expected to be in real-time which allows little room to add to the computational complexity of the algorithms applied at various steps. Therefore, the shadow detection and removal step is supposed to be light and quick to avoid unbearable overhead. This means that any method used for shadow detection in traffic video analytics cannot be computationally expensive. Another observation that is specific to applications such as traffic videos is the fact that most road users, such as vehicles are located between a single environmental light source and the corresponding cast shadows, which makes it possible to apply heuristics based on the light direction and assuming one single region as the shadow for each foreground blob. In addition to the above considerations, we can also note that in most traffic surveillance videos the background region that corresponds to a shadowed region is the road surface which is mostly uniform in texture and color, and therefore, the shaded regions also exhibit similar color and texture values.

Some studies have attempted to benefit from these observations in order to develop computationally light shadow removal algorithms for traffic surveillance videos by applying a combination of low-level features and statistical modeling approaches at pixel and region levels [79, 80]. For example, Hang et al. [78, 81] propose a hierarchical cast shadow detection framework that consists of four steps. After segmenting the foreground by applying the GFM method [34], the foreground pixels are classified into two groups representing moving objects and cast shadows. First, a set of new robust chromatic criteria is presented to detect the candidate shadow pixels in the HSV color-space. The HSV color-space is chosen due to its property of separating the intensity from chromaticity [54]. Note that using precise limits as chromatic criteria as the initial step for extracting candidate shadows instead of only considering intensity attenuation makes this approach much less sensitive to dark objects that have lower brightness than the background, similarly to shadows. These global criteria are based on the assumption of uniformity in the road surface as the source for the lit samples which is a property in a number of applications, such as traffic video analytics. Second, a new shadow region detection method is presented that clusters the candidate shadow pixels into shaded and object regions.

The proposed clustering method in [78] resolves one of the main issues that most shadow detection methods face which is the shadow outlines at the penumbra of shadows that are often classified as foreground. The assumption about the moving objects and their corresponding cast shadows define continuous regions in applications, such as traffic video analytics, helps with

(a)   (b)

FIGURE 4.3
The foreground mask after removing shadows. (a) Original video frame. (b) The results of the Hang and Liu [78] shadow detection method.

clustering each foreground blob into two regions only based on the centroids of the candidate object and shadow pixels obtained at the previous step. Third, a statistical global shadow modeling method is proposed to model the shaded values using a single Gaussian distribution. A fusion of the supposedly shadowed pixels at the two previous steps is considered as the input shadow values which are used to estimate the Gaussian distribution of all shadowed pixels. The choice of a single Gaussian distribution for modeling all shadow pixels globally is based on the uniformity assumption of the road surface across each video frame. Finally, the results of the three shadow detection steps are integrated by calculating a weighted summation of the candidate shadowed pixels. A gray-scale map for shadows is generated and thresholded to derive a shadow-free foreground. Compared to other shadow removal approaches, the method of Hang et al. [78, 81] shows high performance while using little computational resources which makes it applicable in real-world traffic video applications. Figure 4.3 illustrates an example of moving cast shadow removal by applying the method introduced in [78].

4.2.3   Vehicle classification

Vehicle classification is one of the main functionalities of traffic monitoring systems and has broad applications in intelligent transportation systems and smart cities. A high-performance classification system that is able to classify the vehicles into different types accurately, can help the traffic monitoring systems with effective operations and transportation planning. The geometric roadway design is based on the types of vehicles that commonly pass along the road. The estimation of the capacity of highways and the planning for pavement maintenance are affected by the information about the number of different motor-vehicle types utilizing the highway sections. There are numerous vehicle detection and classification systems deployed around the world that use different sensors and are implemented at different locations. Based on the location of the system deployment, vehicle classification systems are generally categorized into three groups which are in-roadway-based, over-roadway-based, and side-roadway-based systems [82]. Vehicle classification in terms of intelligent video analytics falls under the second category as the cameras are the most widely used sensors for this group of systems. Compared to other groups of vehicle classification systems, installation of cameras over the roadways not only covers a wide area and multiple lanes at once, it greatly reduces the cost of construction and maintenance.

The general process of camera-based vehicle classification systems consists of several steps from capturing images from the moving vehicles, feature extraction, and applying classification techniques. For example, Chen et al. [83] use the GMM method along with shadow removal to detect the moving vehicles, track the vehicles by applying the Kalman filter technique, and apply SVM to classify the vehicles into five categories, i.e., cars, vans, buses, motorcycles, and unknown. Unzueta et al. [84] develop a multi-cue background subtraction method that can deal with dynamic changes in the background. Spatial and temporal features are applied in order to generate two-dimensional estimations of each vehicle's silhouette and augment the estimations to three-dimensional vehicle volumes. Vehicles are classified into two-wheels, light vehicles, and heavy vehicles. Karaimer et al. [85] try to improve the classification performance by combining the shape-based classification with Histogram of Oriented Gradient (HOG) feature-based classification techniques. The k-nearest neighbors algorithm (kNN) is applied on shape-based features such as convexity, rectangularity, and elongation, and the results are combined with the classification results of a Support Vector Machine (SVM) applied on HOG features. The combination is based on the sum rules and the product rules looking for the maximum probability values. In this study, vehicles are classified into three types, namely, cars, vans, and motorcycles.

Many studies have used methods based on deep neural networks (DCNNs) in order to extract high-level and effective features for the task of vehicle classification. For example, Huttunen et al. [86] develop a DNN to extract features from car images and align a bounding box around each car. The method is evaluated on a dataset consisting of 6,555 images of four vehicle classes, namely, small cars, trucks, buses, and vans. Convolutional neural networks (CNNs) have also gained a lot of attention in object classification tasks. Dong et al. [87] use vehicle front-view images for classification in a semi-supervised CNN with two stages. First, the effective filter back of CNN is obtained in an unsupervised manner to capture discriminative features. Second, the multitask learning approach [88] is applied to train a Softmax classifier in order to estimate the probability of each vehicle type. This method was evaluated on two datasets containing different vehicle types and road users, such as bus, micro-bus, minivan, sedan, SUV, truck, passenger car, and sedan. Adu-Gyamfi et al. [89] fine-tune a pre-trained DCNN with a domain-specific dataset containing images captured by CCTV cameras at Iowa and Virginia in order to quickly classify the vehicles into 13 types. Faruque et al. [90] apply two representative DCNN-based methods, namely Faster R-CNN [91] and YOLO [92] to classify vehicles in three datasets constructed from videos that are provided

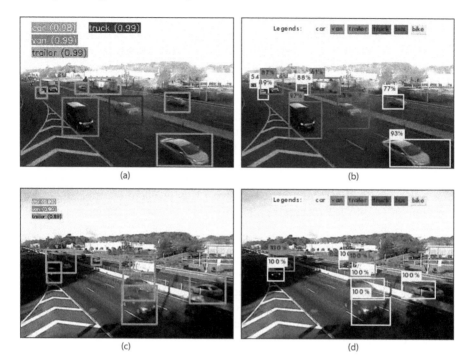

FIGURE 4.4

Example testing results of the YOLO and the Faster R-CNN deep-learning methods using traffic videos that are not seen during training. The frames in (a) and (c) show the vehicle classification results by using the YOLOv3 deep-learning method. The frames in (b) and (d) show the vehicle classification results by using the Faster R-CNN deep-learning method [90].

by the New Jersey Department of Transportation (NJDOT). The vehicles are classified into six types as defined by the Federal Highway Association (FHWA): bike, car, truck, van, bus, and trailer. According to the evaluations, YOLO has performed faster and with higher precision in comparison to Faster RCNN. Figure 4.4 shows samples of vehicle classification using the YOLO and Faster R-CNN methods.

Despite the ability of deep-learning approaches to extract higher-level and more effective features and the improvements in the accuracy of classification tasks, several challenges are faced when applying these methods. Methods based on deep learning require a considerable amount of data which is hard to acquire and not always available due to the need for manual labeling and annotations. The multiple hidden layers in the structure of the neural networks make these methods more computationally expensive and difficult to interpret and configure the parameters. Also, in terms of generalizability, deep-learning methods usually fail to adapt to unseen data in case the data domains have different distributions. Most vision-based vehicle classification systems are trained using relatively small datasets which contain little variability in terms of view perspective, weather conditions, and image resolution. Theagarajan et al. [93] use a large dataset of 786,702 vehicle images from several cameras at 8,000 different locations in the USA and Canada. A DNN-based vehicle classification framework is applied in order to classify the road users into ten categories, namely, cars, bicycles, motorcycles, buses, non-motorized vehicles, pedestrians, pickup trucks, articulated trucks, single-unit trucks, and work vans. However, according to a study on using deep-learning techniques for classification of 300 million images [94], it is shown that the performance has increased logarithmically with increasing the size of the training data. Several studies have attempted to deal with this problem by applying different domain adaptation techniques [95–97]. However, the performance of these methods is not good enough to be applicable in real-world scenarios. Another challenge that is faced by most machine-learning methods in terms of vehicle classification is the similarities among different types of vehicles. Javadi et al. [98] have addressed this issue by applying the fuzzy c-means (FCM) clustering method [99] according to the traffic regulations and the differences in the speed of vehicles with similar dimensions. The vehicles are classified into four different types, namely, private cars, buses, light trailers, and heavy trailers, collected from a real highway video.

To mention another challenge in vehicle classification techniques based on machine-learning algorithms we can refer to the lack of distinctions in viewing different parts of each vehicle. As opposed to the multi-glimpse and visual attention mechanism in human vision system [100] meaning that a human is able to distinguish the key parts of an object from the background and only focus on the most relevant parts of an image. Zhao et al. [101] exploit the visual attention mechanism in order to create a focused image which is in turn passed as input to a DCNN for accurate vehicle classification. In this study, the vehicles are classified into five categories: sedans, vans, trucks, SUVs, and coaches. Another challenging issue in object classification which is more prominent in applications such as traffic videos is the occlusion problem. Especially, in traffic surveillance videos that are captured by a stationary camera, the vehicles pass along the roadway continuously and occlusion often happens at the borders of the video frame or when vehicles are close to each other. Chang et al. [102] assumed vehicles to be convex regions and derived a decomposition optimization model in order to differentiate single vehicles from multiple occluded vehicles. The vehicles are classified into five different types, namely, sedans, vans, SUVs, buses, and trucks by applying a convolutional neural network.

4.2.4   Vehicle tracking

The next significant step after detecting the objects in traffic videos is to associate the locations of the objects at each frame with their corresponding location across multiple consecutive frames in order to track the motion of vehicles and other road users. Vision-based object tracking methods are in general divided into two groups of single-object and multiple-object tracking. In terms of traffic videos, we usually have to deal with multiple road users simultaneously at each frame and therefore, multiple-object tracking (MOT) methods are applied. Object tracking methods are also divided into two groups based on the order of the reference frames used to track objects at each frame. While one group of methods, so-called batch-wise strategies, utilize both previous and future frames to keep track of the objects at the current frame [103,104], another group of tracking methods, so-called online tracking strategies, only depend on the information gained up to the current frame [105]. In the case of traffic video analysis in real-world applications, the processing has to be real-time and there is no prior information about the upcoming frame. Therefore, only online tracking approaches are applicable. Many studies have tried to improve the performance of multiple object tracking methods both in terms of object detection and object association [105–108]. Vehicle tracking in traffic videos has been categorized into a number of different approaches based on features extracted from the data and vehicle representation [109,110] which are briefly reviewed.

4.2.4.1 Region-based tracking

The most-used approaches for tracking multiple vehicles in traffic videos are based on regions, also called blobs, each of which is defined to be a connected component of the image where the pixels of each component have common properties, such as similar intensities, color, texture, and temporal features. In this group of methods each region is associated with one vehicle and is tracked across the frames by using various clues, such as appearance, motion, and geometrical properties. As an example of blob-tracking methods, we can refer to the approach applied by Chang et al. [111] where the foreground blobs at each frame are associated with the closest blobs of the previous frame in terms of the Euclidean distances between blob centroids. Region-based tracking methods are computationally efficient and work well for traffic surveillance videos, especially the videos captured from the highways. Nevertheless, these methods have poor performance in case of traffic congestion, vehicle occlusion, and cluttered background as the regions merge and the vehicles cannot be distinguished.

4.2.4.2 Contour-based tracking

Another group of multiple object tracking methods tends to represent the objects using their contours or so-called boundaries which are updated dynamically at each frame. These methods have a lower computational overload in comparison to the region-based methods as they provide a more efficient description of the vehicles and other road users. However, the same problems of the region-based methods caused by vehicle occlusions and merging of the contours result in drastic drops in the performance of tracking. These sorts of problems are usually dealt with by active track grouping strategies and heuristic update policies [112].

4.2.4.3 Feature-based tracking

This group of tracking methods uses various features to represent the vehicles instead of connected components. Various representative features, such as corners, general shape, lines, and speeded up robust features (SURF) are extracted at each frame and matched over the sequence of frames. The performance of this group of tracking methods relies on the discriminatory power of the selected features. Most feature-based tracking methods [112–115] work well under different illumination conditions, however, the partial vehicle occlusions can have negative effects on them.

4.2.4.4 Model-based tracking

In this group of tracking methods a model is projected onto the image and matched across the frames based on the vehicle motion. In model-based methods, [116–118] the trajectories, models, and the geometrical poses of the vehicles are recoverable and estimated more accurately compared to other approaches. For example, Danescu et al. [116] model the vehicles as three-dimensional cuboids that are generated by grouping the three-dimensional points which are obtained using stereo-vision. The tracking is performed by matching the measurements and the models which are in turn done by intersecting the rectangles in the bird's-eye view and a corner-by-corner matching in the image space. As another example, we can refer to the work of Barth [117] where the vehicles from the oncoming traffic are modeled by a cloud of three-dimensional points under the rigid body assumption. The vehicle models are then updated by using stereo vision and tracking image features using an extended Kalman filter. Model-based methods have higher computational complexity and rely on accurate geometric modeling of the objects and therefore, there is a need for additional heuristic cues at initialization.

4.2.5 Region of interest determination

A region of interest (ROI), is a sample within a dataset identified for a particular purpose [119]. In the case of image and video analytics, the region of interest refers to a subspace of the image or video frame that indicates the region of the main focus. If the region of interest is pre-defined all the video analytics algorithms are performed only on the data points which belong to that region and the other samples are marked unimportant and therefore ignored. Indicating one or multiple regions of the video frame as ROI, not only decreases the computational load due to a lower volume of processed data but reduces the unnecessary incorrect results which in turn improves the performance of video analytics applications. Real-world applications of traffic video analytics are operated in real-time and have extremely limited tolerance for false alarms. Therefore, selecting an accurate and well-defined ROI as a pre-processing step means a great deal to traffic video analytics applications. In the case of traffic surveillance video, the region of interest usually refers to the road area and its proximity as the objects of interest are the road users. In most traffic video applications the ROI is selected manually by human agents which are required to be performed for every video during system installation. However, the manual selection of the ROI limits the flexibility of the vision-based traffic monitoring systems as the ROI-determination has to be repeated in case the viewing distance or perspective changes. For example, in the case of the widely used pan-tilt-zoom (PTZ) cameras [120] the operator can adjust the parameters online at any time and the ROI should be redefined each time the view changes. On the other hand, all the subsequent tasks of the video analytics depend on the initial region of interest as they ignore the data samples of other regions. Therefore, intelligent traffic monitoring systems demand the region of interest be defined and updated automatically and accurately without manual intervention.

Many studies have attempted to determine the region of interest automatically by proposing various approaches which are mostly referred to as road detection, lane detection, road segmentation, road boundary detection, vanishing point detection, or ROI determination. We can categorize these studies into two groups based on the application use-cases. One group of road detection methods tend to segment the road region in traffic surveillance videos [121–130], while another group of methods attempts to perform road segmentation in applications with in-vehicle perception [131–137]. Several techniques and algorithms applied in these studies are applicable in both groups; however, the main intention of the former is to determine the region of interest for traffic surveillance video analytics, while the latter aims to automatically segment the road for applications such as driving assistance systems and self-driving vehicles.

Various features and heuristics have been applied to distinguish the pixels that belong to the road region from the remaining pixels. Santos et al. [121] use the information of gray-amount, texture homogeneity, horizontal line, and traffic motion to construct a feature vector for each superpixel which is in turn fed to a support vector machine (SVM) with the aim of classifying the super-pixels into the road and non-road categories. Helala et al. [123, 138] generate a large number of edges using the contours of superpixel blocks and organize the edges into clusters of co-linearly similar sets. A confidence level is assigned to each cluster and the top-ranked pair of clusters are chosen to represent the road boundaries. Almazan et al. [136] use a spatial prior and estimated the location of the vanishing point and horizontal line in order to detect the road under challenging weather conditions. Cheng et al. [137] propose a road segmentation approach to combine the results of a Gaussian mixture model applied on color features with geometric cues within a Bayesian framework. Temporal features have also been applied for the task of ROI determination in traffic videos [129,130]. Lee et al. [129] extract and accumulate the difference images between two consecutive frames as the locations that contain motion in order to form a road map. A centerline divides the road map into two separate regions each of which represents one of the major traffic directions. In a similar approach, Tsai et al. [130] extract and accumulate the difference images using consecutive frames in order to obtain a road map. The motion vectors are utilized in order to separate the road map into two regions representing the bidirectional traffic. This group of methods relies heavily on the performance of the background subtraction and vehicle tracking techniques.

Still, shadows cast by stationary objects degrade the performance of segmentation methods and many studies have attempted to reduce the shadow effects prior to segmentation. In the case of road segmentation methods, shadows are mostly an issue for applications with in-vehicle perception and many approaches have been proposed to remove or reduce the effects of shadows from the road. Tong et al. [139], calculate an effective projection angle in logarithmic space to extract intrinsic images with weakened shadow effect which is used as the bases for road segmentation. The grayscale intrinsic image contains little to no gradient changes in the shadow edges which reduce the segmentation errors. Li et al. [140] estimate the spatial structure of the road and also extract an intrinsic image based on regression analysis using the color and gradient features in order to segment the road with weakened shadows. Similarly, Wang et al. [141] generate an illumination invariant image and use the mean value of a predefined triangular area as the initial color sample to obtain a number of probability maps. The probability maps are further used to segment the road which is then refined by using the road boundaries to avoid leak segmentation errors.

Deep convolutional neural networks (DCNNs) have also been applied for various segmentation tasks including road segmentation. Li et al. [142] propose a model called bidirectional fusion network (BiFNet) that is constructed from two modules: a dense space transformation module for mutual conversion between the image space and the bird's-eye-view (BEV) space, and a context-base feature fusion module for combining the information from sensors based on corresponding features. Junaid et al. [143] extract several abstract features from the explicitly derived representations of the video frames and feed the features to a shallow convolutional neural network. The supervised methods are limited to the domain of the training data and mostly not applicable in real-world scenarios and embedded systems. Nevertheless, some studies have attempted to apply domain adaptation techniques in order to construct generalizable models. For example, Cheng et al. [144] train a fully convolutional network in order to generalize road segmentation to new illumination conditions and view perspectives. The learned geometric prior is anchored by estimating the road vanishing point and is utilized to extract a number of road regions that are considered to be ground-truth data with the purpose of adapting the network to the target domain.

In the case of traffic surveillance videos, the methods which attempt to benefit from a combination of low-level features and heuristics offer a good compromise among real-time execution, performance, and generalizability, which makes them applicable in real-world scenarios. Gharemannezhad et al. [126] propose a novel Adaptive Bidirectional Detection (ABD) of region-of-interest

(ROI) to segment, the roads with bidirectional traffic flow into two regions of interest automatically. In this method, the results of statistical modeling are fused with the results of feature extraction and spatio-temporal connected component analysis in order to obtain an accurate and adaptive ROI. After applying the GFM [34] method for foreground segmentation, the corresponding RGB values of the background model at the location of pixels in the foreground model are utilized as seed point by the flood-fill operation in an accumulative manner in order to obtain an approximate region that represents the road. Simultaneously a Gaussian mixture model has applied to model the vectors of the Lucas-Kanade optical flow algorithm [145] in order to divide the final road region into two regions of interest representing the two major traffic directions.

In other studies, Ghahremannezhad et al. [125, 128] propose a model for fusing various discriminating features to extract the road region reliably and in real-time. In [128], the initial road samples are obtained in a similar way after localizing the moving vehicles by applying the GFM method. Seven probability maps are generated using the color value differences in RGB and HSV color-spaces, normalized histograms of road and non-road samples, and the results of the flood-fill operation. The generated probability maps are integrated in order to obtain a single road probability value for each location. The final road segmentation is carried away by thresholding the resulting probability map. In [125], after obtaining the initial road samples, a feature vector is constructed for each pixel by integrating features from the RGB and HSV color-spaces and a probability map is generated based on the standardized Euclidean distance between the feature vectors. On the other hand, the flood-fill operation is applied to connect the road components benefiting from the gradient information at the dominant road boundaries. This region is refined by limiting the connected component process and avoiding leak-segmentation errors by using the estimated road boundaries which are in turn calculated by second-degree polynomial curve fitting. The probability map and the estimated region located by the connected component process are integrated in order to derive the final segmented road region. In Figure 4.5 some examples of road segmentation can be seen by applying the method introduced in [125].

4.2.6   Incident detection

Besides collecting useful traffic data one of the main functionalities of traffic monitoring systems is to detect or predict various traffic incidents in real-time in order to take appropriate actions upon discovering an unusual event. Along with various sensors which the road networks are equipped with, surveillance cameras provide useful and continuous information, such as speed, occupancy, and traffic flow. This information can be used for Automatic Incident Detection (AID) in order to notify the operators in the control center who are on the alert for unusual traffic conditions. The faster an incident is detected and verified, the more time is saved in recovery. On the other hand, automatic incident detection improves the safety of the roads and enables more prompt

(a) video frame   (b) ROI

FIGURE 4.5
Examples of road extraction in traffic videos [125].

response to the victims of traffic accidents within the golden hour (the first hour after the occurrence of the accident).

Implementing a coordinated and systematic set of response actions in order to prevent incidents in potentially dangerous situations and handling the incidents quickly after occurrence are the main objectives of traffic incident management. The continuous image streams from traffic surveillance cameras are processed as input data in AID systems with the purpose of recognizing patterns that indicate potential incidents. Each time a potential incident is detected the traffic operators are notified who try to validate the event and take appropriate steps towards incident management. Video-based incident detection algorithms should be able to detect various events, such as stopped vehicles, slow speed, wrong-way driving vehicles, congested traffic lanes, vehicle collisions, and pedestrians in the road. Many studies have attempted to deal with detecting a specific incident or a general anomaly in vision-based traffic monitoring [146, 147]. Here, we review some of the most common traffic incidents and the various approaches attempted to detect the events automatically by applying different algorithms.

4.2.6.1 Stopped vehicle detection

Stopped vehicles in the middle or on the shoulders of roads and highways pose a great deal of danger to the drivers and passengers and are one the main reasons for traffic accidents. According to the AAA Foundation for Traffic Safety [148], approximately 12 percent of interstate highway deaths were due to accidents on the road shoulders. Therefore, it is necessary to quickly notify the authorities if a vehicle has stopped on the highway in an unexpected location. Since manual contact and notification can take several minutes, traffic monitoring systems demand an automatic approach to detect the stopped vehicles. There have been a number of studies with the specific goal of detecting stopped vehicles in traffic surveillance videos. For example, Alpatov et al. [149] apply a simple background subtraction technique for detecting stopped vehicles by subtracting the earliest background estimations from the current background estimation. Several other studies [44, 150–153], which are mostly referred to as abandoned object detection or stationary object detection methods, have also attempted to propose various approaches to detect the objects such as vehicles continuously after they stop moving.

4.2.6.2 Wrong-way vehicle detection

Wrong-way driving (WWD), also known as counterflow driving, refers to the act of driving a motor vehicle against the pre-defined correct direction of traffic. WWD is an illegal driving behavior with high risk and is especially a serious issue when it comes to driving on divided highways and freeways which can result in fatal head-on collisions. Wrong-way driving should be detected and handled instantly in order to avoid endangering the lives of road users. One of the applications of intelligent traffic video analytics is detecting wrong-way vehicles in real-time based on various vehicle detection and tracking techniques. There are a number of studies that have focused on the task of wrong-way vehicle detection [154–156]. Ha et al. [156] implement a system that combines background subtraction with an optical flow algorithm to analyze and learn the motion orientation of each lane and detect the vehicles moving in the wrong direction. Rahman et al. [154] propose an automatic wrong-way vehicle detection system for on-road surveillance cameras. Vehicles are detected by applying the You Only Look Once (YOLO) method and their centroids are tracked in the region of interest (ROI) in order to identify the wrong-way vehicles.

4.2.6.3 Traffic congestion detection

One of the main purposes of traffic management systems is to detect traffic congestion, otherwise known as traffic jams, in order to analyze the information for better planning. Road traffic congestion refers to a condition in transport where the number of vehicles exceeds a certain limit based on the capacity of the road lanes which results in longer trip times and increased vehicular queuing. Many studies have attempted to detect traffic congestion [157–160] in videos which helps the traffic monitoring centers with useful information about traffic jams. Ke et al. [157] present a multi-step method for road congestion detection that consists of foreground density estimation based on gray-level co-occurrence matrix, speed detection based on Lucas-Kanade optical flow algorithm with pyramid implementation, background modeling based on the Gaussian mixture model, and applying a convolutional neural network (CNN) for vehicle detection from the candidate foreground objects. Road congestion is detected in terms of a multidimensional feature space by integrating the information of traffic density, road occupancy, traffic velocity, and traffic flow. Wei et al. [158] proposed an urban road traffic congestion algorithm for real-time applications. In this method, the region of interest (ROI) is selected manually and the difference between the texture features of the congestion image and unobstructed image are taken into account in order to detect vehicle congestion through image grayscale relegation and gray-level co-occurrence matrix calculation. Chakraborty et al. [159] present a traffic congestion detection method that applies a deep convolutional neural network (DCNN) for binary classification of the road traffic condition into two categories of jammed or not jammed.

4.2.6.4   Traffic accident detection

Traffic accidents, also referred to as vehicle collisions or car crashes occur when a vehicle collides with other vehicles, pedestrians, animals, or stationary objects on or around the road. Traffic accidents often result in injuries or fatalities and are estimated to be one of the leading causes of death around the globe. There are several types of traffic accidents, including rear-end, side-impact, head-on collisions, vehicle rollovers, or single-car accidents. Immediate reports of traffic accidents can help to reduce the time delay between the occurrence of an accident and the dispatch of the first responders to the collision location which in turn reduces the mortality rates.

There have been several studies about different approaches addressing the issue of detecting traffic accidents immediately after their occurrence [45, 161–166]. Due to the specificity and detailed motion properties of traffic collisions, they are more difficult and require higher-level features and heuristics to be detected compared to other types of traffic incidents. Yun et al. [161] propose a novel method by modeling the shape of the water surface in a field form, which is called Motion Interaction Field (MIF), using Gaussian kernels. Traffic accidents are detected by utilizing the symmetric properties of the MIF without the need for vehicle tracking. Ijjina et al. [162] propose a framework for vision-based accident detection that applies the Mask R-CNN method for accurate object detection followed by an object tracking method based on the centroids. They define the probability of an accident based on the detected anomalies in vehicle speed and trajectory after an overlap with another vehicle. The framework is evaluated on a dataset with various weather and illumination condition.

Ghahremannezhad et al. [45] propose a novel and real-time single-vehicle traffic accident detection framework which is composed of an automatic region-of-interest detection method, a new traffic direction estimation method, and a first-order logic (FOL) traffic accident detection approach. The traffic region is detected automatically by applying the global foreground modeling (GFM) method and based on the general flow of the traffic. Then, the traffic direction is estimated based on blob-tracking in order to identify the vehicles that make rapid changes of direction. Finally, the traffic accidents are detected using the rules of the first-order logic (FOL) decision-making system. This method is capable of detecting single-vehicle run-off-road crashes in real-time which is one of the most important types of traffic accidents, especially on roads and highways.

4.2.6.5 Traffic anomaly detection

Aside from the studies that are focused on detecting a specific event, there are many studies that attempt to detect abnormal incidents as anomaly [167–171]. Wei et al. [167] consider stopped vehicles as anomalous behavior and introduce a novel method based on the mixture of Gaussian (MOG) background subtraction method in order to obtain the background which contains the stopped vehicles. The Faster R-CNN method is applied on the subtracted background in order to detect the stopped vehicles and report them as anomalies. Tan et al. [169] present a real-time anomaly detection algorithm by improving the efficiency of the optical flow computation with foreground segmentation and spatial sampling and by increasing the robustness of the optical flow algorithm using good feature points selection and forward-backward filtering. They detect anomalies such as pedestrians crossing the road at a wrong location, vehicles entering the side-road at a wrong location, and wrong-way vehicles. Nguyen et al. [170] propose a novel semi-supervised anomaly detection method using Generative Adversarial Network (GAN) which is trained on regular video sequences to predict future frames. The next frame is compared to the predicted ordinary frame in order to determine whether an abnormal event such as a traffic accident or stopped vehicle has occurred.

4.3 Main Challenges in Intelligent Traffic Video Analytics

In order to have a reliable vision-based traffic monitoring system the applied algorithms are supposed to be generalizable, accurate, and responsive, all at the same time. The traffic video streams should be analyzed automatically and in real-time in order to provide the operators at traffic monitoring centers with useful information about the collected traffic data, such as volume, speed, and vehicle types, as well as to detect incidents and create alerts for different events, such as stopped vehicles, slow-speed, wrong-way vehicles, pedestrians, and traffic accidents. The most challenging issue with incident detection algorithms is the need for a good compromise between generalizability and performance. Here, we briefly discuss the main challenges in intelligent traffic video analytics systems.

4.3.1   Performance and reliability

In order to be able to trust an intelligent traffic video analytics system with automatic data processing and incident detection, the operators at the traffic monitoring center tend to set the system at a sensitive level to minimize undetected events. However, configuring the parameters to increase the sensitivity of the system can inevitably result in the occurrences of false alarms. On the other hand, the tolerance for false alarms is limited and a high false alarm rate can be exhausting and reduce the reliability and usability of the traffic video analytics systems. A traffic operation center may house a large number of traffic video feeds simultaneously and an accurate intelligent video analytics system can help lighten the workload of operators to a great extent. Nevertheless, if the intelligent video analytics framework fails to collect sound and reliable data and detect all incidents correctly without too many false alarms, it can become a liability and increase the manual work. Although the operation staff can set different alert criteria for the monitored roads based on the corresponding properties, the system is still supposed to be designed with reliable algorithms that can effectively produce desired outputs under various circumstances. Therefore, the performance of intelligent traffic video analytics systems is a vital aspect and the expectations for accurate and reliable output are generally high in real-world scenarios. The intelligent traffic video analytics systems are deployed based on the existing infrastructures and the provided input video data has relatively low resolution and frame-rate which makes it even more challenging to design effective algorithms.

4.3.2   Flexibility and versatility

Traffic videos are continuously captured during day and night in many locations and can vary in weather condition, illumination, resolution, and frame-rate. Some CCTV (Closed Circuit Television) cameras provide the functionality of adjusting Pan, Tilt, and Zoom (PTZ) for the operator which means a wide range of possible changes in camera viewing angle and distance. All these various possibilities in visual properties among traffic surveillance videos, make it challenging to construct a generalizable framework for performing the tasks of intelligent traffic video analytics. Some factors are in the hands of the human agents who install and operate the traffic monitoring systems and can be chosen in a way that optimizes the applicability of the framework. For example, astute strategies in selecting camera locations along a coverage area, camera position in terms of proximity to the traffic flow, zoom, and tilt levels, and viewing angles, camera settings such as shutter speed and max gain, and the configuration parameters of the video analytics program can have a positive impact on the performance of the systems. However, there are some factors, such as illumination and weather conditions that cannot be controlled by the operators, and the intelligent video analytics system is expected to be able to adapt well in various situations.

4.3.3   Efficiency and responsiveness

One of the main factors in designing intelligent traffic video analytics systems is the processing time which is expected to be short enough to be able to process multiple video frames at each second. Real-time responsiveness is a required property for traffic video analytics applications due to the continuous stream of video frames that should be processed without delay. The computational capacity of the underlying platform, the video quality, and resolution, the number of video frames captured at each time interval, and compliance with the specified cost-efficiency policies are the key points of consideration in defining the limits for the complexity of the designed algorithms. Especially, in the case of embedded and decentralized systems where the video streams are processed locally and on-board in the camera units, the real-time constraints become more prominent. The entire procedure of reading the input video data, pre-processing steps, object detection and classification, object tracking, region of interest determination, and event detection has to be applied for each video frame multiple times during each second. The requirements for real-time performance limit the ability to apply methods and algorithms with high complexity that demand a relatively large amount of resources. Therefore, one of the fundamental challenges in intelligent traffic video analytics is to design effective algorithms, while keeping the complexity to the minimum and as consistent as possible.

4.4   Conclusions

In this chapter, the developments and challenges in intelligent traffic video analytics are reviewed briefly. After a general introduction of the benefits of intelligent traffic video analytics in traffic management systems, the main steps are discussed along with some of the most popular methods applied at each step. First, some of the representative studies about object detection and classification are summarized. Various methods proposed for the tasks of foreground segmentation, shadow removal, and vehicle classification are described that are applied to detect and classify various road users in traffic videos. Then the benefits of defining a region of interest are discussed along with a number of studies that introduce different algorithms to detect the region of interest automatically. Since incident detection is one of the main goals of intelligent traffic video analytics systems besides collecting traffic data, a section is dedicated to providing a brief introduction to the studies that have attempted to detect common traffic events. Afterward, the major challenges in intelligent traffic video analytics are briefly outlined. The imperfection of the current technologies in dealing with various challenging factors in video analytics reveals that additional efforts are needed for further improving the performance and generalization capability of the current traffic video analytics systems, which renders an important field of study with promising research opportunities.

Bibliography

[1] Nela Vlahinić Lenz, Helga Pavlić Skender, and Petra Adelajda Mirković. The macroeconomic effects of transport infrastructure on economic growth: the case of central and eastern eu member states. *Economic research-Ekonomska istraživanja*, 31(1):1953–1964, 2018.

[2] Ashley Auer, Shelley Feese, Stephen Lockwood, and Booz Allen Hamilton. History of intelligent transportation systems. Technical report, United States. Department of Transportation. Intelligent Transportation ..., 2016.

[3] Juan Guerrero-Ibáñez, Sherali Zeadally, and Juan Contreras-Castillo. Sensor technologies for intelligent transportation systems. *Sensors*, 18(4):1212, 2018.

[4] Barbara Barbagli, Gianfranco Manes, Rodolfo Facchini, and Antonio Manes. Acoustic sensor network for vehicle traffic monitoring. In *Proceedings of the 1st International Conference on Advances in Vehicular Systems, Technologies and Applications*, pages 24–29, 2012.

[5] Luz-Elena Y Mimbela, Lawrence A Klein, et al. Summary of vehicle detection and surveillance technologies used in intelligent transportation systems. 2007.

[6] Khairi Abdulrahim and Rosalina Abdul Salam. Traffic surveillance: A review of vision based vehicle detection, recognition and tracking. *International Journal of Applied Engineering Research*, 11(1):713–726, 2016.

[7] Huansheng Song, Haoxiang Liang, Huaiyu Li, Zhe Dai, and Xu Yun. Vision-based vehicle detection and counting system using deep learning in highway scenes. *European Transport Research Review*, 11(1):1–16, 2019.

[8] Ala Mhalla, Thierry Chateau, Sami Gazzah, and Najoua Essoukri Ben Amara. An embedded computer-vision system for multi-object detection in traffic surveillance. *IEEE Transactions on Intelligent Transportation Systems*, 20(11):4006–4018, 2018.

[9] Yi Tan, Yanjun Xu, Subhodev Das, and Ali Chaudhry. Vehicle detection and classification in aerial imagery. In *2018 25th IEEE International Conference on Image Processing (ICIP)*, pages 86–90. IEEE, 2018.

[10] Tomás Rodríguez and Narciso García. An adaptive, real-time, traffic monitoring system. *Machine Vision and Applications*, 21(4):555–576, 2010.

[11] Bin Tian, Qingming Yao, Yuan Gu, Kunfeng Wang, and Ye Li. Video processing techniques for traffic flow monitoring: A survey. In *2011 14th International IEEE Conference on Intelligent Transportation Systems (ITSC)*, pages 1103–1108. IEEE, 2011.

[12] Zhengxia Zou, Zhenwei Shi, Yuhong Guo, and Jieping Ye. Object detection in 20 years: A survey. *arXiv preprint arXiv:1905.05055*, 2019.

[13] KB Saran and G Sreelekha. Traffic video surveillance: Vehicle detection and classification. In *2015 International Conference on Control Communication & Computing India (ICCC)*, pages 516–521. IEEE, 2015.

[14] S Cheung Sen-Ching and Chandrika Kamath. Robust techniques for background subtraction in urban traffic video. In *Visual Communications and Image Processing 2004*, volume 5308, pages 881–892. International Society for Optics and Photonics, 2004.

[15] Belmar Garcia-Garcia, Thierry Bouwmans, and Alberto Jorge Rosales Silva. Background subtraction in real applications: Challenges, current models and future directions. *Computer Science Review*, 35:100204, 2020.

[16] Adi Nurhadiyatna, Wisnu Jatmiko, Benny Hardjono, Ari Wibisono, Ibnu Sina, and Petrus Mursanto. Background subtraction using gaussian mixture model enhanced by hole filling algorithm (gmmhf). In *2013 IEEE International Conference on Systems, Man, and Cybernetics*, pages 4006–4011. IEEE, 2013.

[17] Anuva Chowdhury, Sang-jin Cho, and Ui-Pil Chong. A background subtraction method using color information in the frame averaging process. In *Proceedings of 2011 6th International Forum on Strategic Technology*, volume 2, pages 1275–1279. IEEE, 2011.

[18] Christopher Richard Wren, Ali Azarbayejani, Trevor Darrell, and Alex Paul Pentland. Pfinder: Real-time tracking of the human body. *IEEE Transactions on Pattern Analysis and Machine Intelligence*, 19(7):780–785, 1997.

[19] Alexandre RJ François and Gérard G Medioni. Adaptive color background modeling for real-time segmentation of video streams. In *Proceedings of the International Conference on Imaging Science, Systems, and Technology*, volume 1, pages 227–232. Citeseer, 1999.

[20] Peter Henderson and Matthew Vertescher. An analysis of parallelized motion masking using dual-mode single gaussian models. *arXiv preprint arXiv:1702.05156*, 2017.

[21] Wafa Nebili, Brahim Farou, and Hamid Seridi. Background subtraction using artificial immune recognition system and single gaussian (airs-sg). *Multimedia Tools and Applications*, 79(35):26099–26121, 2020.

[22] Chris Stauffer and W Eric L Grimson. Adaptive background mixture models for real-time tracking. In *Proceedings. 1999 IEEE Computer Society Conference on Computer Vision and Pattern Recognition (Cat. No PR00149)*, volume 2, pages 246–252. IEEE, 1999.

[23] Zoran Zivkovic and Ferdinand Van Der Heijden. Efficient adaptive density estimation per image pixel for the task of background subtraction. *Pattern Recognition Letters*, 27(7):773–780, 2006.

[24] Fida El Baf, Thierry Bouwmans, and Bertrand Vachon. Type-2 fuzzy mixture of gaussians model: application to background modeling. In *International Symposium on Visual Computing*, pages 772–781. Springer, 2008.

[25] Thierry Bouwmans, Fida El Baf, and Bertrand Vachon. Background modeling using mixture of gaussians for foreground detection-a survey. *Recent Patents on Computer Science*, 1(3):219–237, 2008.

[26] Lucia Maddalena and Alfredo Petrosino. A self-organizing approach to background subtraction for visual surveillance applications. *IEEE Transactions on Image Processing*, 17(7):1168–1177, 2008.

[27] Zhenjie Zhao, Thierry Bouwmans, Xuebo Zhang, and Yongchun Fang. A fuzzy background modeling approach for motion detection in dynamic backgrounds. In *International Conference on Multimedia and Signal Processing*, pages 177–185. Springer, 2012.

[28] Chris Stauffer and W. Eric L. Grimson. Learning patterns of activity using real-time tracking. *IEEE Transactions on Pattern Analysis and Machine Intelligence*, 22(8):747–757, 2000.

[29] Zoran Zivkovic. Improved adaptive gaussian mixture model for background subtraction. In *Proceedings of the 17th International Conference on Pattern Recognition, 2004. ICPR 2004.*, volume 2, pages 28–31. IEEE, 2004.

[30] Dar-Shyang Lee. Effective gaussian mixture learning for video background subtraction. *IEEE Transactions on Pattern Analysis and Machine Intelligence*, 27(5):827–832, 2005.

[31] Atsushi Shimada, Daisaku Arita, and Rin-ichiro Taniguchi. Dynamic control of adaptive mixture-of-gaussians background model. In *2006 IEEE International Conference on Video and Signal Based Surveillance*, pages 5–5. IEEE, 2006.

[32] Nuria M Oliver, Barbara Rosario, and Alex P Pentland. A bayesian computer vision system for modeling human interactions. *IEEE Transactions on Pattern Analysis and Machine Intelligence*, 22(8):831–843, 2000.

[33] Shao-Yi Chien, Wei-Kai Chan, Yu-Hsiang Tseng, and Hong-Yuh Chen. Video object segmentation and tracking framework with improved threshold decision and diffusion distance. *IEEE Transactions on Circuits and Systems for Video Technology*, 23(6):921–934, 2013.

[34] Hang Shi and Chengjun Liu. A new foreground segmentation method for video analysis in different color spaces. In *2018 24th International Conference on Pattern Recognition (ICPR)*, pages 2899–2904. IEEE, 2018.

[35] Olivier Barnich and Marc Van Droogenbroeck. Vibe: a powerful random technique to estimate the background in video sequences. In *2009 IEEE International Conference on Acoustics, Speech and Signal Processing*, pages 945–948. IEEE, 2009.

[36] Martin Hofmann, Philipp Tiefenbacher, and Gerhard Rigoll. Background segmentation with feedback: The pixel-based adaptive segmenter. In *2012 IEEE Computer Society Conference on Computer Vision and Pattern Recognition Workshops*, pages 38–43. IEEE, 2012.

[37] Kyungnam Kim, Thanarat H Chalidabhongse, David Harwood, and Larry Davis. Background modeling and subtraction by codebook construction. In *2004 International Conference on Image Processing, 2004. ICIP'04.*, volume 5, pages 3061–3064. IEEE, 2004.

[38] Kyungnam Kim, Thanarat H Chalidabhongse, David Harwood, and Larry Davis. Real-time foreground–background segmentation using codebook model. *Real-time Imaging*, 11(3):172–185, 2005.

[39] Hanzi Wang and David Suter. A consensus-based method for tracking: Modelling background scenario and foreground appearance. *Pattern Recognition*, 40(3):1091–1105, 2007.

[40] Thierry Bouwmans. Traditional and recent approaches in background modeling for foreground detection: An overview. *Computer Science review*, 11:31–66, 2014.

[41] Pakorn KaewTraKulPong and Richard Bowden. An improved adaptive background mixture model for real-time tracking with shadow detection. In *Video-based Surveillance Systems*, pages 135–144. Springer, 2002.

[42] Zezhi Chen and Tim Ellis. A self-adaptive gaussian mixture model. *Computer Vision and Image Understanding*, 122:35–46, 2014.

[43] Kalpana Goyal and Jyoti Singhai. Review of background subtraction methods using gaussian mixture model for video surveillance systems. *Artificial Intelligence Review*, 50(2):241–259, 2018.

[44] Hang Shi and Chengjun Liu. A new global foreground modeling and local background modeling method for video analysis. In *International Conference on Machine Learning and Data Mining in Pattern Recognition*, pages 49–63. Springer, 2018.

[45] Hadi Ghahremannezhad, Hang Shi, and Chengjun Liu. A real time accident detection framework for traffic video analysis. In *the 16th International Conference on Machine Learning and Data Mining*, pages 77–92. IBAI, 2020.

[46] Andrea Prati, Ivana Mikic, Mohan M Trivedi, and Rita Cucchiara. Detecting moving shadows: algorithms and evaluation. *IEEE Transactions on Pattern Analysis and Machine Intelligence*, 25(7):918–923, 2003.

[47] Andres Sanin, Conrad Sanderson, and Brian C Lovell. Shadow detection: A survey and comparative evaluation of recent methods. *Pattern Recognition*, 45(4):1684–1695, 2012.

[48] Rohini Mahajan and Abhijeet Bajpayee. A survey on shadow detection and removal based on single light source. In *2015 IEEE 9th International Conference on Intelligent Systems and Control (ISCO)*, pages 1–5. IEEE, 2015.

[49] Agha Asim Husain, Tanmoy Maity, and Ravindra Kumar Yadav. Vehicle detection in intelligent transport system under a hazy environment: a survey. *IET Image Processing*, 14(1):1–10, 2019.

[50] Rita Cucchiara, Costantino Grana, Massimo Piccardi, and Andrea Prati. Detecting moving objects, ghosts, and shadows in video streams. *IEEE Transactions on Pattern Analysis and Machine Intelligence*, 25(10):1337–1342, 2003.

[51] Chun-Ting Chen, Chung-Yen Su, and Wen-Chung Kao. An enhanced segmentation on vision-based shadow removal for vehicle detection. In *The 2010 International Conference on Green Circuits and Systems*, pages 679–682. IEEE, 2010.

[52] Bangyu Sun and Shutao Li. Moving cast shadow detection of vehicle using combined color models. In *2010 Chinese Conference on Pattern Recognition (CCPR)*, pages 1–5. IEEE, 2010.

[53] Ivan Huerta, Michael B Holte, Thomas B Moeslund, and Jordi Gonzàlez. Chromatic shadow detection and tracking for moving foreground segmentation. *Image and Vision Computing*, 41:42–53, 2015.

[54] Vitor Gomes, Pablo Barcellos, and Jacob Scharcanski. Stochastic shadow detection using a hypergraph partitioning approach. *Pattern Recognition*, 63:30–44, 2017.

[55] Sohail Nadimi and Bir Bhanu. Physical models for moving shadow and object detection in video. *IEEE Transactions on Pattern Analysis and Machine Intelligence*, 26(8):1079–1087, 2004.

[56] Yang Wang. Real-time moving vehicle detection with cast shadow removal in video based on conditional random field. *IEEE Transactions on Circuits and Systems for Video Technology*, 19(3):437–441, 2009.

[57] Andres Sanin, Conrad Sanderson, and Brian C Lovell. Improved shadow removal for robust person tracking in surveillance scenarios. In *2010 20th International Conference on Pattern Recognition*, pages 141–144. IEEE, 2010.

[58] Ruiqi Guo, Qieyun Dai, and Derek Hoiem. Paired regions for shadow detection and removal. *IEEE Transactions on Pattern Analysis and Machine Intelligence*, 35(12):2956–2967, 2012.

[59] Tomas F Yago Vicente, Minh Hoai, and Dimitris Samaras. Leave-one-out kernel optimization for shadow detection and removal. *IEEE Transactions on Pattern Analysis and Machine Intelligence*, 40(3):682–695, 2017.

[60] Jun-Wei Hsieh, Wen-Fong Hu, Chia-Jung Chang, and Yung-Sheng Chen. Shadow elimination for effective moving object detection by gaussian shadow modeling. *Image and Vision Computing*, 21(6):505–516, 2003.

[61] Liu Zhi Fang, Wang Yun Qiong, and You Zhi Sheng. A method to segment moving vehicle cast shadow based on wavelet transform. *Pattern Recognition Letters*, 29(16):2182–2188, 2008.

[62] Chia-Chih Chen and Jake K Aggarwal. Human shadow removal with unknown light source. In *2010 20th International Conference on Pattern Recognition*, pages 2407–2410. IEEE, 2010.

[63] Yupei Wang, Xin Zhao, Yin Li, Xuecai Hu, Kaiqi Huang, and NLPR CRIPAC. Densely cascaded shadow detection network via deeply supervised parallel fusion. In *IJCAI*, pages 1007–1013, 2018.

[64] Lei Zhu, Zijun Deng, Xiaowei Hu, Chi-Wing Fu, Xuemiao Xu, Jing Qin, and Pheng-Ann Heng. Bidirectional feature pyramid network with recurrent attention residual modules for shadow detection. In *Proceedings of the European Conference on Computer Vision (ECCV)*, pages 121–136, 2018.

[65] Hieu Le, Tomas F Yago Vicente, Vu Nguyen, Minh Hoai, and Dimitris Samaras. A+ d net: Training a shadow detector with adversarial shadow attenuation. In *Proceedings of the European Conference on Computer Vision (ECCV)*, pages 662–678, 2018.

[66] Rita Cucchiara, Costantino Grana, Massimo Piccardi, Andrea Prati, and Stefano Sirotti. Improving shadow suppression in moving object detection with hsv color information. In *ITSC 2001. 2001 IEEE Intelligent Transportation Systems. Proceedings (Cat. No. 01TH8585)*, pages 334–339. IEEE, 2001.

[67] Thanarat Horprasert, David Harwood, and Larry S Davis. A statistical approach for real-time robust background subtraction and shadow detection. In *IEEE ICCU*, volume 99, pages 1–19. Citeseer, 1999.

[68] Stephen J McKenna, Sumer Jabri, Zoran Duric, Azriel Rosenfeld, and Harry Wechsler. Tracking groups of people. *Computer Vision and Image Understanding*, 80(1):42–56, 2000.

[69] Nicolas Martel-Brisson and Andre Zaccarin. Learning and removing cast shadows through a multidistribution approach. *IEEE Transactions on Pattern Analysis and Machine Intelligence*, 29(7):1133–1146, 2007.

[70] Fatih Porikli and Jay Thornton. Shadow flow: A recursive method to learn moving cast shadows. In *Tenth IEEE International Conference on Computer Vision (ICCV'05) Volume 1*, volume 1, pages 891–898. IEEE, 2005.

[71] Nicolas Martel-Brisson and André Zaccarin. Kernel-based learning of cast shadows from a physical model of light sources and surfaces for low-level segmentation. In *2008 IEEE Conference on Computer Vision and Pattern Recognition*, pages 1–8. IEEE, 2008.

[72] Jia-Bin Huang and Chu-Song Chen. Moving cast shadow detection using physics-based features. In *2009 IEEE Conference on Computer Vision and Pattern Recognition*, pages 2310–2317. IEEE, 2009.

[73] Jurgen Stander, Roland Mech, and Jörn Ostermann. Detection of moving cast shadows for object segmentation. *IEEE Transactions on Multimedia*, 1(1):65–76, 1999.

[74] Daniel Toth, Ingo Stuke, Andreas Wagner, and Til Aach. Detection of moving shadows using mean shift clustering and a significance test. In *Proceedings of the 17th International Conference on Pattern Recognition, 2004. ICPR 2004.*, volume 4, pages 260–263. IEEE, 2004.

[75] M-T Yang, K-H Lo, C-C Chiang, and W-K Tai. Moving cast shadow detection by exploiting multiple cues. *IET Image Processing*, 2(2):95–104, 2008.

[76] Ariel Amato, Mikhail G Mozerov, Andrew D Bagdanov, and Jordi Gonzalez. Accurate moving cast shadow suppression based on local color constancy detection. *IEEE Transactions on Image Processing*, 20(10):2954–2966, 2011.

[77] Mosin Russell, Ju Jia Zou, and Gu Fang. An evaluation of moving shadow detection techniques. *Computational Visual Media*, 2(3):195–217, 2016.

[78] Hang Shi and Chengjun Liu. A new cast shadow detection method for traffic surveillance video analysis using color and statistical modeling. *Image and Vision Computing*, 94:103863, 2020.

[79] Hadi Ghahremannezhad, Hang Shi, and Chengjun Liu. A new online approach for moving cast shadow suppression in traffic videos. In *2021 IEEE International Intelligent Transportation Systems Conference (ITSC)*, pages 3034–3039. IEEE, 2021.

[80] Hadi Ghahremannezhad, Hang Shi, and Chengjun Liu. Illumination-aware image segmentation for real-time moving cast shadow suppression. In *2022 IEEE International Conference on Imaging Systems and Techniques (IST)*, pages 1–6. IEEE, 2022.

[81] Hang Shi and Chengjun Liu. Moving cast shadow detection in video based on new chromatic criteria and statistical modeling. In *2019 18th IEEE International Conference on Machine Learning and Applications (ICMLA)*, pages 196–201. IEEE, 2019.

[82] Myounggyu Won. Intelligent traffic monitoring systems for vehicle classification: A survey. *IEEE Access*, 8:73340–73358, 2020.

[83] Zezhi Chen, Tim Ellis, and Sergio A Velastin. Vehicle detection, tracking and classification in urban traffic. In *2012 15th International IEEE Conference on Intelligent Transportation Systems*, pages 951–956. IEEE, 2012.

[84] Luis Unzueta, Marcos Nieto, Andoni Cortés, Javier Barandiaran, Oihana Otaegui, and Pedro Sánchez. Adaptive multicue background subtraction for robust vehicle counting and classification. *IEEE Transactions on Intelligent Transportation Systems*, 13(2):527–540, 2011.

[85] Hakki Can Karaimer, Ibrahim Cinaroglu, and Yalin Bastanlar. Combining shape-based and gradient-based classifiers for vehicle classification. In *2015 IEEE 18th International Conference on Intelligent Transportation Systems*, pages 800–805. IEEE, 2015.

[86] Heikki Huttunen, Fatemeh Shokrollahi Yancheshmeh, and Ke Chen. Car type recognition with deep neural networks. In *2016 IEEE Intelligent Vehicles Symposium (IV)*, pages 1115–1120. IEEE, 2016.

[87] Zhen Dong, Yuwei Wu, Mingtao Pei, and Yunde Jia. Vehicle type classification using a semisupervised convolutional neural network. *IEEE Transactions on Intelligent Transportation Systems*, 16(4):2247–2256, 2015.

[88] Abhishek Kumar and Hal Daume III. Learning task grouping and overlap in multi-task learning. *arXiv preprint arXiv:1206.6417*, 2012.

[89] Yaw Okyere Adu-Gyamfi, Sampson Kwasi Asare, Anuj Sharma, and Tienaah Titus. Automated vehicle recognition with deep convolutional neural networks. *Transportation Research Record*, 2645(1):113–122, 2017.

[90] Mohammad O Faruque, Hadi Ghahremannezhad, and Chengjun Liu. Vehicle classification in video using deep learning. In *the 15th International Conference on Machine Learning and Data Mining*, pages 117–131. IBAI, 2019.

[91] Shaoqing Ren, Kaiming He, Ross Girshick, and Jian Sun. Faster r-cnn: Towards real-time object detection with region proposal networks. *Advances in Neural Information Processing Systems*, 28:91–99, 2015.

[92] Joseph Redmon and Ali Farhadi. Yolov3: An incremental improvement. *arXiv preprint arXiv:1804.02767*, 2018.

[93] Rajkumar Theagarajan, Federico Pala, and Bir Bhanu. Eden: Ensemble of deep networks for vehicle classification. In *Proceedings of the Conference on Computer Vision and Pattern Recognition Workshops*, pages 33–40, 2017.

[94] Chen Sun, Abhinav Shrivastava, Saurabh Singh, and Abhinav Gupta. Revisiting unreasonable effectiveness of data in deep learning era. In *Proceedings of the IEEE International Conference on Computer Vision*, pages 843–852, 2017.

[95] Sangrok Lee, Eunsoo Park, Hongsuk Yi, and Sang Hun Lee. Strdan: Synthetic-to-real domain adaptation network for vehicle re-identification. In *Proceedings of the IEEE/CVF Conference on Computer Vision and Pattern Recognition Workshops*, pages 608–609, 2020.

[96] Luca Ciampi, Carlos Santiago, Joao Paulo Costeira, Claudio Gennaro, and Giuseppe Amato. Unsupervised vehicle counting via multiple camera domain adaptation. *arXiv preprint arXiv:2004.09251*, 2020.

[97] Mei Wang and Weihong Deng. Deep visual domain adaptation: A survey. *Neurocomputing*, 312:135–153, 2018.

[98] Saleh Javadi, Muhammad Rameez, Mattias Dahl, and Mats I Pettersson. Vehicle classification based on multiple fuzzy c-means clustering using dimensions and speed features. *Procedia Computer Science*, 126:1344–1350, 2018.

[99] James C Bezdek. *Pattern Recognition with Fuzzy Objective Function Algorithms*. Springer Science & Business Media, 2013.

[100] Ronald A Rensink. The dynamic representation of scenes. *Visual Cognition*, 7(1-3):17–42, 2000.

[101] Dongbin Zhao, Yaran Chen, and Le Lv. Deep reinforcement learning with visual attention for vehicle classification. *IEEE Transactions on Cognitive and Developmental Systems*, 9(4):356–367, 2016.

[102] Jianlong Chang, Lingfeng Wang, Gaofeng Meng, Shiming Xiang, and Chunhong Pan. Vision-based occlusion handling and vehicle classification for traffic surveillance systems. *IEEE Intelligent Transportation Systems Magazine*, 10(2):80–92, 2018.

[103] Siyu Tang, Mykhaylo Andriluka, Bjoern Andres, and Bernt Schiele. Multiple people tracking by lifted multicut and person re-identification. In *Proceedings of the IEEE Conference on Computer Vision and Pattern Recognition*, pages 3539–3548, 2017.

[104] Anton Milan, Stefan Roth, and Konrad Schindler. Continuous energy minimization for multitarget tracking. *IEEE Transactions on Pattern Analysis and Machine Intelligence*, 36(1):58–72, 2013.

[105] Siyu Tang, Bjoern Andres, Miykhaylo Andriluka, and Bernt Schiele. Subgraph decomposition for multi-target tracking. In *Proceedings of the IEEE Conference on Computer Vision and Pattern Recognition*, pages 5033–5041, 2015.

[106] Guillem Brasó and Laura Leal-Taixé. Learning a neural solver for multiple object tracking. In *Proceedings of the IEEE/CVF Conference on Computer Vision and Pattern Recognition*, pages 6247–6257, 2020.

[107] Andrii Maksai, Xinchao Wang, Francois Fleuret, and Pascal Fua. Non-markovian globally consistent multi-object tracking. In *Proceedings of the IEEE International Conference on Computer Vision*, pages 2544–2554, 2017.

[108] Liangliang Ren, Jiwen Lu, Zifeng Wang, Qi Tian, and Jie Zhou. Collaborative deep reinforcement learning for multi-object tracking. In *Proceedings of the European Conference on Computer Vision (ECCV)*, pages 586–602, 2018.

[109] Sokemi Rene Emmanuel Datondji, Yohan Dupuis, Peggy Subirats, and Pascal Vasseur. A survey of vision-based traffic monitoring of road intersections. *IEEE Transactions on Intelligent Transportation Systems*, 17(10):2681–2698, 2016.

[110] Yuqiang Liu, Bin Tian, Songhang Chen, Fenghua Zhu, and Kunfeng Wang. A survey of vision-based vehicle detection and tracking techniques in its. In *Proceedings of 2013 IEEE International Conference on Vehicular Electronics and Safety*, pages 72–77. IEEE, 2013.

[111] Fu Chang, Chun-Jen Chen, and Chi-Jen Lu. A linear-time component-labeling algorithm using contour tracing technique. *Computer Vision and Image Understanding*, 93(2):206–220, 2004.

[112] Nicolas Saunier and Tarek Sayed. A feature-based tracking algorithm for vehicles in intersections. In *The 3rd Canadian Conference on Computer and Robot Vision (CRV'06)*, pages 59–59. IEEE, 2006.

[113] Takashi Furuya and Camillo J Taylor. *Road intersection monitoring from video with large perspective deformation*. PhD thesis, University of Pennsylvania, 2014.

[114] Xiaoxu Ma and W Eric L Grimson. Edge-based rich representation for vehicle classification. In *Tenth IEEE International Conference on Computer Vision (ICCV'05) Volume 1*, volume 2, pages 1185–1192. IEEE, 2005.

[115] Jun-Wei Hsieh, Shih-Hao Yu, Yung-Sheng Chen, and Wen-Fong Hu. Automatic traffic surveillance system for vehicle tracking and classification. *IEEE Transactions on Intelligent Transportation Systems*, 7(2):175–187, 2006.

[116] Radu Danescu, Sergiu Nedevschi, Marc-Michael Meinecke, and Thorsten Graf. Stereovision based vehicle tracking in urban traffic environments. In *2007 IEEE Intelligent Transportation Systems Conference*, pages 400–404. IEEE, 2007.

[117] Alexander Barth. *Vehicle tracking and motion estimation based on stereo vision sequences*. PhD thesis, Inst. für Geodäsie und Geoinformation, 2013.

[118] Artur Ottlik and H-H Nagel. Initialization of model-based vehicle tracking in video sequences of inner-city intersections. *International Journal of Computer Vision*, 80(2):211–225, 2008.

[119] Ron Brinkmann. *The Art and Science of Digital Compositing: Techniques for Visual Effects, Animation and Motion Graphics*. Morgan Kaufmann, 2008.

[120] K-T Song and J-C Tai. Dynamic calibration of pan–tilt–zoom cameras for traffic monitoring. *IEEE Transactions on Systems, Man, and Cybernetics, Part B (Cybernetics)*, 36(5):1091–1103, 2006.

[121] Marcelo Santos, Marcelo Linder, Leizer Schnitman, Urbano Nunes, and Luciano Oliveira. Learning to segment roads for traffic analysis in urban images. In *2013 IEEE Intelligent Vehicles Symposium (IV)*, pages 527–532. IEEE, 2013.

[122] Qing-Jie Kong, Lucidus Zhou, Gang Xiong, and Fenghua Zhu. Automatic road detection for highway surveillance using frequency-domain information. In *Proceedings of 2013 IEEE International Conference on Service Operations and Logistics, and Informatics*, pages 24–28. IEEE, 2013.

[123] Mohamed A Helala, Ken Q Pu, and Faisal Z Qureshi. Road boundary detection in challenging scenarios. In *2012 IEEE Ninth International Conference on Advanced Video and Signal-Based Surveillance*, pages 428–433. IEEE, 2012.

[124] José Melo, Andrew Naftel, Alexandre Bernardino, and José Santos-Victor. Detection and classification of highway lanes using vehicle motion trajectories. *IEEE Transactions on Intelligent Transportation Systems,*, 7(2):188–200, 2006.

[125] Hadi Ghahremannezhad, Hang Shi, and Chenajun Liu. Robust road region extraction in video under various illumination and weather conditions. In *2020 IEEE 4th International Conference on Image Processing, Applications and Systems (IPAS)*, pages 186–191. IEEE, 2020.

[126] Hadi Ghahremannezhad, Hang Shi, and Chengjun Liu. A new adaptive bidirectional region-of-interest detection method for intelligent traffic video analysis. In *2020 IEEE Third International Conference on Artificial Intelligence and Knowledge Engineering (AIKE)*, pages 17–24. IEEE, 2020.

[127] Hang Shi, Hadi Ghahremannezhadand, and Chengjun Liu. A statistical modeling method for road recognition in traffic video analytics. In *2020 11th IEEE International Conference on Cognitive Infocommunications (CogInfoCom)*, pages 000097–000102. IEEE, 2020.

[128] Hadi Ghahremannezhad, Hang Shi, and Chengjun Liu. Automatic road detection in traffic videos. In *2020 IEEE Intl Conf on Parallel & Distributed Processing with Applications, Big Data & Cloud Computing, Sustainable Computing & Communications, Social Computing & Networking (ISPA/BDCloud/SocialCom/SustainCom)*, pages 777–784. IEEE, 2020.

[129] Woochul Lee and Bin Ran. Bidirectional roadway detection for traffic surveillance using online cctv videos. In *2006 IEEE Intelligent Transportation Systems Conference*, pages 1556–1561. IEEE, 2006.

[130] Li-Wu Tsai, Yee-Choy Chean, Chien-Peng Ho, Hui-Zhen Gu, and Suh-Yin Lee. Multi-lane detection and road traffic congestion classification for intelligent transportation system. *Energy Procedia*, 13:3174–3182, 2011.

[131] Chin-Kai Chang, Jiaping Zhao, and Laurent Itti. Deepvp: Deep learning for vanishing point detection on 1 million street view images. In *2018 IEEE International Conference on Robotics and Automation (ICRA)*, pages 4496–4503. IEEE, 2018.

[132] Li Zhang and Er-yong Wu. A road segmentation and road type identification approach based on new-type histogram calculation. In *2009 2nd International Congress on Image and Signal Processing*, pages 1–5. IEEE, 2009.

[133] Taeyoung Kim, Yu-Wing Tai, and Sung-Eui Yoon. Pca based computation of illumination-invariant space for road detection. In *2017 IEEE Winter Conference on Applications of Computer Vision (WACV)*, pages 632–640. IEEE, 2017.

[134] Tobias Kühnl, Franz Kummert, and Jannik Fritsch. Monocular road segmentation using slow feature analysis. In *2011 IEEE Intelligent Vehicles Symposium (IV)*, pages 800–806. IEEE, 2011.

[135] Wook-Sun Shin, Doo-Heon Song, and Chang-Hun Lee. Vehicle classification by road lane detection and model fitting using a surveillance camera. *Journal of Information Processing Systems*, 2(1):52–57, 2006.

[136] Emilio J Almazan, Yiming Qian, and James H Elder. Road segmentation for classification of road weather conditions. In *European Conference on Computer Vision*, pages 96–108. Springer, 2016.

[137] Gong Cheng, Yiming Qian, and James H Elder. Fusing geometry and appearance for road segmentation. In *Proceedings of the IEEE International Conference on Computer Vision Workshops*, pages 166–173, 2017.

[138] Mohamed A Helala, Faisal Z Qureshi, and Ken Q Pu. Automatic parsing of lane and road boundaries in challenging traffic scenes. *Journal of Electronic Imaging*, 24(5):053020, 2015.

[139] Guofeng Tong, Yong Li, Anan Sun, and Yuebin Wang. Shadow effect weakening based on intrinsic image extraction with effective projection of logarithmic domain for road scene. *Signal, Image and Video Processing*, 14(4):683–691, 2020.

[140] Yong Li, Guofeng Tong, Anan Sun, and Weili Ding. Road extraction algorithm based on intrinsic image and vanishing point for unstructured road image. *Robotics and Autonomous Systems*, 109:86–96, 2018.

[141] Ende Wang, Yong Li, Anan Sun, Huashuai Gao, Jingchao Yang, and Zheng Fang. Road detection based on illuminant invariance and quadratic estimation. *Optik*, 185:672–684, 2019.

[142] Haoran Li, Yaran Chen, Qichao Zhang, and Dongbin Zhao. Bifnet: Bidirectional fusion network for road segmentation. *arXiv preprint arXiv:2004.08582*, 2020.

[143] Muhammad Junaid, Mubeen Ghafoor, Ali Hassan, Shehzad Khalid, Syed Ali Tariq, Ghufran Ahmed, and Tehseen Zia. Multi-feature view-based shallow convolutional neural network for road segmentation. *IEEE Access*, 8:36612–36623, 2020.

[144] Gong Cheng, Yue Wang, Yiming Qian, and James H Elder. Geometry-guided adaptation for road segmentation. In *2020 17th Conference on Computer and Robot Vision (CRV)*, pages 46–53. IEEE, 2020.

[145] Bruce D Lucas, Takeo Kanade, et al. An iterative image registration technique with an application to stereo vision. Vancouver, British Columbia, 1981.

[146] Hang Shi, Hadi Ghahremannezhad, and Chengjun Liu. Anomalous driving detection for traffic surveillance video analysis. In *2021 IEEE International Conference on Imaging Systems and Techniques (IST)*, pages 1–6. IEEE, 2021.

[147] Kelathodi Kumaran Santhosh, Debi Prosad Dogra, and Partha Pratim Roy. Anomaly detection in road traffic using visual surveillance: A survey. *ACM Computing Surveys (CSUR)*, 53(6):1–26, 2020.

[148] BC Tefft, LS Arnold, and JG Grabowski. Aaa foundation for traffic safety. *Washington, DC*, 2016.

[149] Boris A Alpatov and Maksim D Ershov. Real-time stopped vehicle detection based on smart camera. In *2017 6th Mediterranean Conference on Embedded Computing (MECO)*, pages 1–4. IEEE, 2017.

[150] A Singh, S Sawan, Madasu Hanmandlu, Vamsi Krishna Madasu, and Brian C Lovell. An abandoned object detection system based on dual background segmentation. In *2009 Sixth IEEE International Conference on Advanced Video and Signal Based Surveillance*, pages 352–357. IEEE, 2009.

[151] Rajesh Kumar Tripathi, Anand Singh Jalal, and Charul Bhatnagar. A framework for abandoned object detection from video surveillance. In

2013 Fourth National Conference on Computer Vision, Pattern Recognition, Image Processing and Graphics (NCVPRIPG), pages 1–4. IEEE, 2013.

[152] Álvaro Bayona, Juan C SanMiguel, and José M Martínez. Stationary foreground detection using background subtraction and temporal difference in video surveillance. In *2010 IEEE International Conference on Image Processing*, pages 4657–4660. IEEE, 2010.

[153] Devadeep Shyam, Alex Kot, and Chinmayee Athalye. Abandoned object detection using pixel-based finite state machine and single shot multibox detector. In *2018 IEEE International Conference on Multimedia and Expo (ICME)*, pages 1–6. IEEE, 2018.

[154] Zillur Rahman, Amit Mazumder Ami, and Muhammad Ahsan Ullah. A real-time wrong-way vehicle detection based on yolo and centroid tracking. In *2020 IEEE Region 10 Symposium (TENSYMP)*, pages 916–920. IEEE, 2020.

[155] Sarah A Simpson et al. Wrong-way vehicle detection: proof of concept. Technical report, Arizona. Dept. of Transportation, 2013.

[156] Synh Viet-Uyen Ha, Long Hoang Pham, Ha Manh Tran, and Phong Ho Thanh. Improved optical flow estimation in wrong way vehicle detection. *Journal of Information Assurance & Security*, 9(7), 2014.

[157] Xiao Ke, Lingfeng Shi, Wenzhong Guo, and Dewang Chen. Multi-dimensional traffic congestion detection based on fusion of visual features and convolutional neural network. *IEEE Transactions on Intelligent Transportation Systems*, 20(6):2157–2170, 2018.

[158] Li Wei and Dai Hong-ying. Real-time road congestion detection based on image texture analysis. *Procedia Engineering*, 137:196–201, 2016.

[159] Pranamesh Chakraborty, Yaw Okyere Adu-Gyamfi, Subhadipto Poddar, Vesal Ahsani, Anuj Sharma, and Soumik Sarkar. Traffic congestion detection from camera images using deep convolution neural networks. *Transportation Research Record*, 2672(45):222–231, 2018.

[160] Jason Kurniawan, Sensa GS Syahra, Chandra K Dewa, et al. Traffic congestion detection: learning from cctv monitoring images using convolutional neural network. *Procedia Computer Science*, 144:291–297, 2018.

[161] Kimin Yun, Hawook Jeong, Kwang Moo Yi, Soo Wan Kim, and Jin Young Choi. Motion interaction field for accident detection in traffic surveillance video. In *2014 22nd International Conference on Pattern Recognition*, pages 3062–3067. IEEE, 2014.

[162] Earnest Paul Ijjina, Dhananjai Chand, Savyasachi Gupta, and K Goutham. Computer vision-based accident detection in traffic surveillance. In *2019 10th International Conference on Computing, Communication and Networking Technologies (ICCCNT)*, pages 1–6. IEEE, 2019.

[163] Siyu Xia, Jian Xiong, Ying Liu, and Gang Li. Vision-based traffic accident detection using matrix approximation. In *2015 10th Asian Control Conference (ASCC)*, pages 1–5. IEEE, 2015.

[164] Fu Jiansheng et al. Vision-based real-time traffic accident detection. In *Proceeding of the 11th World Congress on Intelligent Control and Automation*, pages 1035–1038. IEEE, 2014.

[165] In Jung Lee. An accident detection system on highway using vehicle tracking trace. In *ICTC 2011*, pages 716–721. IEEE, 2011.

[166] S Veni, R Anand, and B Santosh. Road accident detection and severity determination from cctv surveillance. In *Advances in Distributed Computing and Machine Learning*, pages 247–256. Springer, 2021.

[167] JiaYi Wei, JianFei Zhao, YanYun Zhao, and ZhiCheng Zhao. Unsupervised anomaly detection for traffic surveillance based on background modeling. In *Proceedings of the IEEE Conference on Computer Vision and Pattern Recognition Workshops*, pages 129–136, 2018.

[168] Waqas Sultani, Chen Chen, and Mubarak Shah. Real-world anomaly detection in surveillance videos. In *Proceedings of the IEEE Conference on Computer Vision and Pattern Recognition*, pages 6479–6488, 2018.

[169] Hanlin Tan, Yongping Zhai, Yu Liu, and Maojun Zhang. Fast anomaly detection in traffic surveillance video based on robust sparse optical flow. In *2016 IEEE International Conference on Acoustics, Speech and Signal Processing (ICASSP)*, pages 1976–1980. IEEE, 2016.

[170] Khac-Tuan Nguyen, Dat-Thanh Dinh, Minh N Do, and Minh-Triet Tran. Anomaly detection in traffic surveillance videos with gan-based future frame prediction. In *Proceedings of the 2020 International Conference on Multimedia Retrieval*, pages 457–463, 2020.

[171] Keval Doshi and Yasin Yilmaz. Fast unsupervised anomaly detection in traffic videos. In *Proceedings of the IEEE/CVF Conference on Computer Vision and Pattern Recognition Workshops*, pages 624–625, 2020.

5

Live Cell Segmentation and Tracking Techniques

Yibing Wang

Division of Physics, Engineering, Mathematics, and Computer Science, Delaware State University, Dover, Delaware, USA

Onyekachi Williams

Division of Physics, Engineering, Mathematics, and Computer Science, Delaware State University, Dover, Delaware, USA

Nagasoujanya Annasamudram

Division of Physics, Engineering, Mathematics, and Computer Science, Delaware State University, Dover, Delaware, USA

Sokratis Makrogiannis

Division of Physics, Engineering, Mathematics, and Computer Science, Delaware State University, Dover, Delaware, USA

CONTENTS

5.1	Cell Segmentation		130
	5.1.1	Conventional cell segmentation	130
		5.1.1.1 Model-based methods	130
		5.1.1.2 Morphological methods	133
		5.1.1.3 Feature-based methods	133
		5.1.1.4 Clustering methods	134
	5.1.2	CSTQ – joint morphological and level-set-based cell segmentation	135
	5.1.3	Deep learning	138
	5.1.4	Evaluation methods	144
	5.1.5	Discussion	145
5.2	Cell Tracking		146
	5.2.1	Tracking by detection	147
	5.2.2	Tracking by model evolution	149
	5.2.3	Tracking by probabilistic approaches	150
	5.2.4	Tracking by deep learning	151
	5.2.5	CSTQ – tracking by detection using motion prediction	152

DOI: 10.1201/9781003053262-5

5.2.6   Performance evaluation of cell tracking, detection, and
        segmentation 155
5.2.7   Tracking results on live cell image sequences 157
5.2.8   Discussion 158
5.3 Conclusion 159

5.1   Cell Segmentation

Automated image segmentation for cell analysis plays an essential part in the field of image processing and computer vision. The goal of image segmentation is to find the object regions, or the meaningful parts of an object. It separates the images into several distinct regions, which contains useful information. The cell segmentation methods can be categorized into six main categories:

1. Model-based Methods

2. Morphological Methods

3. Feature-based Methods

4. Clustering Methods

5. Machine and Deep Learning Methods

In the past decade, deep learning methods have matched or exceeded performances of state-of-the-art techniques and have become very popular. Hence in our overview of cell segmentation methods we broadly separated them into traditional, and deep learning methods.

5.1.1   Conventional cell segmentation

The conventional methods include model-based, morphological, feature-based and clustering methods. Figure 5.1 shows the publications regarding traditional cell segmentation techniques in the PubMed database. We computed the difference between search results using the key words "cell microscopy image segmentation" and "cell microscopy image segmentation deep learning" to get the traditional cell microscopy image segmentation techniques from 1977 to 2021. The plot shows the publications exponentially increased over the previous decades, which implies that the interest for cell segmentation techniques has been constantly increasing. In this section, we specifically review traditional cell segmentation techniques.

5.1.1.1   Model-based methods

Model-based methods use a mathematical, physical, or other scientific models to partition the image into regions [1–4]. The model often can be formulated

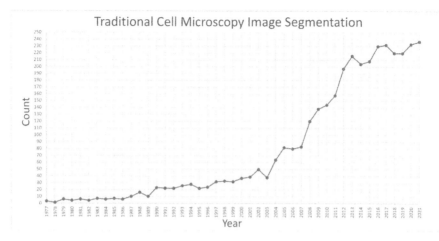

FIGURE 5.1
The number of publications per year on traditional cell segmentation techniques. *Source: PubMed*

by certain Partial Differential Equations (PDEs) to model the surface evolution. The numerical solution corresponds to the stable state of the model that delineates the object boundaries [5]. Active contour models are the most popular physical-based models for object delineation and image segmentation. The active contours can be expressed explicitly by parametric snake models [4], or implicitly by level sets [6].

Parametric active contour models – Snakes

A snake can be explicitly represented by a parametric curve $v(s) = (x(s), y(s))$, $s \in (0, 1)$. The contours will move toward the object boundaries under external and internal forces. Snake model requires the initial contours placed closely enough to the object. Besides, the capture range is also limited. Due to these drawbacks, Xu and Prince [7] proposed a new external force named Gradient Vector Flow (GVF) to cope with the poor convergence for concave boundary. Inspired by GVF [7], Kong et al. [8] presented a cell segmentation technique for 3D fluorescence microscopy images. The grouped voxels are detected from the computed GVF field, then separated into foreground or background voxels by a cell component identification process. Jia et al. [9] also combine a GVF snake model with watershed algorithm for segmenting the nuclei from background as an initial segmentation results. Puttamadegowda et al. [10] presented an automated segmentation method for white blood cells by using two algorithms: fuzzy C means [11] for cell clustering, and snakes [4] to refine the final contours. Of note is that both of these techniques require preprocessing steps, by different filters.

Geometric active contour models-level sets

Level set methods were originally proposed for tracking the motion of propagating curves by Osher and Sethian in 1988 [6]. Level set-based techniques have been widely used for cell segmentation [12–14]. Level set-based methods represent a closed curve C by the zero level set of Lipschitz continuous function $\phi : \Omega \to R$, such that $C = \{x \in \Omega : \phi(x) = 0\}$.

Level set methods can be divided into two main categories: *region-based level set models* [13, 15], and *edge-based level set models* [16, 17]. Edge-based models use the gradient information of the image to stop the curve evolution. Thus, the edge-based models are more sensitive to noise and the curve will cross the boundaries where the edges are weak [18]. However, region-based models utilize region information to control the curve evolution, which can detect the non-defined gradient objects. The Chan-Vese (CV) level set model [12] is a widely used region-based method, that can be considered as a special case of Mumford-Shah model [3]. The CV model is formulated as follows:

$$
\begin{aligned}
F\left(c_1, c_2, C\right) = &\mu \cdot \text{ Length } (C) + \nu \cdot \text{Area(inside } (C)) \\
&+ \lambda_1 \int_{\text{inside } (C)} |I(x,y) - c_1|^2 \, dxdy \\
&+ \lambda_2 \int_{\text{outside } (C)} |I(x,y) - c_2|^2 \, dxdy
\end{aligned}
\tag{5.1}
$$

where $I(x, y)$ is the image, c_1 and c_2 are the constants depending on C, μ, ν, λ_1 and $\lambda_2 \geq 0$ are fixed parameters.

Cell segmentation is a challenging task due to clustered and overlapping cells, and several other issues. Level set is widely used to segment the touching cells in many application over the recent years. Kaur et al. [13] proposed using level set with a curvelet initialization method to segment the nuclei and boundaries of low contrast touching cells. The image data are real cells from Herlev database. In this method, multiplication top hat filter and h-maximum are used for low contrast enhancement. Gharipour et al. [15] introduced a new combination method of Bayesian risk [19] and weighted image patch-based level set with splitting area identification to segment the fluorescence microscopy cell images. The performance of the proposed method shows superiority in detection of touching cells than several state-of-art methods such as DRLSE [16] and LSBR [19]. Furthermore, Al-Dulaimi et al. [20] applied geometric active contours-based level set method for the segmentation of white blood nuclei from cell wall and cytoplasm since it can track the topological changes of cell nuclei. Foreground estimation by means of morphological operations is used before level set evolution for image restoration [21]. In the same year, this method was implemented to HEp-2 cell segmentation that has large complexity and variations in cell densities [22]. To deal with 3D fluorescence microscopy images, Lee et al. [14] presented a 3D region-based active contour model, which is an extension of Chan-Vese [12] method to solve intensity inhomogeneity problem. Considering the influence of noise to an image, Yan et al. [17] proposed an improved edge-based distance regularization level set method which originally proposed by Liet al. [16] for medical image segmentation. The innovation of this method is that they use morphological reconstruction [23, 24] and fuzzy C-means clustering [25] to produce an initial contour. Their approach overcomes the shortcomings presented in the DRLSE method such as sensitivity to the initial level set.

5.1.1.2 Morphological methods

Many morphological cell segmentation techniques employ the watershed transform. The watershed transform interpets a gray level image as a topographic relief. Earlier methods utilized mathematical morphological operations for watershed-based image segmentation. Direct application of watershed segmentation usually leads to over-segmentation due to intensity variations and other issues [18]. To solve this problem, marker-controlled watershed segmentation can be used for cell segmentation [21]. The crucial step in this method is marker selection. In [26], the foreground markers are chosen from the regional maxima of gradient map and background markers are obtained by distance transform. Then, the watershed transform is applied for lactobacillus bacterial cell segmentation. Instead of classifying the blood cell type first, Miao et al. [27] proposed a marker-controlled watershed algorithm which can segment white blood cells (WBC) and red blood cells (RBC) simultaneously. Thus, a classification process is applied after the segmentation step based on a known condition that WBC has nuclei and RBC does not. Tonti et al. [28] proposed a solution of automated HEp-2 cell segmentation in Immunofluorescence (IIF) image using adaptive marker-controlled watershed techniques. This method can segment HEp-2 cells with inhomogeneous intensity levels and staining patterns without any prior knowledge. The parameter setting may be adapted to the contrast and brightness characteristics of the input images. We note that although Table 5.1 shows that the precision and recall values of this method are not excellent, it still yielded an very good performance compared to other techniques. Chang et al. [29] proposed a double thresholding cell segmentation method using morphological opening to separate connected cells. In this method, a global threshold is applied in the pre-processing stage to map the input images into a binary matrix and then an adaptive threshold is presented to separate the cells into connected components.

5.1.1.3 Feature-based methods

Feature-based segmentation techniques aim to detect local features such corners, lines, and edges [21, 30]. However, sole use of feature-based methods normally produces a coarse segmentation. Thus, they are usually combined with other segmentation methods to obtain accurate results. Figure 5.2 shows an overview of the numbers of relevant publications per year. We note that some of the publications also involve deep learning and machine learning methods. For instance, Yuan et al. [31] proposed a blob detection-based method using deep learning RCNN to segment nuclei of time lapse phase images. Pan et al. [32] proposed a bacterial foraging-based edge detection algorithm for cell

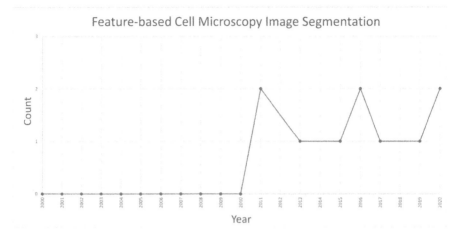

FIGURE 5.2
The number of publications per year using key words "Feature-based cell microscopy image segmentation" on PubMed.

segmentation. This algorithm marketed the tracks of the bacteria based on the pixel density map for nutrient concentration as the detected edges for cells. Karabag et al. [33] introduced an unsupervised algorithm which uses edge detection [34] to separate the nuclei envelope (NE) of HeLa cells from cytoplasm and nucleoplasm for segmentation. A 2D NE surface can be generated from 3D modeling against a spheroid which can provide more biological information of nuclei. For accessing more information from mammalian tissue images, Leonavicius et al. [35] proposed a simplified feature-based nucleus segmentation and probabilistic cytoplasmic detection method. This method can easily deal with 3D noisy cell images and crowded cell images for image analysis. Cao et al. [36] explored a biological morphology feature-based cell segmentation method. In this method, the initial segmentation regions were obtained by seeded active contours [12] and watershed transformation [37]. Next, a refined segmentation result was obtained by a morphological feature constrained k-means clustering method.

5.1.1.4   Clustering methods

Clustering methods aim to group pixels that share the same characteristics into clusters [38]. Zheng et al. [39] applied K-means clustering to extracting foreground region from white blood cell images, and coarse segmentation results were generated by a splitting method. This is a fully automated method without requiring manual parameter tuning. Sharma et al. [40] proposed a novel cell segmentation method based on a bio-inspired clustering technique named as "Teacher-Learner Based Optimization" (TLBO) [41] for cell segmentation. Specifically, it can obtain an optimal cluster for cell segmentation comparing with several traditional methods such as K-means clustering and particle swarm optimization (PSO) [42]. In order to obtain accurate detection and segmentation of white blood cells, Ferdosi et al. [43] introduced a combination method using K-means clustering and morphological operators. In this method, the segmentation results obtained from K-means clustering method [44] may include Non-WBCs. Then the final segmentation results will be refined according to the presence of nuclei. Another K-means clustering-based segmentation method for WBC is presented by Ghane et al. [45]. The cell image was initially segmented by Otsu thresholding method [46] and K-means clustering [47] method for nucleus segmentation. Due to the shortcomings of the traditional fuzzy C-means(FCM) clustering method, Bai et al. [48] proposed an improved FCM method combined with fractional-order velocity-based particle swarm optimization method for cell segmentation which is less sensitive of initialization. Albayrak et al. [49] proposed a superpixel-based clustering method for cell segmentation in histopathological images. The initial superpixels are obtained by simple linear iterative clustering (SLIC) algorithm [50] and the final segmentation results are generated from state-of-art clustering methods such as fuzzy C-means algorithm, k-means algorithm and density-based spatial clustering of application with noise (DBSCAN) algorithm [51].

5.1.2 CSTQ – joint morphological and level-set-based cell segmentation

The purpose of this research is the development of techniques for fully automated cell segmentation and detection on fluorescent microscopy images [53], as part of a Cell Segmentation, Tracking and Quantification (CSTQ) framework. A non-linear diffusion PDE model [54] was employed to detect cell motion in the spatio-temporal domain. A probabilistic edge map was computed by Parzen density estimation method [55] and watershed transformation was applied to delineate moving cell regions on edge map. The final segmentation result was obtained by a temporal linking region-based level set method. The clustered cells were separated by a signed distance transform.

Pre-processing

To reduce the intensity variation from frame to frame, histogram transformation is applied on each three consecutive frames to match the intensity distribution by Equation 5.2 with an intensity distribution learned from whole sequences.

$$P_{3F}(I) = \lim_{N_{total} \to \infty} \frac{N(I)}{N_{total}}, \quad F_{3F}(k) = \int_0^k P_{3F}(I)dI \qquad (5.2)$$

Spatio-temporal diffusion

The initial moving regions are detected by numerically solving a system of non-linear PDE-based diffusion equations on every three consecutive frames.

TABLE 5.1
Summary of traditional cell segmentation methods.

Authors	Dim.	Approach	Remarks	Results	Ref
Kaur et al., 2016	2D	Model-based	Region-based level set	Accuracy > 0.93	[13]
Gharipour et al., 2016	2D	Model-based	Region-based level set	JC > 0.759	[15]
Al-Dulaimi et al., 2016	2D	Model-based	Region-based level set	JD < 0.003	[20]
Lee et al., 2017	3D	Model-based	Region-based level set	Accuracy > 0.8771	[14]
Yan et al., 2020	2D	Model-based	Edge-based level set	DSC > 0.998	[17]
Puttamadegowda et al., 2016	2D	Model-based	Snakes	Accuracy > 0.96	[10]
Kong et al., 2015	3D	Model-based	GVF snakes	JC > 0.95	[8]
Jia et al., 2021	2D	Model-based	GVF snakes	Precision = 0.913	[9]
Shetty et al., 2018	2D	Morphological	Marker-controlled watershed	-	[26]
Miao et al., 2018	2D	Morphological	Marker-controlled watershed	Accuracy:WBC > 0.972 RBC > 0.948	[27]
Tonti et al., 2015	2D	Morphological	Marker-controlled watershed	Precision > 0.655	[28]
Chang et al., 2018	2D	Morphological	Morphological opining	-	[29]
Pan et al., 2017	2D, 3D	Feature-based	Edge detection	F-measure > 0.7119	[32]
Karabag et al., 2019	2D	Feature-based	Edge detection	JC > 0.93	[33]
Leonavicius et al., 2020	3D	Feature-based	Cell feature-based	-	[35]
Cao et al., 2018	2D	Feature-based	Biological morphology feature-based	-	[36]
Zheng et al., 2018	2D	Clustering	K-means clustering	-	[39]
Sharma et al., 2019	2D	Clustering	TLBO-clustering	-	[40]
Ferdosi et al., 2018	2D	Clustering	K-means and morphological operators	Precision > 98.3584	[43]
Ghane et al., 2017	2D	Clustering	K-means	Nucleus: Precision = 0.9607 Cells: Precision=0.9747	[45]
Bai et al., 2018	2D	Clustering	Fuzzy C-means	Recall = 0.9025	[48]
Albayrak et al., 2019	2D	Clustering	Superpixels-based	Precision: SLIC+k-means = 0.675 SLIC+FCM = 0.682 SLIC+DBSCAN = 0.614	[49]
Boukari et al., 2018	2D	Model-based	Region-based level set	SEG > 0.5088	[52]

Figure 5.3 (a) is an diffusion map example. At time points $\tau = \{t-1, t, t+1\}$

$$\frac{\partial I(i,j,\tau,s)}{\partial s} = g(|\nabla_{x,y,t} I(i,j,\tau,s)|) \cdot \Delta_{x,y,t} I(i,j,\tau,s)$$
$$+ \nabla_{x,y,t} g(|\nabla_{x,y,t} I(i,j,\tau,s)|) \cdot \nabla_{x,y,t} I(i,j,\tau,s) \qquad (5.3)$$

Initial condition

$$I(i,j,\tau,0) = I(i,j,\tau) \qquad (5.4)$$

Boundary condition

$$\frac{\partial I}{\partial \vec{n}} = 0 \qquad \text{on } \partial\Omega \times \partial T \times (0, S). \qquad (5.5)$$

Cell delineation – watershed transform

The likelihood of edge occurrence is estimated using parzen density estimation [55]. Next, the edge map can be interpreted as a topographic

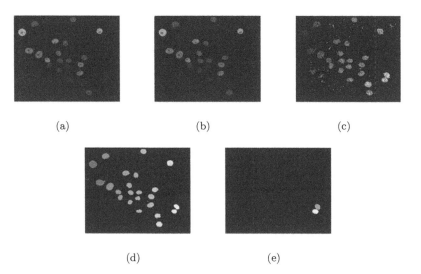

(a) (b) (c)

(d) (e)

FIGURE 5.3
Segmentation results on Fluo-N2DL-GOWT1-01 sequence from the CTC
dataset: (a) input frame, (b) diffused frame, (c) watershed segmentation result, (d) segmentation mask, and (e) ground truth.

surface which is followed by watershed segmentation to divide the edge map
into disjoint regions according to regional minimum. Figure 5.3 (c) displays
an example of watershed segmentation map. The mean intensity and areas are over watershed regions used as criteria for separating the cells from
background.

Temporal linking region-based level set segmentation

A region-based active contour method [12] is applied to refine the delineation results from previous step. For segmenting a time lapse cell images,
we propose a temporal linking approach to generate the cell segmentation results using previous level set contour information according to their temporal
connection:

$$\phi_{n+1}(x, y; 0) = \phi_n(x, y; i_{\text{final}}), \forall(x, y) \in \Omega \tag{5.6}$$

Experimental results

The datasets were obtained from the ISBI Cell Tracking Challenge (CTC)
website *http://celltrackingchallenge.net/*. CSTQ was validated on 2D time
lapse live-cell video sequences. Figures 5.3 (d) and 5.4 (b), show segmentation
results on GOWT1 mouse stem cells, and MSC rat mesenchymal stem cells,
respectively.

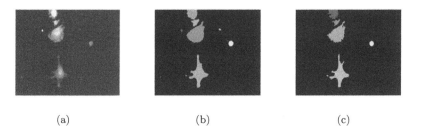

(a)   (b)   (c)

FIGURE 5.4
Segmentation results on Fluo-C2DL-MSC-02 sequence from the CTC dataset:
(a) input frame, (b) segmentation mask, and (c) ground truth.

5.1.3   Deep learning

Deep learning techniques have become the state-of-the-art for medical image segmentation and classification. They have produced remarkable results for efficiency and accuracy [56, 57]. Deep learning has its origins from late seventies to late nineties [56]. Over the years, technological advancements boosted the applications of deep learning in various fields. One such field, is cell image analysis, where deep learning has become a powerful tool for cell segmentation, detection, classification and quantification. Cell image analysis plays a crucial for analyzing and research of cellular events such as mitosis, migration, effects of drugs, cellular immune response and embryonic development [58]. Thereby, deep learning methods aid the automation of cell image analysis techniques of segmentation, quantification and tracking, that were otherwise, long, tedious and required experts. One of the biggest challenges faced in cell image analysis by deep learning is availability of large datasets for training networks. A combination of image processing methods such as data-augmentation [59], contrast enhancement [60], histogram equalization, and other techniques [61] with deep learning have proven to be successful in generating high accuracy and efficiency for cell segmentation. Figure 5.5 represents the number of publications in PubMed using deep learning for cell segmentation in the past 10 years.

Types of neural networks and how they are applied

Artificial neurons form the basis of deep-learning networks also called neural networks. The neurons or nodes are units that have weights and biases and perform transfer functions, followed by activation functions a. The transformation operations by the neurons help to select and transmit information from one node to the other. The transfer function which is also called the learning function is given as

$$a(x) = \sigma(w^T x + b) \tag{5.7}$$

where a is the activation function, $\sigma(.)$ is the element-wise non-linear function, $\mathbf{x}$ is the input, w is the weight, and b is the bias parameter.

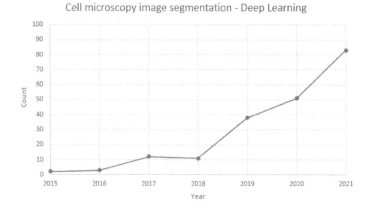

FIGURE 5.5
The number of publications per year on cell segmentation using deep learning.
Source: PubMed

Widely used neural networks include:

1. **Convolutional neural networks**: convolutional neural networks (CNN) employ convolution operations for feature extraction. They usually process digital matrices, that have grid like topology. A fully connected convolutional network is a network, where all the nodes of one layer are connected to all the nodes of the next layer.

2. **Recurrent neural networks**: recurrent neural networks (RNN) are artificial networks that can store past or historic data. They work well with sequential or time-series types of data to predict the future outputs.

3. **Autoencoders and stacked autoencoders**: auto encoders are unsupervised learning models which employ back propagation methods to predict the output. Stacked encoders are essentially several layers of sparse encoders that are similar to autoencoders, but they can learn features automatically from unlabeled data.

Hardware and software used for deep learning

Technological progress in circuit design and manufacturing led to the development of complex and fast digital processing systems. These improvements led to more powerful CPUs and GPUs that have enabled the implementation deep learning algorithms in practical computational times on personal computers. Software libraries and packages like CUDA, TensorFlow, Keras, and Pytorch have been developed exclusively for deep learning algorithms.

Combination of image processing methods and deep learning

Deep learning has found wide-range applications, with combining the methods with other image processing techniques. These hybrid networks give rise to a

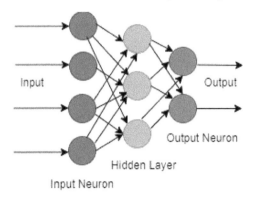

FIGURE 5.6
Structure of an artificial neural network.

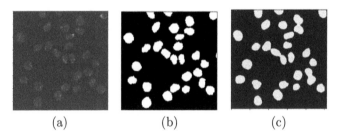

          (a)   (b)   (c)

FIGURE 5.7
Segmentation of Fluo-N2DH-SIM+01 by the method in [62]. (a) input image,
(b) reference frame, and (c) segmented image.

system of networks that can handle pre-processing, post-processing and the
segmentation process. Figure 5.7 presents an output of segmentation of Cell
Tracking Challenge dataset of Fluo-N2DH-SIM+ 01 by a U-net-based method
[62]. This dataset is a simulated two-dimensional video sequence containing
65 frames of fluorescent microscopy images.

Survey of methods

A variety of methods have been applied for cell segmentation using deep learning. Overall, the most popular deep learning method is CNN. The U-net is a
CNN-based encoder-decoder architecture Figure 5.8 proposed by Ronneberger
et al. [63], which is widely used for image segmentation. The encoder performs
downsampling on the image for feature extractions and the decoder performs
upsampling of the image to segment and classify the test images. There exist
skip connections to transfer the spatial information from the encoder network
to decoder network.

Other neural networks, such as autoencoders and recurring neural networks are also popular. Figure 5.9 provides an overview of various deep learning networks that have been applied for cell segmentation from the 2010 to 2021. The trend observed here is an increased use of CNN networks compared to RNNs and autoencoders. The CNNs have an architecture that is more popular due to the properties of feed-forward networks and skip connections that provide spatial coherence.

CNN-based segmentation was proposed in [64], where the model is able to perform cell segmentation for both two-dimensional and three-dimensional images. To achieve this complex operation a micronet has been developed, which has added layers of convolutional network that bypass certain information for better segmentation. Some of the other methods include modified U-nets called as U-net+ [65] which is a network of dense blocks that are fully connected. The dense blocks used in the paper are blocks of convolutional neural networks that perform feature extraction. In the paper proposed by Al-Kofahi et al. [66] for 2-D cell microscopy images, where markers were used for staining the nucleus and the cell borders with different intensities, the authors applied thresholding and watershed to perform the segmentation of the cells from the images. This method combines traditional image processing methods with deep learning. Another hybrid method proposed by Lux et al. [67], follows a similar approach of applying watershed segmentation at the end for segmentation. In this method, the authors used two sets of outputs from the CNN networks to predict cell markers and cell masks. The difference between the markers and masks is used as loss function to optimize the cell prediction.

Similar methods have been proposed by authors like Nikita Moshkov et al. [59], who have compared the performance of two different architectures of U-net and Mask-Region based Convolutional Neural Network (Mask-RCNN). The training and test datasets were augmented, and a loss function was used optimize the prediction between the original datasets and augmented datasets.

The increase of the complexity in cell segmentation from different modalities of images, has been addressed in more recent methods. The authors in [68], propose to use bounding boxes that surround the cells named k-bounding boxes. Then the method predicts probability of cell occurrence in a particular region. It uses two networks to generate the boxes using fully Convolutional Neural Network (fCNN) and the CNN to predict the segmentation. Other region proposal networks like Mask-RCNN have become very popular; Ren et al. [69] propose Mask-RCNN with Non-maximum suppression (NMS) model. In this paper, they apply MASK-RCNN model, where an additional branch is added to predict the object mask in parallel with predicting the bounding box regions. A NMS model has been implemented to calculate the scores of the segmentation to obtain more accurate final segmented mask. Other authors like Hiramatsu et al. [70] have proposed architectures that combine multiple CNNs forming into a gating network and expert network. In this method, the gating network divides the input image into several problems and the expert

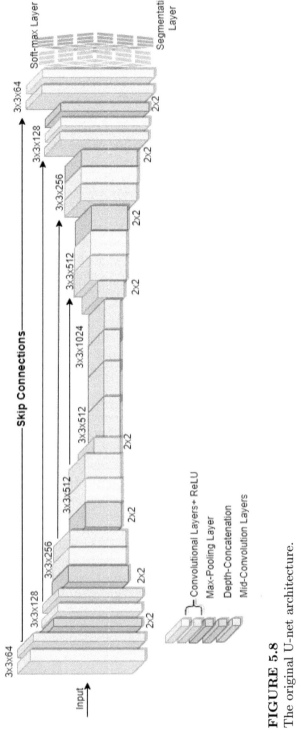

FIGURE 5.8

The original U-net architecture.

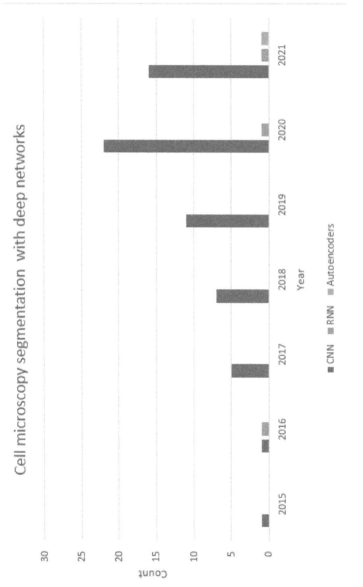

FIGURE 5.9
Timeline of published articles on cell segmentation categorized by network type *Source: PubMed.*

TABLE 5.2

Cell segmentation methods using deep learning techniques.

Authors	Dim.	Approach	Auto (End to End)	Remarks	Ref
Al-Kofahi et al., 2018	2-D	CNN	No	Encoder-Decoder Network predicts nucleus seeds and cytoplasm for cell detection	[66]
Scherr et al., 2020	2-D, 3-D	CNN	No	Applied distance maps to calculate cell borders	[71]
Akram et al., 2016	2-D	fCNN	Yes	Encoder-Decoder network proposes bounding regions to detect and segment cells	[68]
Lux et al., 2020	2-D	CNN	No	Networks predicts cell-markers and masks Watershed is applied for final segmentation	[60]
Su et al., 2015	2-D	Stacked Autoencoder	No	Sparse reconstruction and denoising stacked autoencoders	[72]
Ren et al., 2018	2-D	Mask-RCNN & Stacked Autoencoder	Yes	Mask-RCNN trained on stacked autoencoders for automatic segmentation	[69]
Razaa et al., 2018	2-D	CNN	Yes	Multiple CNNs applied to train and test multiple resolutions of the input	[64]
Moshkov et al., 2020	2-D	Mask-RCNN & CNN	No	Encoder-Decoder network for semantic segmentation and Mask-RCNN for instance segmentation	[59]
Xu et al., 2016	3-D	fCNN	No	Fully connected Encoder-Decoder Network with k-terminal cut	[73]
Long, 2020	2-D	Unet+	No	Light-weighted enhanced version of U-Net called U-Net+ with branch modification of the transfer connections from encoder to decoder	[65]
Payer et al., 2019	2-D	ConvGRU & RNN	No	Recurrent stacked hour glass U-net network with convolutional gated recurrent units to transfer between split connections cosine embedded loss function	[74]
Ronnerberger et al., 2015	2-D	CNN	No	Original U-net architecture Encoder-Decoder network	[63]
Hiramatsu et al., 2018	2-D	CNN	No	Integrating multiple CNNs called multiple gating network and multiple expert networks	[70]

method works on one specific problem. This approach improves segmentation accuracy. We list cell segmentation techniques by deep learning in Table 5.2.

5.1.4 Evaluation methods

The following measures are regularly used for performance evaluation of cell segmentation algorithms and are also reported in Table 5.1.

1.

$$Accuracy = \frac{TN + TP}{TP + FP + TN + FN} \qquad (5.8)$$

2.

$$Precision = \frac{TP}{FP + FN} \qquad (5.9)$$

where TP is True Positive, FP is False Positive, TN is True Negative and FN is False Negative.

3.

$$Jaccard(JC) = \left| \frac{R_s \cap R_{Ref}}{R_s \cup R_{Ref}} \right| . \qquad (5.10)$$

JC is from 0 to 1. A bigger value represents a better result.

4. JD: Jaccard Distance Error when equals to 0 is the best

5.

$$F_\beta = \frac{(\beta^2 + 1) PR}{\beta^2 P + R} \qquad (5.11)$$

where P is precision, R is recall and β is usually set to 1. F-measures close to 1 which indicates a better edge detection.

6. SEG: It measures the Jaccard similarity index and the method sets to zero if the indices of cells for which $|R_s \cap R_{Ref}| \leq 0.5 \cdot |R_{Ref}|$

5.1.5 Discussion

In the past few decades, many segmentation approaches have been proposed for cell or nuclei segmentation in microscopy images. One of the major challenges is the segmentation of overlapping or touching cells such as in [13], [15]. Due to the large variability among the cell images including fluorescence microscopy, bright-field microscopy, phrase contrast microscopy [30], it is also a difficult task to make the method applicable for many kinds of image modalities. Among these traditional techniques, each of them has their own strength and weakness.

The watershed transform techniques tend to produce over-segmentation, but are straightforward to implement [18]. Level-set methods cost more computational time for getting the curve to converge, but are flexible to topological changes. For dealing with these problems, more recent methods do not singly use one segmentation technique, but combine several techniques to learn from each other which greatly improves the accuracy of segmentation results. For example, Yan et al. [17] combined model-based level set method with clustering model fuzzy C-means. Besides, Cao et al. [36] applied feature-based model, morphological model watershed and active contours to refine segmentation results. However, the computation time will increase at the same time because of the complexity of algorithm. Therefore, a robust and generalized cell segmentation technique is needed and many researchers are making effort for it. The evaluation results from summarized proposed methods are based on different data dimensions, cell types, and evaluation methods. Thus, it is hard to conclude which method is superior to others.

The literature in recent years suggests that deep learning methods are successful in cell segmentation using various imaging modalities like microscopy, histopathology, H&E imaging, Differential Interference Contrast (DIC), and others. One of the positive attributes of applying deep learning methods is the segmentation speed and accuracy. Another advantage is the ability for supervised or unsupervised learning, and the combination with traditional image processing methods. The development of fast computing technologies, enabled the application of deep learning methods to large-scale datasets. The biggest drawback faced by deep learning algorithms is availability of labeled datasets to enable the network to learn. The large computational requirements in the training stage, also reduce the scope of expansion of the deep learning algorithms. In conclusion, it can be summarized that deep learning algorithms have advanced the field of medical image segmentation, and cell segmentation in particular, by improving its accuracy, and increasing the degree of automation, thereby progressing further from traditional methods that used manually engineered image processing steps.

5.2   Cell Tracking

Cell tracking is a key process for studying the complex behavior of cells and biological processes [75]. Tracking of cells by manual means is tedious, subject to human error, high costs, and may not be reproducible. Automated live cell tracking aims to address challenges encountered in manual tracking. The capability to automatically track cells and analyze cell events in time-lapse microscopy images is quite significant [76, 77].

Biomedical image analysis techniques can be used for identifying cell events such as migration, division, collision, or death. These analyses are applicable to drug development and medical therapy for diseases. However, automated techniques are frequently challenged by low signal-to-noise ratio, varying densities of cells in the field of view, topological complexities of cell shapes, and cell behaviors. Generalization to other modalities such as phase enhancing microscopy techniques, or differential interference contrast microscopy may also be complicated.

Hence, a cell tracking method that is used for a particular dataset may not be effective for another dataset. This is because of the differences of imaging technologies, and variation of cells characteristics, properties and behavior [78]. Also, there are specific problems related to the tracking stage, such as poor detection and association of objects. According to Meijering in [78], two major problems were summarized, which are the recognition of relevant objects, and their separation from the background in every frame, that is the segmentation step, and secondly the association of segmented objects from frame to frame, and making connections, that is the linking step. Many researchers have also incorporated the segmentation approach to track cells after detection. Cell linking has posed to be a problem most of the times. An approach to solving the association problem [78] is to link every segmented cell in any given frame to the nearest (spatial distance or difference in intensity, volume, orientation, and other features) cell in the next frame.

A number of cell tracking algorithms have been proposed by many researchers to help address challenges experienced in tracking of cells as they migrate, leave the field of view, collide, and divide. Figure 5.10 shows examples of cell events that may occur in a time-lapse sequence. The proposed methods in the literature can be generally classified as tracking by detection, tracking by model evolution, and tracking by probabilistic approaches. Deep learning techniques have also been used for cell tracking via deep convolutional neural networks, convolutional siamese networks, and other techniques. Hence they can be considered to be a separate category. Next, we will survey the trend and usefulness of these tracking methods over the years. We will describe methods that have produced good and reliable tracking results.

5.2.1   Tracking by detection

Generally these techniques entail detection of cells based on texture, intensity, or gradient in all frames, then association of the detected cells between frames [79]. The accuracy of detection, matching strategy and segmentation, affect the performance of tracking [80, 81]. These approaches perform well on sequences of low resolution, or low signal-to-noise ratio.

Some researchers have combined tracking by detection with segmentation, while others have solely used the detection stage. Combining the motion features with the structural and appearance features of the cells has been used to track time lapse microscopy images in [82]. This was done by introducing a cell detection method based on h-maxima transformation to reduce segmentation error, then fitting an ellipse to the nucleus shape. Combination of motion features of skewness and displacement with structural features (such as color compatibility, area overlap and deformation) was performed, in order to find the correct correspondences between the detected cells. Also, a template-matching-based backward tracking procedure was employed to recover any discontinuities in a cell trajectory that may occur due to the segmentation errors, or the presence of a mitosis.

Some have integrated the detection method with other probabilistic approaches to detect cell collisions. The tracking algorithm named Augmented Interaction Multiple Models Particle Filter (AIMMPF), was proposed to deal with all interacting cases occurring among cells as seen in [83]. The authors proposed a general framework for multi-cell tracking, especially when cell collision and division occur, by first defining three typical events to characterize interaction modes among cells (cell independence, cell collision, and cell

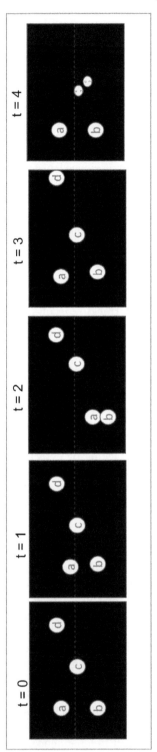

FIGURE 5.10

Illustration of cell events in a time-lapse sequence. Cells are represented as "*a, b, c, d*". At time $t = 2$, cell '*a*' and '*b*' collide; cell '*d*' migrates and leaves the field of view (disappearance), at time $t = 4$. Also, at time $t = 4$ cell c undergoes division to form cells '*c1*' and '*c2*'.

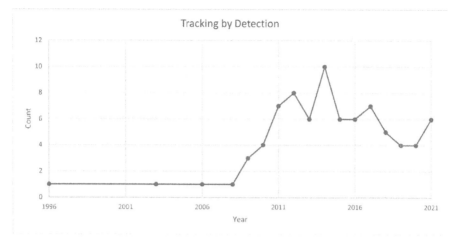

FIGURE 5.11
The number of published articles per year on cell tracking by detection. *Source: PubMed.*

division). Then, the evolving model relevant to each event is described and a tracking algorithm using particle filters on augmented interacting multiple models is proposed for identifying spatially adjacent cells with varying size. In order to reduce the ambiguity of correspondences and establish the trajectories of cells, both structural and dynamic cell features were used to manage the data association problem. This method proved successful when compared to yjr Interaction Multiple Models Particle Filter (IMMPF) that was not augmented. We present the trend of usage by researchers of cell tracking by detection based on the number of published articles per year in Figure 5.11.

5.2.2 Tracking by model evolution

Tracking by model evolution can be described as exploring the relationship of objects between neighboring frames [84]. We can classify the tracking by model evolution into parametric and non-parametric. According to [85], the techniques using parametric active contour models can produce good estimates of cell morphology, but need to be adapted to handle cell-cell contacts and mitosis at increased computational cost. The authors in [86] obtain the edge of an object in the frame by evolving the contour of the former frame. According to Michael et. al. in [87], active contour models attracted much attention for general image segmentation which also can minimize an energy functional of a level-set, in which energy may be minimized according to shape and appearance criteria. There are challenges, which could be due to incorrect segmentation as a result of low contrast edges and noise, or sensitivity of the model to the initial position [88]. The trend of usage of model evolution techniques for cell tracking is displayed in Figure 5.12.

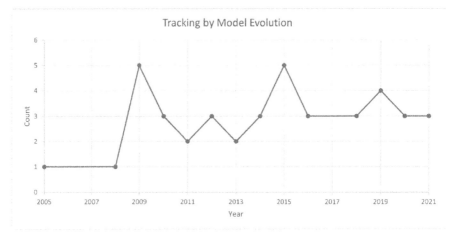

FIGURE 5.12
The number of published articles per year on cell tracking by model evolution.
Source: PubMed

5.2.3   Tracking by probabilistic approaches

These methods make use of Bayesian filtering to construct a motion evolution
model and use the Markov process in techniques such as particle filtering to
track multiple cells. These techniques model the location, speed and acceleration of the cells [89]. They are useful for solving tracking problems, when
there is existence of noise [90].

   The authors in [85], describe a multi-tracking target system that combines
bottom-up and top-down image analysis by integrating multiple collaborative
modules. The lower level consists of a cell detector, a fast geometric active contour tracker, and an interacting multiple models (IMM) motion filter adapted
to the biological context. The higher level consists of two trajectory management modules, called the track compiler and the track linker. By performing
online parameter adaptation, the IMM filter enhances the tracking of varying cell dynamics. Another method was proposed in [91] to detect and track
multiple moving biological spot-like particles showing different kinds of dynamics in image sequences acquired through multidimensional fluorescence
microscopy. The main stages of this method are: a detection stage performed
by a three-dimensional (3-D) undecimated wavelet transform, and prediction
of each detected particle's future state in the next frame. This is accomplished
by using an interacting multiple model (IMM) encompassing multiple biolog-
ically realistic movement types. Tracks are constructed, thereafter, by a data
association algorithm based on the maximization of likelihood of each IMM.
The last stage consists of updating the IMM filters, in order to compute final
estimations of the present image and to improve predictions for the next image. The trend of published articles using probabilistic models for cell tracking
is shown in Figure 5.13.

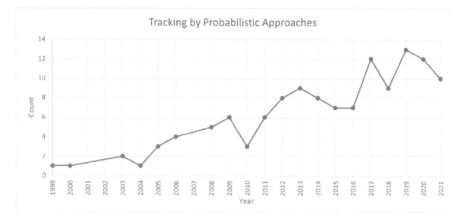

FIGURE 5.13
The number of published articles per year on cell tracking by probabilistic approaches. *Source: PubMed.*

5.2.4 Tracking by deep learning

Deep learning techniques have been widely employed for cell tracking. These methods were motivated by the need for automated pipelines, which could segment and track cell nuclei with little, or no user intervention, and would greatly increase data analysis throughput, as well as the productivity of biologists [92]. Most published works utilize Convolutional Neural Networks for tracking. In [92], the authors followed the approach used in [93] to access the complex features of deep convolutional neural networks, in addition to the real-time speeds of kernelized correlation filters for tracking. The authors applied hierarchical visual feature tracking in two modes. In the first mode, they use the existing DeepCell segmentation network for cell segmentation and tracking of feature maps. In the second mode, they build and train a CNN, in order to identify cell nuclei, which the network uses to generate tracking feature maps. This tracking system consists of three components: CellTrack, DeepCell and RecogNet.

Performance comparisons of cell trackers have shown that CNNs produce competitive cell detection accuracy compared to other contemporary methods [94]. In addition, competitive tracking performances have been reported for both 2D [95–99], and 3D time lapse microscopy data [100–102]. The emergence of machine learning concepts has broadly increased the usage and diversification of the deep networks for cell tracking. This trend is presented in Figure 5.14.

We summarize representative cell tracking methods of the literature in Table 5.3.

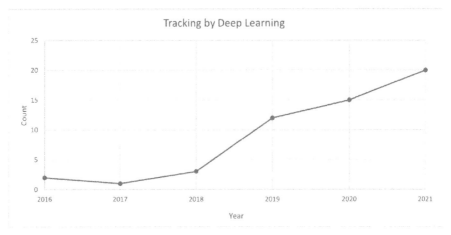

FIGURE 5.14
The number of published articles per year on cell tracking by deep learning.
Source: PubMed

5.2.5   CSTQ – tracking by detection using motion prediction

This is an automated cell tracking method that performs motion prediction
and minimization of a global probabilistic function for each set of cell tracks
[52]. Cell events are also identified by back tracking the cell track stack and
forming new tracks in other to determine a partition of the complete track
set. This can be categorized as a tracking by detection approach. This method
follows a segmentation stage that produces cell label maps for each frame. The
flow chart that outlines the main stages of this method is shown in figure 5.15.

Computation of features, moving cell detection and cell separation

In this phase, the method computes cell properties related to intensity, shape,
and size, to be used for probabilistic cell matching and quantification. It com-
poses a pattern vector that corresponds to each cell. Afterward, it removes all
non-cell detected objects to reduce false positives.

Motion model estimation

This technique estimates the cell motion by use of a variational multi-scale
optical flow technique. The velocity of each pixel between two consecutive
frames is estimated at times t and $t+\delta t$. This stage employs the Combined Lo-
cal/Global Optical Flow Method (CLGOF) approach to estimate the optical
flow. It uses calculus of variations to create a system of differential equations
to model the displacement field. An example of optical flow estimation and
warping on a real time-lapse sequence is shown in Figure 5.16. The motion
field is then applied to the current central frame to produce the warped cells.

TABLE 5.3

Summary of cell tracking techniques.

Author	Tracking Method	Principle	Dim.	Auto	Produces Quantifi-cation	Remark	Ref
Lu et al., 2015	Detection	Multicell Tracking (AIMMPF)	2D	Yes	Yes	Based on cell independence, collision and detection	[83]
Jackson et al., 2016	Deep Learning	Convolutional Neural Network	2D	Yes	Yes	Flexible for different cell types	[92]
Sugawara et al., 2022	Deep learning	Efficient learning using sparse human annotations for nuclear tracking (ELEPHANT)	3D	Yes	Yes	produces an interface that integrates cell track annotations, deep learning, prediction and proof reading	[106]
Jean-Baptiste et al., 2020	Deep learning	(DeLTA) Deep Learning for time lapse analysis)	2D	Yes	Yes	Produces pipeline for segmentation, tracking, and lineage reconstruction	[107]
Li et al., 2008	Model Evolution/filtering	Interacting multiple model (IMM) motion filtering, cell detector, active contour tracker	2D	Yes	Yes	Combine bottom up, and top down image analysis	[85]
Loffler et al., 2021	Detection	Graph based cell tracking algorithm	2D/3D	Yes	Yes	Handled over and under segmentation errors, false negatives	[110]
Dzyubachyk et al., 2010	Model Evolution/filtering	Level set based cell tracking	2D/3D	Yes	Yes	Reduces tracking time	[105]
Bertinetto et al., 2016	Deep Learning	Fully Convolutional Siamese Networks	2D	Yes	Yes	Trained Siamese Network to locate an exemplar image within a larger search image	[109]
Sixta et al., 2020	Detection	Temporal Feedback	2D	Yes	Yes	Able to estimte shapes of the tracked objects	[108]
Akber et al., 2011	Detection	Combine Motion and Topological Features of the Cells	2D	Yes	Yes	Detection based on h-maxima transformation	[82]
Yang et al., 2005	Model Evolution/filtering	Temporal Context	2D	Yes	Yes	Level Set Determine Cell Trajectories	[80]
Moller et al., 2013	Model Evolution/filtering	Normal Velocities	2D	Semi-Auto	Yes	Complete Cell Tracking Framework for obtaining Cell Centroids, Contour of each Cell enabling shape, area of the Cell calculation	[87]
Boukari et al., 2018	Detection	Motion Prediction	2D	Auto	Yes	Uses also minimization of a global probabilistic function for each set of cell tracks	[52]

Cell matching

This stage performs bi-frame cell matching. It searches for the maximum likelihood match for each cell of the current frame among all the cells of the previous warped frame. A reject option is introduced when a newly appearing cell does not match the previous frame.

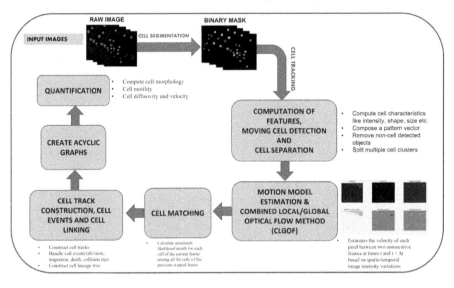

FIGURE 5.15
Flowchart of the CSTQ-MIVIC cell tracking method highlighting the main stages.

Cell track construction, cell events, and cell linking

In this last stage, the cell linked lists are constructed for representation of cell tracks. The lists are back-traced to detect overlapping tracks, and to identify and handle the cell events. Each track is the set of identified cell states in the sequence. After finding all cell events, a cell lineage tree is constructed to store and visualize the cell events. Tracking results are also represented by acyclic-oriented graphs. Figure 5.17 illustrates the formation of cell tracks over a time interval.

Quantification stage

This stage computes cell characteristics, which are broadly categorized into structural (area, perimeter, major and minor principal axes, circularity, eccentricity, convexity), motility (each cell total distance, cell lifetime, distance between the start and the end points in a trajectory), diffusivity, and velocity. Fourier descriptors, Independent Component Analysis (ICA), and Principal Component Analysis (PCA) may also be computed to analyze the structure and appearance of cells. A widely used dynamic cell property is the Mean Squared Displacement (MSD). This is an advanced measure of diffusivity computed by the second-order moment of displacement as a function of time point difference:

$$MSD(n) = \frac{1}{N-n} \sum_{i=1}^{N-n} d^2(\omega_i, \omega_{i+n}) \qquad (5.12)$$

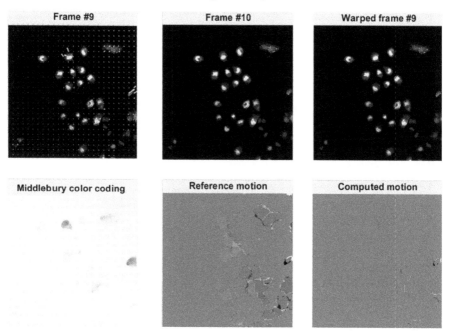

FIGURE 5.16
First row: estimation of optical flow (left) between frames 9 (left) and 10 (middle) of Fluo-C2DL-Huh701 dataset. Second row: Middlebury color coding of computed optical flow (left), the difference between cell indicator functions of reference cell masks of frames 9 and 10 (middle), and the difference between cell masks of frame 10 and warped frame 9 using the computed optical flow (right).

where $\omega_i = (x_i, y_i)$ is the centroid of a cell at time point i, N is the total trajectory lifetime, and n is the interval for computation of MSD.

5.2.6 Performance evaluation of cell tracking, detection, and segmentation

Cell tracks can be mathematically represented by acyclic oriented graphs. The graph vertices describe the spatio-temporal locations of individual cells, and the edges represent temporal relationships between cells. Such a representation stores the information of all identified cellular events, such as migration, division, death, and transit through the field of view. The increasing number of cell tracking algorithms has warranted the development of tracking performance evaluation techniques [103] that enable comparisons. Similar to segmentation evaluation, manually segmented and tracked cells are used as the reference standard, or ground truth (GT), for evaluation of automated

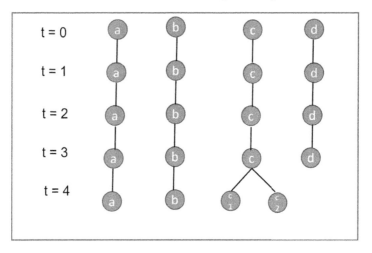

FIGURE 5.17
Example of cell event identification, and formation of cell lineages.

tracking techniques. The TRA measure [103, 104] that we describe next, is among the more widely used measures.

TRA measure

We can understand the tracking accuracy (TRA) of a particular method as how accurately each given object has been identified and followed in successive frames. It is based on comparison of acyclic oriented graphs representing the time development of objects in both the gold reference tracking annotation and each tested method. Numerically, TRA is defined as a normalized Acyclic Oriented Graph Matching (AOGM) measure:

$$TRA = 1 - min(AOGM, AOGM_0)/AOGM_0$$

where $AOGM_0$ is the $AOGM$ value required for creating the reference graph from scratch (i.e., it is the $AOGM$ value for empty tracking results). TRA always falls in the [0,1] interval, with higher values corresponding to better tracking performance.

DET measure

Detection accuracy expresses how accurately each given object has been identified and it is based on comparison of the nodes of acyclic oriented graphs representing objects in both the GT and the tested method. It is defined numerically as a normalized Acyclic Oriented Graph Matching (AOGM-D) measure for detection:

$$DET = 1 - min(AOGM - D, AOGM - D_0)/AOGM - D_0$$

where $AOGM - D$ is the cost of transforming a set of nodes provided by the participant into the set of GT nodes; $AOGM - D_0$ is the cost of creating the set of GT nodes from scratch (i.e., it is $AOGM - D$ for empty detection results). DET always falls in the [0,1] interval, with higher values corresponding to better detection performance.

SEG measure

The SEG measure, which expresses the segmentation accuracy (SEG) of the methods, denotes how well the segmented regions of the cells match the actual cell or nucleus boundaries. It is measured by comparing the segmented objects with the reference annotation of selected frames (2D) and/or image planes (in the 3D cases). It is based on the Jaccard similarity index numerically, where the mean of the J indices of all reference objects over a frame sequence yields the SEG measure. It takes values in the [0, 1] interval, where 1 implies perfect match and 0 implies no match

$$J(S, R) = \frac{|R \cap S|}{|R \cup S|}, \tag{5.13}$$

where R denotes the set of pixels belonging to a reference object and S denotes the set of pixels belonging to its matching segmented object. A reference object R and a segmented object S are considered matching, if and only if, the following condition holds true

$$|R \cap S| > 0.5 \cdot |R|.$$

5.2.7   Tracking results on live cell image sequences

Our goal is to report tracking results produced by representative methods of the literature. The level set-based tracking used by Dzyubachyk in [105] produced very good tracking accuracy on four datasets as shown in Table 5.4. Also, the adaptive interacting multiple model (IMM) motion filter proposed in [85] produces a better result when compared to the Kalman filter as shown Table 5.5. Deep learning methods, such as ELEPHANT [106] produce very good tracking results, which outperform other methods in detection and linking accuracy as shown in Table 5.6. We also ranked the CSTQ method in comparison with the tracking methods published in the ISBI Cell tracking Challenge website as shown in Table 5.7. CSTQ achieved top 10 TRA rates for two sequences and TRA of at least 0.88 for another two test sequences. Example of a lineage tree produced from the CSTQ method, after performing tracking is shown in Figure 5.18. The tracking stage produces global linked cell label maps, cell linking trajectories, and graphs that store the identified cell events.

TABLE 5.4

Tracking performance of the level-set-based tracking method in [105].

Datasets	Precision	Recall	Sequences	False Division
Hoechst	99.7%	99.8%	11	2
H2B-GFP	100.0%	100.0%	21	4
RADI8-YFP	97.1%	93.8%	4	1
PCNA-GFP	99.1%	98.5%	5	2

TABLE 5.5

Tracking performance of the IMM and compared with the Kalman filter tracking method in [85].

Sequence	Kalman	IMM	IMM track linking
Tracking	Validity		
A1	86.4%	88.9%	92.6%
A2	88.2%	82.2%	92.5%
B1	76.1%	79.3%	88.0%
B2	76.9%	80.3%	86.3%
B3	77.8%	79.8%	87.5%
B4	74.1%	77.8%	86.1%
C1	81.0%	84.3%	90.9%
Division	Tracking ratio		
A1	100%	100%	100%
A2	N/A%	N/A%	N/A%
B1	78.2%	78.2%	85.5%
B2	78.8%	80.8%	86.5%
B3	75.0%	77.1%	85.4%
B4	77.2%	82.5%	87.7%
C1	76.2%	78.6%	88.1%

TABLE 5.6

Cell tracking results produced by deep learning methods [106].

Methods	TRA	Detection
ELEPHANT	0.975	0.979
KTH-SE	0.945	0.959
KIT-Sch-GE	0.886	0.930

5.2.8 Discussion

Cell tracking methods have made significant performance gains, driven by the introduction of deep learning methods. A good tracking algorithm should effectively address cell events, such as mitosis, collision, apoptosis, division, cells leaving or entering the field of view, and new cells, or cells entering a field

TABLE 5.7
Ranking of the TRA measure of CSTQ [52].

Datasets	SEG	TRA	TRA RANK
Fluo-C2DL-Huh7	0.476	0.74	7
Fluo-C2DL-MSC	0.645	0.757	4
Fluo-N2DH-SIM+	0.715	0.887	26
Fluo-N2DL-HeLa	0.832	0.954	24

of view. The most complex stage of cell tracking is the association of cells to form tracks. Generally, as the cell number increases, cell linking becomes more complicated.

Frequent causes of errors are incorrect cell detection, wrong cell labeling, erroneous detection of cell division, and incorrect cell matching. Using the augmented interacting multiple models particle filter tracking algorithm proposed in [83], both structural and dynamic cell features are used to manage the data association problem to reduce the ambiguity of correspondence and establish trajectories of interested cells. Table 5.4 shows the tracking performance by level set based tracking method [105]. False detections occurred during cell divisions, rare cell events in the environment of an existing cell, in the presence of dead cells, or when a new cell enters the field of view. In Table 5.5, we made a comparison between the standard Kalman motion filter and a constant velocity motion model. The IMM motion filter was tested with four motion models: random walk, constant velocity, constant acceleration, and constant speed circular turn [91].

5.3 Conclusion

Automated cell tracking is an interesting and significant application of video analytics. Significant advancements of artificial intelligence techniques have driven improvements in cell tracking as well. Challenges still remain because of the variability of cell types, imaging methods, and image quality limitations. However, several techniques have produced promising cell detection and tracking accuracy rates that underscore their applicability to diagnosis techniques, drug development, and personalized medicine.

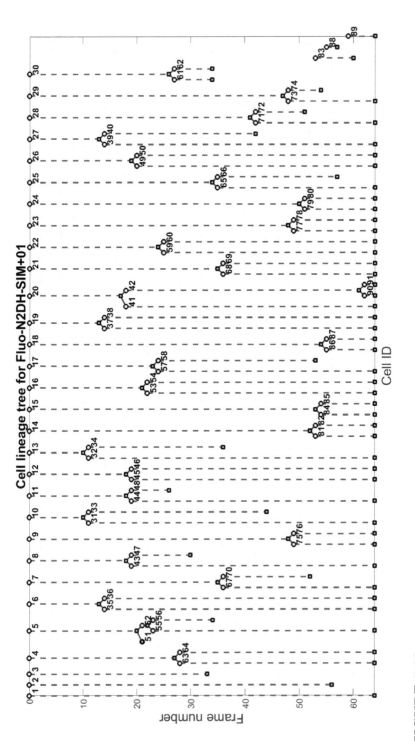

FIGURE 5.18
Lineage tree construction of Fluo-N2DH-SIM+01 dataset by CSTQ, where the cells are labeled by numbers. The dashed lines show migrating cells over time, and the thick lines show cell divisions. The yellow circle nodes denote new cells, or cells entering the field of view. The orange square nodes denote cell death, or cells exiting the field of view.

Acknowledgment

This study was supported by the Army Research Office of the United States Department of Defense under award#W911NF2010095.

Bibliography

[1] Yuri Boykov, Olga Veksler, and Ramin Zabih. Fast approximate energy minimization via graph cuts. *IEEE Transactions on Pattern Analysis and Machine Intelligence*, 23(11):1222–1239, 2001.

[2] Vicent Caselles, Ron Kimmel, and Guillermo Sapiro. Geodesic active contours. *International Journal of Computer Vision*, 22(1):61–79, 1997.

[3] David Bryant Mumford and Jayant Shah. Optimal approximations by piecewise smooth functions and associated variational problems. *Communications Pure and Applied Mathematics*, 1989.

[4] Michael Kass, Andrew Witkin, and Demetri Terzopoulos. Snakes: Active contour models. *International Journal of Computer Vision*, 1(4):321–331, 1988.

[5] Fatima Zerari Boukari. *Mathematical Methods and Algorithms for Identification, Tracking, and Quantitative Analysis in Cellular Bioimaging.* Delaware State University, 2017.

[6] Stanley Osher and James A Sethian. Fronts propagating with curvature-dependent speed: Algorithms based on hamilton-jacobi formulations. *Journal of Computational Physics*, 79(1):12–49, 1988.

[7] Xu, Chenyang, and Jerry L. Prince. Snakes, shapes, and gradient vector flow. *IEEE Transactions on Image Processing*, 7(3):359–369, 1998.

[8] Jun Kong, Fusheng Wang, George Teodoro, Yanhui Liang, Yangyang Zhu, Carol Tucker-Burden, and Daniel J Brat. Automated cell segmentation with 3D fluorescence microscopy images. In *2015 IEEE 12th International Symposium on Biomedical Imaging (ISBI)*, pages 1212–1215. IEEE, 2015.

[9] Dongyao Jia, Chuanwang Zhang, Nengkai Wu, Zhigang Guo, and Hairui Ge. Multi-layer segmentation framework for cell nuclei using improved gvf snake model, watershed, and ellipse fitting. *Biomedical Signal Processing and Control*, 67:102516, 2021.

[10] J Puttamadegowda and SC Prasannakumar. White blood cell sementation using fuzzy c means and snake. In *2016 International Conference on Computation System and Information Technology for Sustainable Solutions (CSITSS)*, pages 47–52. IEEE, 2016.

[11] Congcong Zhang, Xiaoyan Xiao, Xiaomei Li, Ying-Jie Chen, Wu Zhen, Jun Chang, Chengyun Zheng, and Zhi Liu. White blood cell segmentation by color-space-based k-means clustering. *Sensors*, 14(9):16128–16147, 2014.

[12] Tony F Chan and Luminita A Vese. Active contours without edges. *IEEE Transactions on Image Processing*, 10(2):266–277, 2001.

[13] Sarabpreet Kaur and JS Sahambi. Curvelet initialized level set cell segmentation for touching cells in low contrast images. *Computerized Medical Imaging and Graphics*, 49:46–57, 2016.

[14] Soonam Lee, Paul Salama, Kenneth W Dunn, and Edward J Delp. Segmentation of fluorescence microscopy images using three dimensional active contours with inhomogeneity correction. In *2017 IEEE 14th International Symposium on Biomedical Imaging (ISBI 2017)*, pages 709–713. IEEE, 2017.

[15] Amin Gharipour and Alan Wee-Chung Liew. Segmentation of cell nuclei in fluorescence microscopy images: An integrated framework using level set segmentation and touching-cell splitting. *Pattern Recognition*, 58:1–11, 2016.

[16] Chunming Li, Chenyang Xu, Changfeng Gui, and Martin D Fox. Distance regularized level set evolution and its application to image segmentation. *IEEE Transactions on Image Processing*, 19(12):3243–3254, 2010.

[17] Xiaoxiao Yan and Nongliang Sun. Improved fcm based distance regularization level set algorithm for image segmentation. In *2020 IEEE 3rd International Conference of Safe Production and Informatization (IICSPI)*, pages 278–281. IEEE, 2020.

[18] Fuyong Xing and Lin Yang. Robust nucleus/cell detection and segmentation in digital pathology and microscopy images: a comprehensive review. *IEEE Reviews in Biomedical Engineering*, 9:234–263, 2016.

[19] Yao-Tien Chen. A level set method based on the bayesian risk for medical image segmentation. *Pattern Recognition*, 43(11):3699–3711, 2010.

[20] Khamael Al-Dulaimi, Inmaculada Tomeo-Reyes, Jasmine Banks, and Vinod Chandran. White blood cell nuclei segmentation using level set methods and geometric active contours. In *2016 International Conference on Digital Image Computing: Techniques and Applications (DICTA)*, pages 1–7. IEEE, 2016.

[21] Rafael C Gonzalez. *Digital Image Processing*. Pearson education India, 2018.

[22] AL-Dulaimi Khamael, Jasmine Banks, Inmaculada Tomeo-Reyes, and Vinod Chandran. Automatic segmentation of hep-2 cell fluorescence microscope images using level set method via geometric active contours. In *2016 23rd International Conference on Pattern Recognition (ICPR)*, pages 81–83. IEEE, 2016.

[23] Luc Vincent. Morphological grayscale reconstruction in image analysis: applications and efficient algorithms. *IEEE Transactions on Image Processing*, 2(2):176–201, 1993.

[24] Laurent Najman and Michel Schmitt. Geodesic saliency of watershed contours and hierarchical segmentation. *IEEE Transactions on Pattern Analysis and Machine Intelligence*, 18(12):1163–1173, 1996.

[25] James C Bezdek, Robert Ehrlich, and William Full. Fcm: The fuzzy c-means clustering algorithm. *Computers & Geosciences*, 10(2-3):191–203, 1984.

[26] Mangala Shetty and R Balasubramani. Lactobacillus bacterial cell segmentation based on marker controlled watershed method. In *2018 International Conference on Electrical, Electronics, Communication, Computer, and Optimization Techniques (ICEECCOT)*, pages 56–59. IEEE, 2018.

[27] Huisi Miao and Changyan Xiao. Simultaneous segmentation of leukocyte and erythrocyte in microscopic images using a marker-controlled watershed algorithm. *Computational and Mathematical Methods in Medicine*, 2018, 2018.

[28] Simone Tonti, Santa Di Cataldo, Andrea Bottino, and Elisa Ficarra. An automated approach to the segmentation of hep-2 cells for the indirect immunofluorescence ana test. *Computerized Medical Imaging and Graphics*, 40:62–69, 2015.

[29] Chieh-Sheng Chang, Jian-Jiun Ding, Yueh-Feng Wu, and Sung-Jan Lin. Cell segmentation algorithm using double thresholding with morphology-based techniques. In *2018 IEEE International Conference on Consumer Electronics-Taiwan (ICCE-TW)*, pages 1–5. IEEE, 2018.

[30] Erik Meijering. Cell segmentation: 50 years down the road [life sciences]. *IEEE Signal Processing Magazine*, 29(5):140–145, 2012.

[31] Pengyu Yuan, Ali Rezvan, Xiaoyang Li, Navin Varadarajan, and Hien Van Nguyen. Phasetime: Deep learning approach to detect nuclei in time lapse phase images. *Journal of Clinical Medicine*, 8(8):1159, 2019.

[32] Yongsheng Pan, Yong Xia, Tao Zhou, and Michael Fulham. Cell image segmentation using bacterial foraging optimization. *Applied Soft Computing*, 58:770–782, 2017.

[33] Cefa Karabağ, Martin L Jones, Christopher J Peddie, Anne E Weston, Lucy M Collinson, and Constantino Carlos Reyes-Aldasoro. Segmentation and modelling of the nuclear envelope of hela cells imaged with serial block face scanning electron microscopy. *Journal of Imaging*, 5(9):75, 2019.

[34] John Canny. A computational approach to edge detection. *IEEE Transactions on Pattern Analysis and Machine Intelligence*, 8(6):679–698, 1986.

[35] Karolis Leonavicius, Christophe Royer, Antonio MA Miranda, Richard CV Tyser, Annemarie Kip, and Shankar Srinivas. Spatial protein analysis in developing tissues: a sampling-based image processing approach. *Philosophical Transactions of the Royal Society B*, 375(1809):20190560, 2020.

[36] Jianfeng Cao, Zhongying Zhao, and Hong Yan. Accurate cell segmentation based on biological morphology features. In *2018 IEEE International Conference on Systems, Man, and Cybernetics (SMC)*, pages 3380–3383. IEEE, 2018.

[37] Amalka Pinidiyaarachchi and Carolina Wählby. Seeded watersheds for combined segmentation and tracking of cells. In *International Conference on Image Analysis and Processing*, pages 336–343. Springer, 2005.

[38] Scott E Umbaugh. *Digital Image Processing and Analysis: Applications with MATLAB® and CVIPtools*. CRC press, 2017.

[39] Xin Zheng, Yong Wang, Guoyou Wang, and Jianguo Liu. Fast and robust segmentation of white blood cell images by self-supervised learning. *Micron*, 107:55–71, 2018.

[40] Mukta Sharma and Mahua Bhattacharya. Segmentation of ca3 hippocampal region of rat brain cells images based on bio-inspired clustering technique. In *2019 IEEE International Conference on Bioinformatics and Biomedicine (BIBM)*, pages 2438–2445. IEEE, 2019.

[41] Suresh Chandra Satapathy and Anima Naik. Data clustering based on teaching-learning-based optimization. In *International Conference on Swarm, Evolutionary, and Memetic Computing*, pages 148–156. Springer, 2011.

[42] DW Van der Merwe and Andries Petrus Engelbrecht. Data clustering using particle swarm optimization. In *The 2003 Congress on Evolutionary Computation, 2003. CEC'03.*, volume 1, pages 215–220. IEEE, 2003.

[43] Bilkis Jamal Ferdosi, Sharmilee Nowshin, Farzana Ahmed Sabera, et al. White blood cell detection and segmentation from fluorescent images

with an improved algorithm using k-means clustering and morphological operators. In *2018 4th International Conference on Electrical Engineering and Information & Communication Technology (iCEEiCT)*, pages 566–570. IEEE, 2018.

[44] Tapas Kanungo, David M Mount, Nathan S Netanyahu, Christine D Piatko, Ruth Silverman, and Angela Y Wu. An efficient k-means clustering algorithm: Analysis and implementation. *IEEE Transactions on Pattern Analysis and Machine Intelligence*, 24(7):881–892, 2002.

[45] Narjes Ghane, Alireza Vard, Ardeshir Talebi, and Pardis Nematollahy. Segmentation of white blood cells from microscopic images using a novel combination of k-means clustering and modified watershed algorithm. *Journal of Medical Signals and Sensors*, 7(2):92, 2017.

[46] Nobuyuki Otsu. A threshold selection method from gray-level histograms. *IEEE Transactions on Systems, Man, and Cybernetics*, 9(1):62–66, 1979.

[47] Nishchal K Verma, Abhishek Roy, and Shantaram Vasikarla. Medical image segmentation using improved mountain clustering technique version-2. In *2010 Seventh International Conference on Information Technology: New Generations*, pages 156–161. IEEE, 2010.

[48] Xiangzhi Bai, Chuxiong Sun, and Changming Sun. Cell segmentation based on fopso combined with shape information improved intuitionistic fcm. *IEEE Journal of Biomedical and Health Informatics*, 23(1):449–459, 2018.

[49] Abdulkadir Albayrak and Gokhan Bilgin. Automatic cell segmentation in histopathological images via two-staged superpixel-based algorithms. *Medical & Biological Engineering & Computing*, 57(3):653–665, 2019.

[50] Radhakrishna Achanta, Appu Shaji, Kevin Smith, Aurelien Lucchi, Pascal Fua, and Sabine Süsstrunk. Slic superpixels compared to state-of-the-art superpixel methods. *IEEE Transactions on Pattern Analysis and Machine Intelligence*, 34(11):2274–2282, 2012.

[51] Martin Ester, Hans-Peter Kriegel, Jörg Sander, Xiaowei Xu, et al. A density-based algorithm for discovering clusters in large spatial databases with noise. In *kdd*, volume 96, pages 226–231, 1996.

[52] Fatima Boukari and Sokratis Makrogiannis. Automated cell tracking using motion prediction-based matching and event handling. *IEEE/ACM Transactions on Computational Biology and Bioinformatics*, 17(3):959–971, 2018.

[53] Fatima Boukari and Sokratis Makrogiannis. Joint level-set and spatiotemporal motion detection for cell segmentation. *BMC Medical Genomics*, 9(2):179–194, 2016.

[54] Pietro Perona and Jitendra Malik. Scale-space and edge detection using anisotropic diffusion. *IEEE Transactions on Pattern Analysis and Machine Intelligence*, 12(7):629–639, 1990.

[55] Emanuel Parzen. On estimation of a probability density function and mode. *The Annals of Mathematical Statistics*, 33(3):1065–1076, 1962.

[56] Geert Litjens, Thijis Kooi, Babak Ehteshami, Arnaud Arindra Adiyoso Setio, Francesco Ciompi, Mohsen Ghafoorian, Jeroen A.W.M. van der Laak, Bram van Ginneken, and Clara I. Sánchez. A survey on deep learning in medical image analysis. *Medical Image Analysis*, 42:60 – 88, 2017.

[57] Fuyong Xing, Yuanpu Xie, Hai Su, Fujun Liu, and Lin Yang. Deep learning in microscopy image analysis: A survey. *IEEE Transactions on Neural Networks and Learning Systems*, pages 1–19, 2018.

[58] Fatima Boukari and Sokratis Makrogiannis. A joint level-set and spatio-temporal motion detection for cell segmentation. *BMC Medical Genomics*, 2016.

[59] Nikita Moshkov, Botond Mathe, Attila Kertesz-Farkas, Reka Hollandi and Peter Horvath. Test-time augmentation for deep learning-based cell segmentation on microscopy images. *Scientific Reports volume*, 10, 2020.

[60] Filip Lux and Petr Matula. Cell segmentation by combining marker-controlled watershed and deep learning. *ArXiv*, abs/2004.01607, 2020.

[61] Fidel A.Guerrero. Peña, Pedro D. Marrero Fernandez, Paul T. Tar, Tsang I. Ren, Elliot M. Meyerowitz, Alexandre Cunha. J regularization improves imbalanced multiclass segmentation. In *2020 IEEE 17th International Symposium on Biomedical Imaging (ISBI)*, pages 1–5, 2020.

[62] Nikhil Pandey. Kaggle Data science Bowl -2018keras u-net starter - lb 0.277 93ff3b. `https://www.kaggle.com/nikhilpandey360/keras-u-net-starter-lb-0-277-93ff3b`. Accessed: 2022-03-03.

[63] Olaf Ronneberger, Philipp Fischer, and Thomas Brox. U-net: Convolutional networks for biomedical image segmentation. In Nassir Navab, Hornegger Joachim, Wells William M., and Alejandro F. Frangi, editors, *Medical Image Computing and Computer-Assisted Intervention – MICCAI 2015*, pages 234–241, Cham, 2015. Springer International Publishing.

[64] Shan e Ahmed Raza, Linda Cheung, Muhammad Shaban, Simon Graham, David Epstein, Stella Pelengaris, Michael Khan, Nasir M. Rajpoot. Micro-net: A unified model for segmentation of various objects in microscopy images. *Medical Image Analysis*, 52, 04 2018.

[65] Feixiao Long. Microscopy cell nuclei segmentation with enhanced u-net. *BMC Bioinformatics*, 8, 2020.

[66] Yousef Al-Kofahi, Alla Zaltsman, Robert Graves, Will Marshall and Mirabela Rusu. A deep learning-based algorithm for 2-D cell segmentation in microscopy images. *BMC Bioinformatics*, 19, 2018.

[67] Filip Lux and Petr Matula. DIC image segmentation of dense cell populations by combining deep learning and watershed. In *2019 IEEE 16th International Symposium on Biomedical Imaging (ISBI 2019)*, pages 236–239, 2019.

[68] Saad Ullah Akram, Juho Kannala, Lauri Eklund, and Janne Heikkila. Cell segmentation proposal network for microscopy image analysis. In Gustavo Carneiro, Diana Mateus, Loïc Peter, Andrew Bradley, João Manuel R. S. Tavares, Vasileios Belagiannis, João Paulo Papa, Jacinto C. Nascimento, Marco Loog, Zhi Lu, Jaime S. Cardoso, and Julien Cornebise, editors, *Deep Learning and Data Labeling for Medical Applications*, pages 21–29. Springer International Publishing, 2016.

[69] Xuhua Ren, Sihang Zhou, Dinggang Shen, and Qian Wang. Mask-RCNN for cell instance segmentation. *IEEE Transactions on Medical Imaging* ·, 2020.

[70] Yuki Hiramatsu, Kazuhiro Hotta, Ayako Imanishi, Michiyuki Matsuda, and Kenta Terai. Cell image segmentation by integrating multiple cnns. In *2018 IEEE/CVF Conference on Computer Vision and Pattern Recognition Workshops (CVPRW)*, pages 2286–22866, 2018.

[71] Tim Scherr, Katharina Loffler, Moritz Bohland, and Ralf Mikut. Cell segmentation and tracking using cnn-based distance predictions and a graph-based matching strategy. *PL*, 12, 2020.

[72] Ha Su, Xiangfei Kong, Yuanpu Xie, Shaoting Zhang, and Lin Yang. Robust cell detection and segmentation in histopathological images using sparse reconstruction and stacked denoising autoencoders. *Medical Image Computing and Computer Assisted Intervention Society*, 2015.

[73] Lin Yang, Yizhe Zhang, Ian H. Guldner, Siyuan Zhang, and Danny Ziyi Chen. 3D segmentation of glial cells using fully convolutional networks and k-terminal cut. In *MICCAI*, 20F16.

[74] Christian Payer, Darko Stern, Marlies Feiner, Horst Bischof and Martin Urschler. Segmenting and tracking cell instances with cosine embeddings and recurrent hourglass networks. *Medical Image Analysis*, 57:106–119, 2019.

[75] Revathi Ananthakrishnan and Allen Ehrlicher. The forces behind cell movement. *International Journal of Biological Sciences*, 3(5):303, 2007.

[76] Erik Meijering, Oleh Dzyubachyk, Ihor Smal, and Wiggert A van Cappellen. Tracking in cell and developmental biology. In *Seminars in Cell & Developmental Biology*, volume 20, pages 894–902. Elsevier, 2009.

[77] Christophe Zimmer, Bo Zhang, Alexandre Dufour, Ayméric Thébaud, Sylvain Berlemont, Vannary Meas-Yedid, and Jean-Christophe Olivo-Marin. On the digital trail of mobile cells. *IEEE Signal Processing Magazine*, 23(3):54–62, 2006.

[78] Erik Meijering, Oleh Dzyubachyk, and Ihor Smal. Methods for cell and particle tracking. *Methods in Enzymology*, 504:183–200, 2012.

[79] HP Ng, SH Ong, KWC Foong, Poh-Sun Goh, and WL Nowinski. Medical image segmentation using k-means clustering and improved watershed algorithm. In *2006 IEEE Southwest Symposium on Image Analysis and Interpretation*, pages 61–65. IEEE, 2006.

[80] Xiaodong Yang, Houqiang Li, Xiaobo Zhou, and Stephen Wong. Automated segmentation and tracking of cells in time-lapse microscopy using watershed and mean shift. In *2005 International Symposium on Intelligent Signal Processing and Communication Systems*, pages 533–536. IEEE, 2005.

[81] Dirk Padfield, Jens Rittscher, and Badrinath Roysam. Spatio-temporal cell segmentation and tracking for automated screening. In *2008 5th IEEE International Symposium on Biomedical Imaging: From Nano to Macro*, pages 376–379. IEEE, 2008.

[82] M Ali Akber Dewan, M Omair Ahmad, and MNS Swamy. Tracking biological cells in time-lapse microscopy: An adaptive technique combining motion and topological features. *IEEE Transactions on Biomedical Engineering*, 58(6):1637–1647, 2011.

[83] Mingli Lu, Benlian Xu, Andong Sheng, Zhengqiang Jiang, Liping Wang, Peiyi Zhu, and Jian Shi. A novel multiobject tracking approach in the presence of collision and division. *Computational and Mathematical Methods in Medicine*, 2015, 2015.

[84] Thomas A Nketia, Heba Sailem, Gustavo Rohde, Raghu Machiraju, and Jens Rittscher. Analysis of live cell images: Methods, tools and opportunities. *Methods*, 115:65–79, 2017.

[85] Kang Li, Eric D Miller, Mei Chen, Takeo Kanade, Lee E Weiss, and Phil G Campbell. Cell population tracking and lineage construction with spatiotemporal context. *Medical Image Analysis*, 12(5):546–566, 2008.

[86] Neda Emami, Zahra Sedaei, and Reza Ferdousi. Computerized cell tracking: current methods, tools and challenges. *Visual Informatics*, 5(1):1–13, 2021.

[87] Michael Möller, Martin Burger, Peter Dieterich, and Albrecht Schwab. A framework for automated cell tracking in phase contrast microscopic videos based on normal velocities. *Journal of Visual Communication and Image Representation*, 25(2):396–409, 2014.

[88] Yali Huang and Zhiwen Liu. Segmentation and tracking of lymphocytes based on modified active contour models in phase contrast microscopy images. *Computational and Mathematical Methods in Medicine*, 2015, 2015.

[89] Bruno Meunier, Brigitte Picard, Thierry Astruc, and Roland Labas. Development of image analysis tool for the classification of muscle fibre type using immunohistochemical staining. *Histochemistry and Cell Biology*, 134(3):307–317, 2010.

[90] Apurva S Samdurkar, Shailesh D Kamble, Nileshsingh V Thakur, and Akshay S Patharkar. Overview of object detection and tracking based on block matching techniques. In *RICE*, pages 313–319, 2017.

[91] Auguste Genovesio, Tim Liedl, Valentina Emiliani, Wolfgang J Parak, Maité Coppey-Moisan, and J-C Olivo-Marin. Multiple particle tracking in 3-D+ t microscopy: method and application to the tracking of endocytosed quantum dots. *IEEE Transactions on Image Processing*, 15(5):1062–1070, 2006.

[92] Anton Jackson-Smith. Cell tracking using convolutional neural networks. Technical report, Stanford CS Class CS231n, 2016.

[93] Chao Ma, Jia-Bin Huang, Xiaokang Yang, and Ming-Hsuan Yang. Hierarchical convolutional features for visual tracking. In *Proceedings of the IEEE International Conference on Computer Vision*, pages 3074–3082, 2015.

[94] Vladimír Ulman, Martin Maška, Klas EG Magnusson, Olaf Ronneberger, Carsten Haubold, Nathalie Harder, Pavel Matula, Petr Matula, David Svoboda, Miroslav Radojevic, et al. An objective comparison of cell-tracking algorithms. *Nature Methods*, 14(12):1141–1152, 2017.

[95] Runkai Zhu, Dong Sui, Hong Qin, and Aimin Hao. An extended type cell detection and counting method based on fcn. In *2017 IEEE 17th International Conference on Bioinformatics and Bioengineering (BIBE)*, pages 51–56. IEEE, 2017.

[96] Weidi Xie, J Alison Noble, and Andrew Zisserman. Microscopy cell counting and detection with fully convolutional regression networks. *Computer Methods in Biomechanics and Biomedical Engineering: Imaging & Visualization*, 6(3):283–292, 2018.

[97] Yuanpu Xie, Fuyong Xing, Xiaoshuang Shi, Xiangfei Kong, Hai Su, and Lin Yang. Efficient and robust cell detection: A structured regression approach. *Medical Image Analysis*, 44:245–254, 2018.

[98] Xipeng Pan, Dengxian Yang, Lingqiao Li, Zhenbing Liu, Huihua Yang, Zhiwei Cao, Yubei He, Zhen Ma, and Yiyi Chen. Cell detection in pathology and microscopy images with multi-scale fully convolutional neural networks. *World Wide Web*, 21(6):1721–1743, 2018.

[99] Juan C Caicedo, Allen Goodman, Kyle W Karhohs, Beth A Cimini, Jeanelle Ackerman, Marzieh Haghighi, CherKeng Heng, Tim Becker, Minh Doan, Claire McQuin, et al. Nucleus segmentation across imaging experiments: the 2018 data science bowl. *Nature Methods*, 16(12):1247–1253, 2019.

[100] David Joon Ho, Shuo Han, Chichen Fu, Paul Salama, Kenneth W Dunn, and Edward J Delp. Center-extraction-based three dimensional nuclei instance segmentation of fluorescence microscopy images. In *2019 IEEE EMBS International Conference on Biomedical & Health Informatics (BHI)*, pages 1–4. IEEE, 2019.

[101] Edouard A Hay and Raghuveer Parthasarathy. Performance of convolutional neural networks for identification of bacteria in 3D microscopy datasets. *PLoS computational Biology*, 14(12):e1006628, 2018.

[102] Kenneth W Dunn, Chichen Fu, David Joon Ho, Soonam Lee, Shuo Han, Paul Salama, and Edward J Delp. Deepsynth: Three-dimensional nuclear segmentation of biological images using neural networks trained with synthetic data. *Scientific Reports*, 9(1):1–15, 2019.

[103] Pavel Matula, Martin Maška, Dmitry V Sorokin, Petr Matula, Carlos Ortiz-de Solórzano, and Michal Kozubek. Cell tracking accuracy measurement based on comparison of acyclic oriented graphs. *PLoS one*, 10(12):e0144959, 2015.

[104] Martin Maška, Vladimír Ulman, David Svoboda, Pavel Matula, Petr Matula, Cristina Ederra, Ainhoa Urbiola, Tomás España, Subramanian Venkatesan, Deepak MW Balak, et al. A benchmark for comparison of cell tracking algorithms. *Bioinformatics*, 30(11):1609–1617, 2014.

[105] Oleh Dzyubachyk, Wiggert A Van Cappellen, Jeroen Essers, Wiro J Niessen, and Erik Meijering. Advanced level-set-based cell tracking in time-lapse fluorescence microscopy. *IEEE Transactions on Medical Imaging*, 29(3):852–867, 2010.

[106] Ko Sugawara, Çağrı Çevrim, and Michalis Averof. Tracking cell lineages in 3D by incremental deep learning. *Elife*, 11:e69380, 2022.

[107] Jean-Baptiste Lugagne, Haonan Lin and Mary J Dunlop. DeLTA: Automated cell segmentation, tracking, and lineage reconstruction using deep learning. *PLos Computational Biology*, 16(4):e1007673, 2020.

[108] Tomas Sixta, Jlahul Cao, Jochen Seebach, Hans Schnlttler and Boris Flach. Coupling cell detection and tracking by temporal feedback. *Machine Vision and Applications*, 31(24), 2020.

[109] Luca Bertinetto, Jack Valmadre, Joao F. Henriques, Andrea Vedaldi and Philip H.S. Torr. Fully-convolutional siamese networks for object tracking. European Conference on Computer Vision (ECCV), pages 850-865, *Computer Vision*, 2016.

[110] Katharina Loffler, Tim Scherr and Ralf Mikut. A graph-based cell tracking algorithm with few manually tunable parameters and automated segmentation error correction. *PLoS ONE* 16(9): e0249257, 2021.

6

Quantum Image Analysis – Status and Perspectives

Konstantinos Tziridis

MLV Research Group, Department of Computer Science, International Hellenic University, Kavala, Greece

Theofanis Kalampokas

MLV Research Group, Department of Computer Science, International Hellenic University, Kavala, Greece

George A. Papakostas

MLV Research Group, Department of Computer Science, International Hellenic University, Kavala, Greece

CONTENTS

6.1   Introduction 174
6.2   Quantum Computation – Fundamentals 176
     6.2.1   Qubits 176
     6.2.2   Quantum gates 178
          6.2.2.1   Hadamard gate 178
          6.2.2.2   Pauli gates 179
          6.2.2.3   Not gate 179
          6.2.2.4   Phase gate 179
          6.2.2.5   Identity gate 179
          6.2.2.6   CNOT gate 179
     6.2.3   Quantum parallelism 180
     6.2.4   Entanglement 180
6.3   Quantum Computer Vision 180
     6.3.1   Quantum image representation 181
          6.3.1.1   Quantum image processing 190
          6.3.1.2   Feature extraction 193
          6.3.1.3   Decision (quantum model selection training
                   and evaluation) 194
     6.3.2   Tools and hardware 202
          6.3.2.1   Quantum computing tools 202
          6.3.2.2   Quantum hardware 205
6.4   Conclusions 207

DOI: 10.1201/9781003053262-6

6.1   Introduction

Since the first announcement of the words Quantum Computing, tremendous effort has been given by research communities to elaborate on this domain from both hardware and software sides. Based on PitchBook the funding for quantum computing in U.S. has passed $300 million from 2015 to 2020 which is a significant rise considering that Quantum Computing exists from 1979 to 1980. Despite the economic events that point out the significance of this domain there are other facts, studies, and theories that supported the importance of Quantum Computing and they consist of the history that covers the topics about quantum computing importance.

The formation of Quantum Computing domain started with Paul Benioff's publication [1] in 1980 in which he proposed the theoretical model of the quantum computer and how could it be built. Before him, there were other researchers like Roman S. Ingraden who pointed out that classical information theory cannot fully describe quantum information theory and thus vice versa should be applied [2]. In 1980 Yuri Manin proposed a basic idea for quantum computing [3]. Next in 1981 Richard Feynman proved that it is impossible to simulate quantum problems on classical computers from the point that volume of data would outpoint the available resources and thus it is important to build quantum computers that will be adapted under quantum mechanics and physics laws [4], then the capacity of information that could be stored in a quantum state outperforms classical computer capabilities. The next important contribution was made in 1982 [5] where Tsui noticed that in a very low temperature of a matter quantum entanglement could be observed from the microscope. A big step came from Bennett in 1993 who proved that with the exploitation of entanglement state among qubits it is possible to send information at a far distance and it is proposed as Quantum Teleportation [6]. Next in 1994, Peter Shor proposed a factoring algorithm [7] that exploits quantum mechanics and can calculate the prime factors of any N-digit number in a time order of $(\log(N))^3$ where any algorithm of the same kind in a classical computer would have taken just $\log(N)$ faster. In the year 1995 Peter Shor and A. Steaneoffered proposed the quantum error-correction codes in order to encode and decode information in a quantum state with the smallest possible error in probability sampling. In 1998 another significant breakthrough in quantum computing came from Lov K. Grover who proposed Grover's [8] algorithm as a search database method. In this research, it is given the example that a search of a specific item in a database that contains N items would take N time order by a classical computer whereas by quantum computer with Grover's algorithm would take $\sqrt{N}$ due to the exploitation of quantum parallelism and the probability representation by the square of the amplitudes in the block-sphere. With this contribution in 2001, IBM manufactured the first quantum device and since then quantum computing has become a topic of interest by the research community and the technology industry with the presentation of the first commercial quantum computer by D-Wave in 2010.

Since then a race has been started around quantum computers in order to increase their capabilities from more qubits that translate to more space in order to encode information, to more stable circuits and gates with less noise. From the quantum computing side, new domains of software engineering started to exploit quantum mechanics and quantum computers in order to solve problems that classical computers are considered less efficient.

The main focus of this chapter is image analysis including image processing and computer vision algorithms. The applications and the algorithms that form the previously referred domains are computationally expensive since they have to process a huge amount of pixels. Image as a data representation is considered one of the richest data sources and heavy at the same way and thus it would be a great challenge to exploit the quantum computing advantages.

Quantum image processing was first introduced in 1997 by Vlasov [9] who exploited quantum mechanics properties to implement an image recognition system. In 2003, Schützhold proposed a quantum algorithm [10] that searches and extracts patterns in binary images. On the same page Beach, Lomot and Cohen proved that [11] quantum image processing might be advantaged by classical quantum algorithms like Grover's, through quantum image processing application that detects posture. Since then quantum image processing domain has met a significant rise with the evolution of quantum machine learning. In this domain, a variety of streams have been started by the research community where some deal with the applicability of classical machine-learning methods to data from quantum systems, other deal with the expansion of classical machine-learning models in quantum computers in order to manipulate quantum states as inputs and outputs. The highest interest has been given to the formation of models and learning procedures that bridge quantum and classical devices under one hybrid process. From the quantum side, the heaviest computationally subroutine of the learning process is implemented under quantum computing properties and the rest of the process hosted in a classical computer. Under this scheme, many machine-learning and deep-learning models have been proposed like QSVM and QGAN accordingly. The models that have been proposed are mainly applied in classical data that have been encoded in a quantum state. With this trend, quantum image processing and analysis domains have focused their interest on different topics around data representation methods of classical data in a quantum computer, feature extraction methods of classical data in a quantum computer, new quantum processing methods, and addressing image data problems to quantum computing properties. Even with this effort quantum computer vision domain is not clearly formed since there are some limitations that do not let many applications and methods execute under quantum conditions.

To conclude it is crystal clear that quantum computers give a significant computational speed-up and expand the capabilities in the amount of data structures that can be implemented under quantum states which might solve some problems in the Computer Vision domain in the near future.

Despite the advantages of quantum computing over classical computer problem solving there are some disadvantages that do not exist in classical environments and they put a barrier between research and commercial level of quantum image processing and quantum computer vision domains. According to the no-cloning principle where a quantum state could not be replicated to another quantum state it is not clear how an image could be replicated, which is a fundamental mechanism for image processing algorithms. Another difficulty is the measurement of the manipulated image in a quantum state which may fall in a collapsing state. Despite the quantum computing limitation, there are some hardware limitations also that will be discussed in the presented paper like the low number of available qubits, the presence of noise during the measurement of probabilities by quantum states and many others.

The purpose of this chapter is to present an analysis of quantum computing research work with the main focus on quantum image processing and quantum computer vision. To cover most topics a presentation will be done in other domains of quantum computing like tools and hardware. The rest of this chapter is organized as follows. Section 6.2 the basic components of quantum circuits will present, with an addition of some quantum computing fundamentals like quantum parallelism, entanglement, superposition, and quantum teleportation. In Section 6.3 a brief review will be presented regarding research on quantum computer vision algorithms from the preprocessing to the evaluation with a presentation also on quantum tools and hardware. Finally, in Section 6.4 the conclusions will highlight the state of quantum computers at the current stage, which affects the applications of the quantum algorithms proposed in real quantum computers.

6.2   Quantum Computation – Fundamentals

In this section, the basic concepts of quantum computing will be presented and analyzed like quantum bits, quantum gates, quantum circuits and some basic characteristics of quantum computers that rely on quantum mechanics laws and theories. A fundamental rule of quantum computers is that they process on an atomic level where the properties of atomic particles are having different characteristics. Some of these characteristics are quantum parallelism, entanglement, superposition, and they will be analyzed in this section. Based on these concepts, the entire quantum computing theory has been invented and expanded until today.

6.2.1   Qubits

The fundamental unit for information representation and processing in classical computers is the bit that can exist in one state taking two possible values 0 or 1. On this concept, the entire computing theories and achievements are formed and it is a belief that with a similar approach quantum computing can rise with its own unique characteristics. Like bits in classical computers the basic unit for the information process in a quantum computer is the qubit and like the bit can exist in one of the two states mentioned above:

$$0 \rightarrow |0\rangle, 1 \rightarrow |1\rangle, \tag{6.1}$$

As shown in Equation 6.1 the only similarity between bit and qubit is that at the end they collapse in one of the two states zero or one. The state of a qubit as presented in Equation 6.1 is contained in the notation $|\rangle$ and it is called state vector or a ket from the Dirac notation. Qubits are two-dimensional vectors and they exist in Hilbert space. Hilbert space is an infinite-dimensional space and it covers all complex numbers C in order to satisfy the condition of completeness. Quantum computers exploit Hilbert space characteristics and thus they have the ability to grow in size structure exponentially. Based on this a Hilbert space with two-dimension can be called a qubit and the most general normalized state can be expressed as:

$$\alpha|0\rangle + \beta|1\rangle, \tag{6.2}$$

In Equation 6.2 α and β are complex numbers and they must satisfy $|\alpha|^2 + |\beta|^2 = 1$. Where $|\alpha|^2$ and $|\beta|^2$ are given the ability to obtain a probability measurement of the qubit state in $|0\rangle$ or $|1\rangle$ accordingly based on Born's rule. Despite the two states that a qubit can collapse there is another position that it can be found and it's called superposition where the qubit can exist in a linear combination of the states $|0\rangle$ or $|1\rangle$ and mathematically can be represented as:

$$|\psi\rangle = \alpha|0\rangle + \beta|1\rangle, \tag{6.3}$$

In Equation 6.3 $|\psi\rangle$ represents a wave function and it can be correlated with the decomposition of a musical note to its sub-component frequencies. This wave function represents all the possible values that a qubit can have in a superposition state. A fact about quantum computers is that when a measurement is applied the quantum system is disturbed and thus the superposition of a qubit collapse in one of the two states. In Equations 6.2 and 6.3 α and β are not encoded only the probabilities of the state, but also the relative phase and amplitude of the qubit state which has physical significance. For physicists in Equation 6.3, α and β can represent the spin of an atom around the axis (like a photon polarize). Based on this the representation of a qubit can be formed in a three-dimensional space called Bloch-sphere as presented in Figure 6.1.

Bloch-sphere is the visual representation of a qubit where the relative phase ϕ and amplitude θ are presented. A basic rule of quantum mechanics and physics is that at an atomic level atomic particles behave like a wave signal, even if a single atom has its own geometrical and physical properties. And thus in quantum computing, it is achievable to encode a huge amount of information only in a single qubit. Qubits and superposition states can be

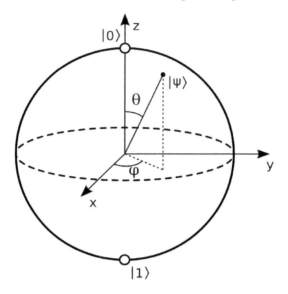

FIGURE 6.1
Bloch-sphere

found in a quantum computer under different quantum phenomena such as trapped ions, superconductors, or photons which will be analyzed in the next sections. To conclude this is the basic unit of information process in a quantum computer and it gave the capacity to overcome the amount of information that can be encoded in a classical computer.

6.2.2   Quantum gates

The computation model of a quantum computer is usually the quantum circuit. Quantum circuits consist of qubits and quantum gates where a sequence of quantum gates can perform complex operations in a quantum computer. In general, quantum gates achieve logical transformations through unitary transformations of quantum states and they can be represented in the form of a matrix. The most used single-qubit gates are Hadamard, Pauli-X, Y, and Z, NOT, Phase, Identity, and CNOT.

6.2.2.1   Hadamard gate

One of the common quantum gates is Hadamard and it can bring a qubit in a superposition state. In the form of a matrix, it can be represented as

$$H = \frac{1}{\sqrt{2}} \begin{vmatrix} 1 & 1 \\ 1 & -1 \end{vmatrix}, \tag{6.4}$$

A Hadamard gate can bring an n-qubit system on an all basis equally in a superposition.

6.2.2.2 Pauli gates

Pauli gates consist of three gates that apply rotation on one of the three axis X, Y, and Z of a qubit in a Bloch-sphere. In a matrix form, Pauli or Rotation gates can be represented as

$$X = \begin{vmatrix} 0 & 1 \\ 1 & 0 \end{vmatrix}, Y = \begin{vmatrix} 0 & -i \\ i & 0 \end{vmatrix}, Z = \begin{vmatrix} 1 & 0 \\ 0 & -1 \end{vmatrix}, \tag{6.5}$$

Based on Equation 6.5, a qubit is rotated by π radians along any of the above axis.

6.2.2.3 Not gate

The NOT quantum gate is identical with the classical NOT gate and it changes the quantum state at the exact opposite and in a form of a matrix can be represented as

$$NOT = \begin{vmatrix} 0 & 1 \\ 1 & 0 \end{vmatrix}, \tag{6.6}$$

6.2.2.4 Phase gate

Phase gate changes the phase of a qubit along to Z-axis as represented in Bloch-sphere but it leaves the probability of basis states of the qubit unchanged. In a form of a matrix, it can be represented as

$$P(\phi) = \begin{vmatrix} 1 & 0 \\ 0 & e^{i\phi} \end{vmatrix}, \tag{6.7}$$

In Equation 6.7, when $\phi = \theta$ then phase gate is similar to Pauli Z gate.

6.2.2.5 Identity gate

The Identity quantum gate is a simple gate that leaves unchanged the state of a qubit and from the computing point of view, it can be used to represent the none or do-nothing operation. In a form of a matrix, it is represented as

$$I = \begin{vmatrix} 1 & 0 \\ 0 & 1 \end{vmatrix}, \tag{6.8}$$

6.2.2.6 CNOT gate

The CNOT quantum gate is a gate that acts on two qubits where on the first qubit it performs control that decides whether the NOT operation should be applied on the second qubit or not. In the form of a matrix, it is represented as:

$$CNOT = \begin{vmatrix} 1 & 0 & 0 & 0 \\ 0 & 1 & 0 & 0 \\ 0 & 0 & 0 & 1 \\ 0 & 0 & 1 & 0 \end{vmatrix}, \tag{6.9}$$

The CNOT gate is a very useful gate since it can simulate the operation of if-else statements.

These are the most common quantum gates that can be used in sequences to conduct qubit's manipulation and information processing in the form of a quantum circuit. The number of gates in a quantum circuit represents its depth and the more the depth of the circuit grows the more noise is produced from the reason that quantum gates are not fully stable.

6.2.3 Quantum parallelism

A basic difference between quantum computers and classical computers is that classical computers can execute one computational path at a time while quantum computers can execute multiple computational paths at a time. This can happen based on the ability of the quantum memory register to exist in a superposition of base states. Its individual state of superposition can be thought of as a single argument to a function. A function performed in a superposition is thus performed on each state of the superposition. Since the number of possible states is 2^N where N is the number of qubits, it is feasible to perform one operation on a quantum computer that would take an exponential number of operations on a classical computer.

6.2.4 Entanglement

Entanglement is the fundamental characteristic of quantum mechanics and it is what gave quantum computing a strong advantage. From the quantum mechanics point of view at an atomic level, the particles are connected to each other. Thus in quantum computing a qubit is susceptible to any kind of change with regard to all other qubits. This gives the capability to quantum computers to form 2^n classical states with only N qubits since in an entangled state the information is encoded not in each qubit but in the relationship among them. This gave different characteristics to quantum computers from classical ones.

These are the two fundamental characteristics of the qubits that exist in quantum computers and they are the reason for all computation advantages that a quantum computer has. Another reason that parallelism and entanglement are so important is that based on no-cloning theory a quantum gate cannot operate on more than one circuit executions and it must end the circuit execution and then reuse it. To conclude, these are the basic concepts of quantum computing based on which quantum hardware and software are using.

6.3 Quantum Computer Vision

Quantum Computer Vision (QCV) is an emerging field of research, which can lead to orders of magnitude speed increase, compared to its classical counterparts. Although the current state is far from this point due to technical

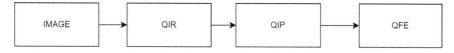

FIGURE 6.2
Pipeline of Quantum Image Analysis

challenges regarding Quantum Computers. Despite the challenges, numerous efforts have been made in order to create the Quantum equivalent of popular computer vision algorithms, which are mainly tested in simulations. In order to realize QCV with the already known algorithms it is essential to highlight a pipeline for all the necessary operations as shown in Figure 6.2. As a starting point for QCV is the conversion of the classical image in a data structure described with quantum states. Next, quantum versions of computer vision algorithms are necessary in order to leverage the parallel computational speedup quantum computers can offer. Another important step for a complete pipeline is the feature extraction step which is widely used in classical computer vision for various tasks. The final required operation is the retrieval of the classical image derived from the previous processing steps.

6.3.1 Quantum image representation

In order to develop algorithms for quantum computers, it is necessary to encode both data and operations using quantum states and gates. Inspired by classical computers quantum image representation (QIR) consists of two parts that capture information about position and color for every pixel in an image.

For this task, several methodologies have been proposed. In this review, we will briefly present the available QIRs found in the literature.

- Flexible Representation for Quantum Images – FRQI

FRQI [12] captures the information about colors and the corresponding positions. FRQI expression is represented in Equation 6.10, where $\otimes$ is the tensor product notation, $|0\rangle$ and $|1\rangle$ are 2-D computational basis quantum states, $|i\rangle$, $i = 0, 1, \ldots, 2^{2n} - 1$ are $2^{2n} - D$ computational basis quantum states and $\theta = (\theta_0, \theta_1, \ldots, \theta_2^{2n} - 1)$ is the vector of angles encoding colors. The two parts of FRQI representation are $|c_i\rangle|$, which encodes the information about colors and $|i\rangle|$ that encodes positional information:

$$|I(\theta)\rangle = \frac{1}{2^n} \sum_{i=0}^{2^{2n}-1} |c_i\rangle \otimes |i\rangle, \qquad (6.10)$$

$$|c_i\rangle = \cos\theta_i|0\rangle + \sin\theta_i|1\rangle, \qquad (6.11)$$

where $\theta_i \in [0, \frac{\pi}{2}], i = 0, 1, \ldots, 2^{2n} - 1$

• Multi-Channel Representation for Quantum Images – MCRQI

MCRQI [13] was inspired from FRQI targeting RBG α images. It is achieved by using three qubits to encode the image's color information as well as it's transparency. MCRQI expression is presented in Equation 6.12, where $|c_{RGB\alpha}^i\rangle$ carries the color information and α channel and is defined in Equation 6.13, where $\theta_{Ri}, \theta_{Gi}, \theta_{Bi}, \theta_{\alpha i} \in [0, \frac{\pi}{2}]$ are vectors encoding the RGB and α channel respectively; $\otimes$ is the tensor product notation; $|000\rangle, \ldots, |111\rangle$ are 8-D computational basis quantum states and $|i\rangle$ for $i = 0, 1, \ldots, 2^{2n} - 1$ are $|2^{2n} - D\rangle$ computational basis states:

$$|I(\theta)\rangle = \frac{1}{2^n + 1} \sum_{i=0}^{2^{2n}-1} |c_{RGB\alpha}^i\rangle \otimes |i\rangle, \tag{6.12}$$

$$\begin{aligned}|c_{RGB\alpha}^i\rangle = {} & \cos\theta_{Ri}|000\rangle + cos\theta_{Gi}|001\rangle \\ & + cos\theta_{Bi}|010\rangle + cos\theta_{\alpha i}|011\rangle \\ & + sin\theta_{Ri}|100\rangle + sin\theta_{Gi}|101\rangle \\ & + sin\theta_{Bi}|110\rangle + \sin\theta_{\alpha i}|111\rangle.\end{aligned} \tag{6.13}$$

• Novel Enhanced Quantum Representation – NEQR

In 2013, through analysis of the FRQI model, NEQR quantum image representation method was proposed to overcome FRQI's drawbacks [14]. To do so, NEQR model uses two entangled qubit sequences to store the grayscale and position information, and stores the image in the superposition of the two qubit sequences. For a gray range of 2^q in an image, the binary sequence $C_{YX}^0 C_{YX}^1 \ldots C_{YX}^{q-2} C_{YX}^{q-1}$ encodes the gray-scale value $f(Y, X)$ of the corresponding pixel (Y, X) as in Equation 6.14:

$$f(Y, X) = C_{YX}^0 C_{YX}^1 \ldots C_{YX}^{q-2} C_{YX}^{q-1}, C_{YX}^k \in [0, 1], f(Y, X) \in [0, 2^q - 1], \tag{6.14}$$

The complete representation of a quantum image of dimensions $2^n \times 2^n$ is presented in Equation 6.15:

$$I\rangle = \frac{1}{2^n} \sum_{Y=0}^{2^n-1} \sum_{X=0}^{2^n-1} |f(Y,X)\rangle|YX\rangle = \frac{2}{2^n} \sum_{Y=0}^{2^n-1} \sum_{X=0}^{2^n-1} \overset{q-1}{\underset{i=0}{\otimes}} |C_{YX}^i|YX\rangle, \tag{6.15}$$

• Colored Quantum Image Representation – CQIR

CQIR [15] is expressed in Equation 6.16:

$$|I\rangle = |C\rangle_m \otimes |P\rangle_{2n} = \frac{1}{2^n} \sum_{i=0}^{2^{2n}-1} \sum_{j=0}^{2^m-1} a_{ij}|j\rangle|i\rangle, \tag{6.16}$$

Pixel positions are encoded in register $|P\rangle$ using $2n$ qubits. Register $|P\rangle$ is in the form $|y\rangle|x\rangle$ where y and x encode the row and column of the image, respectively. Color is represented using $m = \log_2 L$ qubits encoding L colors (gray levels) of the image, and α_{ij} determines the color of the pixel at position i via superposition state of all possible colors.

- Quantum Image Representation for Log-Polar Images – QUALPI

QUALPI [16] model was presented inspired by FRQI. Instead of storing the representation in Cartesian coordinates, the authors proposed storing the representation in Log-polar coordinates, which helps in many complex affine transformations, like rotation and scaling due to the irreversible interpolations required. The QUALPI representation method is defined in Equation 6.17, where $g(\rho, \theta)$ represents the gray scale value of the corresponding pixel. With a gray range of 2^q the gray scale values can be encoded by binary sequence $C_0 C_1 \ldots C_{q-2} C_{q-1}$ as shown in Equation 6.18. The basis state consists of the tensor product of three qubit sequences, where all the information of a pixel includes the gray scale value, the log-radius position, and the angular position:

$$I\rangle = \frac{1}{\sqrt{2^{m+n}}} \sum_{\rho=0}^{2^m-1} \sum_{\theta=0}^{2^n-1} (|g(\rho, \theta)\rangle \otimes |\theta\rangle), \tag{6.17}$$

$$g(\rho, \theta) = C_0 C_1 \ldots C_{q-2} C_{q-1}, g(\rho, \theta) \in [0, 2^q - 1], \tag{6.18}$$

- Quantum States for M Colors and Quantum States for N Coordinates position and QSMC&QSNC

In [17], QSMC&QSNC model was introduced. The model leverages two sets of quantum states QSMC&QSNC for image storing. QSMC represents colors and QSNC represents the coordinates of the pixels in the image. To store both color information and position of the image in a quantum state the authors' mapped m different colors and n coordinate position information into the same number of angle values. The expression for QSMC&QSNC is shown in Equation 6.19, where $|QSMC_i\rangle = \cos\phi_i|0\rangle + \sin\phi_i|1\rangle$, $|QSNC_i\rangle = \cos\theta_i|0\rangle + \sin\theta_i|1\rangle$ and $\phi_i, \theta_i \in [0, \frac{\pi}{2}], i = 0, 1, \ldots 2^{2n} - 1$:

$$|I\rangle = \frac{1}{2^n} \sum_{i=0}^{2^{2n}-1} |QSMC_i\rangle \otimes |QSNC_i\rangle, \tag{6.19}$$

- Normal Arbitrary Quantum Superposition State – NAQSS

NAQSS [18] model was proposed due to the difficulties in processing multi-dimensional color images, like three-dimensional images. The problems are realized in the storage of large images as well as in poor accuracy of image segmentation for some images for content-based image searching. The NAQSS is an $(n + 1)$-qubit quantum representation, which is used to represent multi-dimensional color images. Among the bits, the first n are used to represent the coordinates of the 2^n pixels and the remaining bit represents the segmentation information of the image. NAQSS corresponds to the color and angle of an image, mapping the color information to certain values on the interval $[0, \frac{\pi}{2}]$. NAQSS is represented in Equation 6.20, where $|x_i\rangle = \cos\gamma_i|0\rangle + \sin\gamma_i|1\rangle$ and is used to represent the segmentation information of an image. The locations of the pixels are represented by the coordinates $|\nu_1\rangle\nu_2\rangle \ldots \nu_\kappa\rangle$:

$$|I\rangle = \sum_{i=0}^{2^n-1} \theta_i|\nu_1\rangle\nu_2\rangle \ldots \nu_\kappa\rangle \otimes |x_i\rangle, \qquad (6.20)$$

- MCQI

MCQI [19] quantum image representation was proposed for color image processing, while maintaining color information in a normalized state. MCQI representation is shown in Equation 6.21, where $|C^i_{RGB\alpha}$ is used for color encoding and is defined in Equation 6.22, where $\{\theta^i_R, \theta^i_G \theta^i_B\} \in [0, \frac{\pi}{2}$ and represents the gray values of red, green, and blue channels, respectively:

$$I\rangle = \frac{1}{2^{n+1}} \sum_{i=0}^{2^{2n}-1} |C_RGB\alpha^i\rangle|i\rangle, \qquad (6.21)$$

$$\begin{aligned}|C^i_{RGB\alpha}\rangle = {}& \cos\theta^i_R|000\rangle + \cos\theta^i_G|001\rangle + \cos\theta^i_B|010\rangle + \cos\theta_\alpha|011\rangle \\ & + \sin\theta^i_R|100\rangle + \sin\theta^i_G|101\rangle + \sin\theta^i_B|110\rangle + \sin\theta_\alpha|111\rangle, \quad (6.22)\end{aligned}$$

- Simple Quantum Representation – SQR

SQR [20] model was introduced targeting infrared images and its characteristic regarding the reflectance of infrared radiation energy. SQR was inspired by Qubit Lattice representation for color images, but instead of encoding the color values using angle parameters, a probability of project measurement is used to store radiation energy values for each pixel. The SQR representation model is shown in Equation 6.23, where $|\phi_{ij}\rangle = \cos\theta_{ij}|0\rangle + \sin\theta_{ij}|1\rangle$:

$$|I\rangle = |\phi_{ij}\rangle, \qquad (6.23)$$

- Improved Novel Enhanced Quantum Representation – INEQR

In [21], INEQR model was proposed as an improvement for NEQR for the realization of the image scaling using nearest neighbor interpolation on quantum images. NEQR can process quantum images of size $2^n \times 2^n$. However, when the scaling ratios are not equal for the two dimensions, the size of the scaled image will not have the form of $2^n \times 2^n$. An INEQR quantum image can be expressed as shown in Equation 6.24, where

$$|YX\rangle = |Y\rangle|X\rangle = |y_0 y_1 \ldots y_{n_1-1}\rangle|x_0 x_1 \ldots x_{n_2-1}\rangle, x_i, y_i \in \{0,1\}$$

$$|I\rangle = \frac{1}{2^{\frac{n_1+n_2}{2}}} \sum_{Y=0}^{2^{n_1}-1} \sum_{X=0}^{2^{n_2}-1} |f(Y,X)\rangle|YX\rangle$$

$$= \frac{1}{2^{\frac{n_1+n_2}{2}}} \sum_{Y=0}^{2^{n_1}-1} \sum_{X=0}^{2^{n_2}-1} \overset{q-1}{\underset{i=0}{\otimes}} C_Y^i X\rangle|YX\rangle, \tag{6.24}$$

- Generalized Quantum Image Representation – GQIR

GQIR [21] quantum image representation model can store arbitrary $H \times W$ quantum images. GQIR is able to overcome the problem that the NEQR representation method is only able to represent square images of size $2^n \times 2^n$ making the method compatible with image scaling operations. Both position and color information is captured into normalized quantum states. A GQIR image is shown in Equation 6.25, where $|YX\rangle$ represents the location information and $|C_{YX}\rangle$ is the color information at the corresponding position:

$$|I\rangle = \frac{1}{\sqrt{2}^{h+w}} \left(\sum_{Y=0}^{H-1} \sum_{X=0}^{W-1} \otimes_{i-0}^{q-1} |C_{YX}^i|YX\rangle) \right) \tag{6.25}$$

$$|YX\rangle = |Y\rangle|X\rangle = |y_0 y_1 \ldots y_{h-1}\rangle|x_0 x_1 \ldots x_{w-1}\rangle, y_i, x_i \in \{0,1\}$$
$$|C_{YX}\rangle = |C_{YX}^0 C_{YX}^1 \ldots C_{YX}^{q-1}\rangle, C_i^{YX} \in \{0,1\}, \tag{6.26}$$

- NCQI

NCQI [22] quantum image representation model, inspired by NEQR, uses the basis state of qubit sequence to store the RGB values of each pixel. All pixels are stored in a normalized superposition state and can be operated simultaneously. Compared with previous multi-channel representations, NCQI achieved quadratic speedup in the preparation of the quantum image. The quantum image representation for NCQI is shown in Equation 6.27, where $c(y,x)$ represents the pixel value at the corresponding coordinate:

$$|I\rangle = \frac{1}{2^n} \sum_{y=0}^{2^{2n}-1} \sum_{x=0}^{2^{2n}-1} |c(y,x)\rangle \otimes |yx\rangle, \tag{6.27}$$

- Digital RGB Multi-Channel Representation for Quantum Colored Images – QMCR

In [23], QMCR model was suggested based on the RGB model of the classical computers and derived by extending the gray-scale information in NEQR to color representation. Two entangled qubit sequences are used to encode pixel positions and colors. More specifically one qubit sequence is used to encode pixel position and the other qubit sequence is used to encode the red, green, and blue channels' values of the corresponding pixel at the same time. Equation 6.28 shows the representation of a $2^n \times 2^n$ QMCR image with grayscale range for each color channel 2^q, where $|C_{RGByx}\rangle = |R_{yx}\rangle|G_{yx}\rangle|B_{yx}\rangle$, $|R_{yx}\rangle = |r_{yx}^{q-1}r_{yx}^{q-2}\ldots r_{yx}^0$, $|G_{yx}\rangle = |g_{yx}^{q-1}g_{yx}^{q-2}\ldots g_{yx}^0$, $|B_{yx}\rangle = |b_{yx}^{q-1}b_{yx}^{q-2}\ldots b_{yx}^0$:

$$|I\rangle = \frac{1}{2^n}\sum_{y=0}^{2^n-1}\sum_{x=0}^{2^n-1}|C_{RGByx} \otimes |yx\rangle, \tag{6.28}$$

- Quantum Representation Model for Multiple Images – QRMMI

In [24], QRMMI model was proposed for the representation of multiple images. For a digital image with size $2^n \times 2^n$, then the color information $f_J(Y, X)$ at pixel position (Y, X) of the image J can be represented as a binary sequence as shown in Equation 6.29, where $|JYX\rangle = |J\rangle|Y\rangle|X\rangle = |j_{t-1}j_{t-2}\cdots j_0\rangle|y_{n-1}y_{n-2}\cdots y_0\rangle|x_{n-1}x_{n-2}\cdots x_0\rangle$. $|J\rangle$ and $|YX\rangle$ denote the sequence number of the Jth image and the position information, respectively:

$$f_J(Y, X) = C_{JYX}^{q-1}C_{JYX}^{q-2}C_{JYX}^1C_{JYX}^0, \tag{6.29}$$

- Quantum Representation of Multi-Wavelength Images – QRMW

In [25], QRMW model was proposed, which leverages three separate register qubit sequences to store the position, wavelength, and color value for each pixel. The whole image is saved in a superposition of three-qubit sequences. The expression of a QRMW image is shown in Equation 6.30, where λ and y, x represent channel and position information, respectively:

$$|I\rangle = \frac{1}{\sqrt{2^{b+n+m}}}\sum_{\lambda=0}^{2^{b-1}}\sum_{y=0}^{2^{n-1}}\sum_{x=0}^{2^{m-1}}|f(\lambda, y, x)\rangle \otimes |\lambda\rangle \otimes |yx\rangle, \tag{6.30}$$

- Generalized Novel Enhanced Quantum Representation - GNEQR

GNEQR [26] was proposed from a synthesis of four representations and is presented in Equation 6.31, where $|x\rangle = |i_n\ldots i_{k+1}\rangle$, $|y\rangle = |i_k\ldots i_1\rangle$ and $i_1,\ldots i_k,\ldots i_n \in \{0,1\}$. $|x\rangle$ and $|y\rangle$ represent the X and Y axis, respectively. $f(x, y)$ is the pixel value at the coordinate (x, y), while $f(x, y) \in C_m\ldots C_m$ is a color set equal to $C_m = \{0, 1, \ldots 2^m - 1\}$

$$|\Psi_{G2}^m\rangle = \frac{1}{\sqrt{2^n}}\sum_{x=0}^{2^{n-k}-1}\sum_{y=0}^{2^k-1}|f(x, y)\rangle|x\rangle|y\rangle, \tag{6.31}$$

- Optimized Quantum Representation for Color Images – OCQR

OCQR [27] makes full use of the quantum superposition characteristic to store the RGB values for every pixel. Based on OCQR the model uses three-dimensional quantum sequences to store the color quantum image, one sequence represents the channel value, the other indicates the channel index and the last one stores position information. Equation 6.32 shows the detailed representation model, where $|r_{yx}\rangle = |r_{yx}^{q-1} \ldots r_{yx}^{0}\rangle$, $|g_{yx}\rangle = |g_{yx}^{q-1} \ldots g_{yx}^{0}\rangle$, $|b_{yx}\rangle = |b_{yx}^{q-1} \ldots b_{yx}^{0}\rangle$, and $|s_{yx}\rangle = |0^{q-1} \ldots 0^{0}\rangle$. ch_index is the index of the channel and red, green, and blue channels are represented by $|00\rangle$, $|01\rangle$, and $|10\rangle$, respectively. $|11\rangle$ represents the free channel index, which can be used to store other pixel information, like transparency:

$$I\rangle = \frac{1}{2^{n+1}} \sum \sum |c(x,y)\rangle |ch\_index\rangle |yx\rangle$$

$$= \frac{1}{2^{n+1}} \sum_{y=0}^{2^n-1} \sum_{x=0}^{2^n-1} (|r_{yx}\rangle \otimes |00\rangle + |g_{yx}\rangle \otimes |01\rangle$$

$$+ |b_{yx}\rangle \otimes |10\rangle + |s_{yx}\rangle \otimes |11\rangle)|yx\rangle, \tag{6.32}$$

- Bitplane Representation of Quantum Images – BRQI

BRQI [28] was proposed in order to improve the storage capacity of QIRs. To do so the authors proposed a bitplane representation of quantum images leveraging the GNEQR representation model. For a grayscale image, each bitplane's representation using BRQI method is shown in Equation 6.34, where j represents the jth bitplane, $j = 0, 1 \ldots, 7$, $m = 1$ and $g(x,y) \in C_1 = \{0,1\}$. For the representation of the eight bitplanes via one state, Equation 6.35 is used:

$$|\Psi_m^j\rangle = \frac{1}{\sqrt{2^n}} \sum_{x=0}^{2^{n-k}} \sum_{y=0}^{2^k-1} |g(x,y)\rangle |x\rangle |y\rangle, \tag{6.33}$$

$$|\Psi_B^8\rangle = \frac{1}{\sqrt{2^3}} \sum_{i=0}^{2^{3-1}-1} |\Psi_m^l\rangle |l\rangle, \tag{6.34}$$

$$= \frac{1}{\sqrt{2^{n+3}}} \sum_{l=0}^{2^3-1} \sum_{x=0}^{2^{n-k}-1} \sum_{y=0}^{2^k-1} |g(x,y)\rangle |x\rangle |y\rangle |l\rangle, \tag{6.35}$$

- Order-encoded Quantum Image Model – OQIM

OQIM [29] uses the basis state of a qubit sequence to store the ascending order of each pixel according to their gray values' magnitude. For color and position, amplitude probability of qubits are used. OQIM model is more flexible and better suited for histogram-related tasks like specification, equalization, as well as other similar image enhancement methods like luminance correction. The OQIM model is presented in Equation 6.36, where $cp_i\rangle = \cos\theta_i|00\rangle + \sin\theta_i|10\rangle + \cos\phi_i|01\rangle + \sin\phi_i|11\rangle$ represents both coordinates and colors and $|i\rangle$ encodes the sorted position with the basis state of a qubit sequence:

$$|I\rangle = \frac{1}{2^{n+\frac{1}{2}}} \sum_{i=0}^{2^{2n}-1} |cp_i\rangle \otimes |i\rangle, \tag{6.36}$$

- Improved Flexible Representation of Quantum Images – IFRQI

IFRQI [30] was based on FRQI and NEQR and uses p qubits to store the grayscale value of every pixel of $2p$-bit-deep image. In order to incorporate the advantage of the superposition principle, IFRQI uses two entangled qubit sequences. The first sequence consists of p qubits that represent the grayscale information of each $2p$-bit pixel and the second one is the same as the one in the FRQI model. The IFRQI model is defined in Equation 6.37, where $\alpha_{i,\kappa} = \cos\theta_\kappa$, $\beta_{i,\kappa} = \sin\theta_\kappa$ for all $0 \leq \kappa \leq p-1$

$$|I_q\rangle = \frac{1}{2^n} \sum_{i=0}^{2^{2n}-1} \overset{\kappa=p-1}{\underset{\kappa=0}{\otimes}} (\alpha_{i,\kappa}|0\rangle + \beta_{i,k}|1\rangle) \otimes |i\rangle, \tag{6.37}$$

- Double Quantum Color Images Representation – QRCI

QRCI [31] model and inspired by the NCQI model and stores a color image into two entangled qubit sequences, where the three color channels (R,G,B) information is stored into the first sequence, while the corresponding bit-plane information and position information are stored in the second qubit sequence. For an RGB image of size $2^n \times 2^n$, the color information is encoded as shown in Equation 6.38, where L is the bitplane, $R_{LYX}, G_{LYX}, B_{LYX} \in \{0,1\}$. QRCI representation is shown in Equation 6.39, where $C_L(Y,X)$ represents the color information for the corresponding pixel coordinate (Y,X). $|LYX\rangle$ can be expressed as in Equation 6.40, where L and $|YX\rangle$ represent bitplane and location information, respectively:

$$C_L(Y,X) = R_{LYX}G_{LYX}B_{LYX}, \tag{6.38}$$

$$I\rangle = \frac{1}{\sqrt{2^{2n+3}}} \sum_{L=0}^{2^3-1} \sum_{Y=0}^{2^n-1} \sum_{X=0}^{2^n-1} |C_L(Y,X)\rangle \otimes |LYX\rangle$$

$$= \frac{1}{\sqrt{2^{2n+3}}} \sum_{L=0}^{2^3-1} \sum_{Y=0}^{2^n-1} \sum_{X=0}^{2^n-1} |R_{LYX}G_{LYX}B_{LYX}\rangle \otimes |LYX\rangle, \tag{6.39}$$

$$|LYX\rangle = |L\rangle|Y\rangle|X\rangle = |L_2L_1L_0\rangle|Y_{n-1}Y_{n-2}\ldots Y_0\rangle|X_{n-1}X_{n-2}\ldots X_0\rangle, \tag{6.40}$$

- Quantum Block Image Representation – QBIR

QBIR [32] quantum image representation model encodes pixel gray values and position information of image blocks into two entangled qubit sequences. It was proposed in order to perform encryption on quantum images. The expression for the representation of a QBIR image $|I\rangle$ is defined in Equation 6.41, where $|jtyx\rangle$ represents position information. $|jt\rangle$ stores the position of blocks, while $|yx\rangle$ the positions of the pixels in each block:

$$|I\rangle = \frac{1}{2^2}\frac{1}{2^2}\sum_{j=0}^{2^w-1}\sum_{t=0}^{2^w-1}\sum_{y=0}^{n-w}\sum_{x=0}^{n-w}|C(j,t,y,x)\rangle \otimes |yx\rangle$$

$$= \frac{1}{2^n}\sum_{j=0}^{2^w-1}\sum_{t=0}^{2^w-1}\sum_{y=0}^{n-w}\sum_{x=0}^{n-w}|c_{jtyx}^{q-1}\cdots c_{jtyx}^{0}\rangle|jtyx\rangle \otimes |jt\rangle \otimes |yx\rangle, \qquad (6.41)$$

- Quantum Representation of Indexed Images – QIIR

In [33], QIIR consists of a quantum data matrix and a quantum palette matrix. Each one of these two matrixes is represented by the basic states of qubit sequence for information storing. The representation of the data matrix is shown in Equation 6.42, where $|YX\rangle = |Y_{n-1}Y_{n-2}\ldots Y_0 X_{n-1}X_{n-2}\ldots X_0\rangle$ is the two-dimensional position coordinate of each pixel, $\{Y_i\}_{i=0}^{n-1} \in \{0,1\}$, $\{X_i\}_{i=0}^{n-1} \in \{0,1\}$; $|I_{YX}\rangle = |I_{YX}^{q-1}I_{YX}^{q-2}\ldots I_{YX}^1 I_{YX}^0\rangle$ is the pixel value, which is also the index into the palette matrix, $\{I_{yx}^i\}_{i=0}^{q-1} \in \{0,1\}$. The representation of the palette matrix is shown in Equation 6.43, where $|j\rangle = |j_{q-1}j_{q-2}\ldots j_0\rangle$ is the index in the palette matrix, $\{j_i\}_{i=0}^{q-1} \in \{0,1\}$; $|C_j\rangle = |C_j^{23}C_j^{22}\ldots C_j^0\rangle$, $\{|C_i^j\rangle\}_{i=16}^{23}, \{|C_i^j\rangle\}_{i=8}^{15}, \{|C_i^j\rangle\}_{i=0}^{7}$, are the red, green, and blue component values in a single color, respectively, $\{|C_j^i\rangle\}_{i=0}^{23} \in \{0,1\}$:

$$|Q_{Data}\rangle = \frac{1}{2^n}\sum_{Y=0}^{2^n-1}\sum_{X=0}^{2^n-1}|I_{YX}\rangle \otimes |YX\rangle, \qquad (6.42)$$

$$|Q_{Map}\rangle = \frac{1}{\sqrt{2^q}}\sum_{j=0}^{2^q-1}|C_j\rangle \otimes |j\rangle, \qquad (6.43)$$

- Double Quantum Color Images Representation – DQRCI

In [34], DQRCI model was proposed based on QRCI. With DQRCI two color digital images can be stored into a quantum superposition state. The expression for the DQRCI model is shown in Equation 6.44, where $C_L^1(Y,X)$ and $C_L^2(Y,X)$ represent the color information of pixel $|YX\rangle$ in bitplane $|L\rangle$, in two images, respectively:

$$|D\rangle = \frac{1}{\sqrt{2^{2n+3}}}\sum_{L=0}^{2^3-1}\sum_{Y=0}^{2^n-1}\sum_{X=0}^{2^n-1}|C_L(Y,X)\rangle \otimes |L\rangle \otimes |YX\rangle$$

$$= \frac{1}{\sqrt{2^{2n+3}}}\sum_{L=0}^{2^3-1}\sum_{Y=0}^{2^n-1}\sum_{X=0}^{2^n-1}|C_L^1(Y,X)C_L^2(Y,X)\rangle \otimes |L\rangle \otimes |YX\rangle$$

$$= \frac{1}{\sqrt{2^{2n+3}}}\sum_{L=0}^{2^3-1}\sum_{Y=0}^{2^n-1}\sum_{X=0}^{2^n-1}|R_{LYX}^1 G_{LYX}^1 B_{LYX}^1 R_{LYX}^2 G_{LYX}^2 B_{LYX}^2\rangle$$

$$\times \otimes|L\rangle \otimes |YX\rangle, \qquad (6.44)$$

- Quantum Hue, Saturation and Lightness – QHSL

In [35], QHSL model was proposed. Hue and saturation are encoded and stored as amplitude angles of a qubit. As for lightness, it is stored as a qubit sequence of length q. The complete expression of the QHSL model is expressed in Equation 6.45, where $\theta \in [0, \pi]$ indicates saturation, $\phi \in [0, 2\pi]$ indicates hue and $L^0 L^1 \ldots L^{q-1} \in [0, 2^q - 1]$, where $L^i \in \{0, 1\}$ denotes lightness:

$$|HSL\rangle = |HS\rangle \otimes |L\rangle = \left(\cos\frac{\theta}{2}|0\rangle + e^{i\phi}\sin\frac{\theta}{2}|1\rangle \right) \overset{q-1}{\underset{i=0}{\otimes}} |L^i\rangle, \qquad (6.45)$$

6.3.1.1 Quantum image processing

Quantum image processing is necessary in order to get benefit from the speedup that quantum computers are capable to deliver. There is a plethora of image processing algorithms for classical computers, although not all of them have their quantum equivalents. The available QIP algorithms in the literature mainly focus on tasks like geometric and morphological transformations, edge detection, segmentation, similarity analysis, matching, compression, encryption/decryption, scrambling, denoizing, and watermarking.

- Geometric and morphological transformations

The first application of a quantum geometric transformation was proposed in [36], the authors studied the problem of image translation. More specifically two types of translation circuits were suggested; entire translation and cyclic translation. As a representation method, NEQR was preferred. In [37], algorithms for both global and local translation based on the FRQI representation model were designed. The global translation was implemented by using adder modulo N and for local translation, Gray code is also used. Another geometric transformation is image scaling. Image scaling is a basic and widely used operation in image processing. Image processing gives the ability to adjust the size of digital images. In order to achieve image scaling it is necessary to use interpolation methods to create new pixels in the case when scaling up is needed, or to delete redundant pixels in scale when scaling down is required. In [21], the authors proposed a quantum image scaling algorithm based on the INEQR representation model using nearest-neighbor as the interpolation method. In [38], the authors proposed quantum circuits for nearest-neighbor interpolation for FRQI and NEQR representation models using quantum rotation gates and Control-Not gates. Morphological transformations are simple operations based on the shape of an image, like erosion and dilation, that have been studied in quantum experiments. In [39], the authors based on set operators proposed quantum versions of erosion and dilation algorithms. In [40], the authors proposed dilation and erosion quantum circuits for quantum binary and grayscale images based on the NEQR representation model. In [26], the authors proposed several quantum morphological operations for binary and grayscale quantum images. More specifically erosion and dilation were proposed for both types of images. Moreover, for quantum binary images noise removal, boundary extraction, and skeleton extraction were proposed. For the case of quantum grayscale images edge detection, image enhancement, and texture segmentation were proposed. In [41], the authors proposed a quantum morphological gradient operation that derived from the combination of quantum erosion and dilation morphological operations.

- Edge detection

Edge detection plays a crucial role in several image processing pipelines, which is used to identify the boundaries, also referred to as edges of objects, or regions within an image. In [42], the authors proposed two quantum image processing algorithms based on quantum measurements including quantum edge detection and quantum adaptive median filtering. In [43], the authors proposed a quantum implementation of the Prewitt operator for edge detection producing far better results than existing conventional methods, like Sobel and Canny edge detection. In [44], the authors based on the FRQI representation model, proposed a quantum variation of the Sobel image edge detection algorithm. In [45], the authors proposed a quantum edge detection for medical images using Shannon entropy, producing far better results than conventional methods like Canny and Sobel edge detection methods. In [46], the authors proposed an implementation of a highly efficient quantum algorithm for edge detection independent of the image size by applying Hadamard gates. In [47] proposed a quantum particle swarm optimization to tackle the problem of edge detection tested on images from the Berkeley database producing continuous and clean edges compared with conventional methods. In another study the same year [48], the authors performed edge detection on quantum images based on the NEQR representation model and on the classical Sobel operator and applied X-shift and Y-shift transformation. In another study [49], the authors proposed quantum edge detection for the NEQR representation model based on an improved Prewitt operator that combines the non-maximum suppression method and adaptive threshold value method. Non-maximal suppression is used to refine the edges, while adaptive thresholding can reduce the misjudgment of edge points. In the same year [50], the authors proposed edge detection based on four directional Sobel operators for NEQR quantum images resulting in much richer edges than conventional methods, extending the work later on proposing the usage of improved Sobel mask based on NEQR for purpose of edge detection. In [51], the authors proposed a quantum implementation for the Marr-Hildreth edge detection algorithm for NEQR quantum images for the first time.

- Segmentation

Image segmentation is a process, which results in the separation of the foreground on one or more objects from the background in an image. Various methods and tools have been proposed to perform segmentation in an image. In [52–55], clustering methods have been used for this problem. Thresholding is another popular method for image segmentation used in [51, 56–62], with several methodologies that include optimization methods, like genetic algorithm [61], particle swarm optimization [56], marine predator algorithm [62], leveraging the speedup from parallel calculations of quantum computers. Other methods include optimization algorithms combined with fuzzy level sets [63].

- Similarity analysis and image matching

Similarity analysis of two or more images quantifies the degree of similarity between intensity patterns in the images. Image matching techniques on the other hand try to find the existence of a pattern within a source image. Quantum computing could help make these processes faster, so there have been proposed methodologies for the topics. In [64–67] similarity analysis methodologies were presented for several kinds of images including grayscale and color.

- Image compression

Image compression is an image processing method, that aims to reduce the size in storage requirements for an image without degrading the quality of the image to an acceptable level. Reducing the required disk space for storing images is an essential tool, due to the increased data requirements of the current digital era. Quantum methodologies for image compression have been proposed in the literature leveraging various mechanisms. More specifically in [68] image compression was achieved by using a genetic algorithm combined with a quantum neural network. Another learning-based proposal [69] tackled the problem with a quantum version of a back-propagation neural network. Based on a quantum variation of OMP (Orthogonal Matching Pursuit) of CP (compressive sensing) theory [70], combined with a quantum version of particle swarm optimization another quantum implementation for image compression was proposed. By leveraging Grover's quantum search algorithm, fractal image compression was proposed for quantum computers [71]. A quantum implementation for the widely used JPEG compression method was suggested in [72].

- Image encryption/decryption and scrambling

Image encryption as a processing method results in a corrupted image. It's a necessary task, in several cases, like secure data transfer. In encryption scenarios, a recovery mechanism is essential. On the other hand image scrambling is a technique for image content hiding, but without providing a recovery mechanism. Quantum implementations for image encryption/decryption leverage tools like QFT [73] for gray images, extending the application to color images [74]. Another encryption scheme for color images was suggested using multiple discrete chaotic maps [75]. Image scrambling can be considered a less complicated task, due to the fact that a process for reversing the scrambling is not required. Several studies have been conducted about quantum implementations of image scrambling that leveraged quantum bit-plane gray code [76]. Another proposal involves a block-based image scrambling scheme based on GNEQR [26].

- Image denoising and filtering

Image denoising is the task of removing noise from images. Image filtering refers to changing the appearance of an image by modifying the values of the pixels. Studies for image denoising include quantum implementations of self-organizing neural network [77], quantum image filtering with a Boolean mean filter that works exclusively with computational basis states [78], quantum median filtering in spatial domain [79]. As for filtering quantum circuits for Fourier transform have been studied [15].

- Image watermarking

Watermarking is the process of hiding digital information in a carrier signal. Several use cases require watermarking such as the protection of confidential information or to claim authorship. Quantum-based approaches for the problem of watermarking include transforms like QWT [80] and QFT [81]. Another implementation was proposed using circuits consisting of operators such as flip, coordinate-swap, and orthogonal rotation [82]. An adaptive watermarking approach was proposed by using the weighted QPSO algorithm [83].

6.3.1.2 Feature extraction

Feature extraction is a very crucial topic in image processing. There is a direct association between the performance of the algorithms and the features used as input to the algorithms. As a result, fast and efficient feature extraction methods are required. Classical implementations of algorithms for feature extraction are already efficient. Quantum computers are capable of delivering better performance compared to their classical counterparts, due to the exponential speedup in parallelism that a quantum computer can offer. Many quantum implementations have been used in the literature derived from popular classical feature extraction methodologies like Fourier Transform (FT), Wavelet Transform (WT), Discrete Cosine Transform (DCT), Cosine Transform (CT). QFT has been used in many quantum image processing related to encryption/decryption [73], where a dual-phase encryption/decryption scheme was proposed based on QFT. More applications include filtering [15], compression [18]. Quantum version of Wavelet transform also is deployed as a feature extraction mechanism for tasks such as compression [18] and watermarking [80]. QDCT has been used for image compression [84], while QCT was used for image watermarking [85].

6.3.1.3 Decision (quantum model selection training and evaluation)

As it is well known in machine learning and computer vision implementation schemes after the feature extraction methods follow the model selection and the formation of model output in order to solve a specific problem like image classification or segmentation. In this section, high attention is given to the model selection and the model decision in the quantum computer vision domain. Classical machine-learning domains have adopted a variety of models to exploit image data for analysis. This evolutionary step gave the capability to create new applications like image classification, object recognition and detection, image segmentation, image compression, image generation and many more. With computer vision applications there are many challenges that emerged for machine-learning models like real-time prediction for object detection and tracking or more complex model architectures in order to generalize in higher image resolutions and bigger data distributions. Another challenge is the time that takes for a model to generalize in this huge amount of data. A proof of these challenges can be the evolution of computer vision methods and models historically.

At the beginning of this domain, statistical feature extraction methods have been applied to image data and then classical models have been used to recognize patterns like SVM, KNN, or tree-based models. During time, CNN models started to take place at some point with the beginning of Y. LeCun in 1989 [86] with a computer vision application of image handwritten recognition. Despite the fact that CNN is an old model scheme it has been evolved more than any kind of model in machine-learning and deep-learning domain offering many applications but not the solution to the above-mentioned problems. Until today these breakthroughs are facing the same challenges since the data generated are increased in size and amount which means that more computational resources are needed for the models in order to process them and the formation of more complex model architectures tends to be hard. Based on these challenges machine-learning and computer vision research communities started to exploit quantum computing and quantum computers in order to overcome the limitations of speed, resources, and complexity. As mentioned in the beginning there is a variety of approaches to adapt computer vision and machine-learning models and methods to quantum computing properties and quantum computer limitations. The most generalized categories of quantum machine-learning models as proposed by [87, 88] are quantum-quantum (QQ) where the data and the algorithms are based on quantum properties, quantum-classical (QC) where quantum algorithms are used for pattern recognition over classical data, classical-quantum (CQ) where classical machine-learning models are implemented in order to execute in quantum devices and finally classical-classical where classical machine-learning algorithms and methods are inspired by quantum computing properties like entanglement or super-position. From the above four categories, quantum computer vision models exist mainly in quantum-classical and classical-quantum categories since they demand a huge amount of resources for the image data and the model architecture and so they cannot be fully supported by quantum computers yet. To continue, quantum machine-learning and deep-learning models are presented and analyzed with a highlight on their achievements over classical versions.

Quantum machine-learning models

PQK

Many quantum machine-learning models have been adopted for quantum computer vision problems since they can achieve generalization in much higher feature dimensions than classical and suffer from over-fitting as the amount of parameters rises. Despite classical ML models that have been proposed in a quantum scheme there are new algorithms based on quantum mechanics that can be applied for pattern recognition. In [89] the PQK method was proposed from projected quantum kernels with applications in the MNIST dataset for image classification purposes. The method exploits data encoding in the form of a kernel under quantum properties in order to achieve higher speed and better feature representation since the quantum kernels are extracted based on Hilbert space. The produced quantum kernels are used in the classical model SVC and performed better in comparison to classical kernels.

QKNN

The k-nearest neighbors (KNN) is one of the most known supervised machine-learning models. In order to classify an unseen data sample, it looks at k instances in the training set that is close to the prediction datum. Then it chooses the class that appears most often in the labels of these nearest neighbors as the predicted label. This is the most general procedure of the KNN model and its advantage is that it is non-parametric and makes no assumptions about the data distribution. The problem with these model schemes is that when it comes to predict a large data set the computation demands in order to extract k-nearest neighbors are very high. Based on this limitation QKNN is proposed. To achieve classification QKNN exploits a technique [90] that calculates overlapping between two quantum states in an entanglement, which acts as a similarity measure analogs to Euclidean distance. This quantum distance is calculated through the SWAP-test method and gave the characteristic and the speed to QKNN to classify encoded data in a quantum state very fast. Thus QKNN implementation could overcome the limitation of the classical scheme. QKNN model is proposed in classical-quantum scheme [91] for image classification in known datasets. The images are processed in the

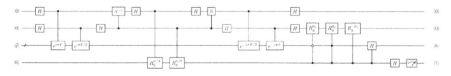

FIGURE 6.3
QSVM quantum circuit [95]

classical computer for feature extraction and then with the quantum computing properties, the discrimination based on nearest neighbors takes place in
order to measure image similarities. Since it is proposed QKNN has been used
in computer vision applications with some additions like [92] where QKNN
has been proposed with K-L transformation of the input data in the classical
computer and then in a quantum state. The addition of the transformation
gave QKNN a significant performance boost in comparison with the classical
contribution paper in image classification problems with known datasets like
(MNIST and Caltech).

QSVM

The next most common machine-learning model with applications in
the quantum computer vision domain is quantum support vector machines
(QSVM). Support vector machine (SVM) [93] is a popular supervised machine-
learning model with applications in classification and regression. The basic
concept of the model is to determine the optimal hyperplane that separates
data instances in order to be distinguishable with a maximum margin between
them. Since some data instances may not be strictly separable, slack variables
are proposed that provide a soft margin by allowing some data to violate the
margin condition. This formula is known as the maximum margin classifier
and requires the data to be linearly separable. To overcome this requirement
SVMs exploit kernels, which transform data in a higher-dimensional feature
space where non-separable data samples in the original space now are separable in the higher dimension. The problem with SVMs is that as the data
size grows the kernel computations tend to be computationally expensive from
time and memory. To overcome this limitation in [94] quantum support vector
machine (Figure 6.3) is proposed with the claim that the evaluation of the inner product which is the inner product of the feature map can be done faster
on a quantum computer. Also, QSVM shows its supremacy over classical from
the point that feature maps are represented as a superposition, which transforms the quadratic programming scheme of classical SVM to the solution of
a linear equation system which is much faster on a quantum computer. Below
a visual representation of the QSVM quantum circuit is presented.

With these advantages, QSVM has proven very useful in training with
image data [96, 97] where QSVM has been used for handwritten recognition
systems and outperforms classical SVM in the matter of evaluation performance and speed.

From the aspects of quantum machine-learning models, there are more models that could be presented like quantum decision trees but since they have not yet been applied or adopted in the quantum computer vision domain it is not given any interest. An important notice based on the above is that quantum machine-learning models' evolution is following the same pattern with classical machine-learning models evolution as mentioned above. With this progress, it is clear that quantum machine-learning models have unlocked new capabilities for the computer vision domain to adopt quantum computing properties from the perspective of methods and applications.

Quantum deep learning

Since the conversion of the first invented machine-learning methods and models in their quantum version is covered, a review of quantum deep-learning models QNN and its variations are presented with an interest in contributions that correlated with the quantum computer vision domain like previous. In order to reach a quantum neural network scheme, it is important to present the fundamental concepts of the classical scheme for the reason of understanding the improvement that has been achieved.

QNN

Neural networks physically exist inside the brain and they equip living things like humans with a variety of abilities like generalization which is a brain functionality. The introduction of artificial neural networks came in 1943 [90] presenting the basic model of artificial neurons for learning problem purposes and the most completed scheme has introduced with the name perceptron in 1957 [98]. In an abstract view, artificial neurons can be characterized as real-valued functions, which is parameterized by a vector of non-negative weights. These weights are the connections between artificial neurons and their values are representing the strength between each connection. These two components with an activation function that controls the neuron outputs form the basic scheme of an artificial neural network. For the sake of correct terminology, the perceptron is a special category of artificial neurons with a specific activation function and training rules. During the evolution of artificial neural networks, many different approaches have been proposed from different activation functions, and a variety of architectures where each of them tend to have limitations and new challenges. In their quantum version, neural networks tend to overcome some of their limitations like speed, performance, and simplicity. The mainstream model scheme of quantum neural networks has been proposed in [99] using a variational quantum circuit, which is very similar to ANN models and it can be found with the name variational quantum algorithm. This scheme is the most promising for execution on real quantum hardware as presented in [100]. VQA are circuits that consist of three blocks, the data encoding in a quantum state, the variational ansatz, and the classical update of the learnable parameters. This algorithm is a hybrid

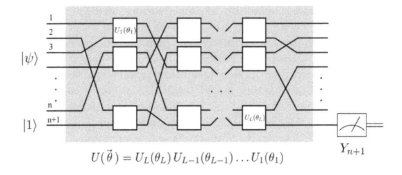

$$U(\vec{\theta}) = U_L(\theta_L)\, U_{L-1}(\theta_{L-1}) \dots U_1(\theta_1)$$

FIGURE 6.4
QNN quantum circuit [102]

classical-quantum form since the calculation of the objective function that optimizes the quantum learning parameters is executed in a classical computer. VQA algorithm is a quantum circuit of the Parametrized Quantum Circuit (PQC) category. This procedure gave the base to form the QNN since it is the same concept as the classical neural networks training procedure. Thus QNN can be characterized as a PQC and in a sequence of stacked circuits, a QNN model can be formed. A limitation of this QNN model scheme is that the updated value of a neuron in the model cannot be transferred to every neuron in the following layer for the reason of non-cloning theory in quantum mechanics. Thus a classical computer is used to calculate the parameters and update the quantum neural model or quantum neural model layer based on an objective function. Another limitation is the divergence between non-linear properties of classical CNN based on their activation functions and the linear unitary dynamics of quantum computing. A solution to this problem is the construction of quantum circuits that approximate or fit nonlinear functions like RUS architecture [101]. Below a visual representation of the QNN quantum circuit is presented in Figure 6.4.

Even though QNN models gave some advantages over their classical versions related to speed and accuracy, as presented in [103] where QNN model is compared with classical models like 2D CNN in the classification of CT scan images of COVID-19 patients. The QNN experiments were executed on a real quantum hardware D-Wave on a 2041 qubit system. The results have shown a significant accuracy boost of QNN over the classical version models. From the matter of speed, it is mentioned that the training time of the QNN model was 52 min and the training time of 2D CNN in a K80 GPU was 1h 30min. This study is a confirmation of the above research breakthroughs around the quantum computer vision domain.

The evolution of classical neural networks was the introduction of convolutional neural networks and their variations where pioneer was Y. LeCun [86]. CNN model scheme contains convolution layers and pooling layers. In convolution layers, the neurons are connected locally with each other and as a

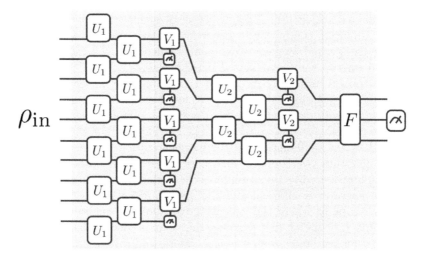

FIGURE 6.5
QCNN quantum circuit [104]

sequence, they can be characterized as an abstraction feature mechanism. In other words, convolution layers extract hidden features by linear combinations between surrounding pixels. Pooling layers are used to reduce the dimensions of the feature maps produced from the convolution layers in order to avoid over-fitting. The training procedure of the above-presented model is based on the gradient descent method in order to minimize the loss between the produced labels for some data by the CNN with the real labels. CNN models have mainly been developed to process image data in a huge numbers and resolutions. For many years there have been many contributions in various topics around CNN like the training algorithms (supervised, un-supervised, and semi-supervised), new architectures (Darknet, VGG, and DeepMind), new functionalities in the convolution layers, new models schemes like Generative Artificial Networks or Boltzmann Machines. Similar to classical CNN quantum CNN as proposed by Cong [104] in 2019 exploits the PQC procedure where the convolutional layer is represented as a quasi-local unitary operation on the input state and it is repeated several times within the model structure. The quantum pooling layer is applied by performing measurements that break the superposition states and thus the dimensionality reduction is succeeded where the nearby qubits that are not measured are applied in a unitary rotation. Another approach to the quantum pooling layer is by using CNOT gates to achieve dimensionality reduction since from the two-qubits one is fall to the zero state. Finally, when the dimension of the quantum system is small, the fully connected layer anticipates classification results. A visual representation of the QCNN quantum circuit is presented in Figure 6.5.

Another artificial neural network classical scheme that had a significant impact on quantum computing is Generative Adversarial Networks (GANs). GANs are categorized as generative models where the generative process is distributed on two separated neural networks generator and discriminator, which are trained iteratively in order for the generator to learn to produce high-quality data samples with the aim to fool the discriminator that tries to classify the generator output as fake or real. The quantum version of GANs consists of two QNNs, an architecture scheme that was not the only one applied. In [105] the combinations of the implementation schemes around classical and quantum GAN model take place where when both the generator and discriminator are in a quantum form the performance is higher. Although when the generator is classical and the discriminator is quantum they will never reach a point where the generator can fool the quantum discriminator since the probability distributions that can be accessed by a quantum computer are much larger than classical. Interesting results have been shown based on the scheme where the generator is quantum and the discriminator is classical. In this scheme, the generated data can produce real data faster. The problem with the hybrid forms of QGAN is that a computational speed limitation is taking place since there is a part of the process that will encode and decode data. The performance of QGANs applications has proven superior to classical as presented in [106] where classical and quantum GAN in image data generation using MNIST dataset were compared. The advantages of QGAN are tested in [107] where a quantum adversarial defense mechanism is proposed for various adversarial attacks. Based on the results the quantum GAN model could achieve an accuracy of over 98% despite the attacks that occurred. Finally, in [108] QGAN with a Boltzmann Machine (BM) for image generation in multiple datasets were combined, where the quantum-associated adversarial network (QAAN) performed better than classical model schemes like DCGAN and others.

On the same principle of generative models, a lot of interest has been shown in Boltzmann machine neural networks due to the fact that they have simpler structures than GANs and are very close to the Ising model which is expressed under quantum mechanics. It is formed as a neural network with the aim to learn a set of weights in a way that results in an approximation of a target probability distribution which is implicitly given by training data. Boltzmann Machine neural networks consist of visible layer neurons and hidden layer neurons which are connected with many forms (fully, restricted); for each connection, a joint probability is associated and with a weight coefficient of each connection, the BM neural network is structured. In general BM networks could be characterized as undirected graphs. The training process consists of methods like gradient descent to optimize the weight values in order to maximize the probabilities that reproduce the statistics of observed data. Changing the classical nodes with qubits and the classical energy function with a quantum Hamiltonian, the Quantum Boltzmann Machine (QBM) is introduced in [109] with the capability to generate quantum and classical distributions. It has been applied on a real quantum computer D-Wave [110]

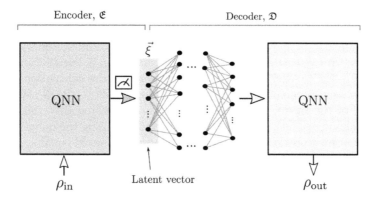

FIGURE 6.6
Quantum autoencoder [109]

for the purpose of image generation and classification using known datasets of MNIST digits and fashion. The results have shown that QBM performed better in comparison with its classical version in both classification and generation experiments. Another contribution that verifies the superiority of QBM over classical was reported in [111]. Quantum and classical models of BM are compared in image classification and reconstruction in D-Wave quantum computer where QBM has proven superior to both experiments.

Another very popular generative model is the variational autoencoder and they have been used in many applications in the computer vision domain. The goal is to learn a complex data distribution like images in order to sample and generate new data based on the given distribution. The limitations of this model are the same as the rest of machine and deep-learning models, especially in computer vision tasks. Quantum autoencoders have been proposed under a quantum-classical model scheme where an encoder QNN takes a quantum state from Hilbert space to a subset of real vector space, and a decoder QNN performs the inverse operation. The classical part is the minimization and the update of the learning parameters of the model. The above concept is presented in Figure 6.6 below:

As it is presented in [112] QVAE performed well in image generation of MNIST dataset, where the most important part is that it is executed in real quantum hardware of D-Wave.

Another interesting topic around quantum neural networks is their trainability. Training is the process to find optimal models or parameters that minimize the cost function derived from learning problems. The capabilities of training routines that have been proposed by the research community around quantum deep learning justify the advantages of quantum computing properties and introduce new challenges at the same time. Starting with the challenges there is an often phenomenon that occurs while training quantum neural network which is called Barren Plateau and it appears when the amount of qubits and circuit depth is comparatively large, making QNN un-trainable with the objective function to become very flat. A major cause of this phenomenon is that entanglement tends to spread information across the whole quantum system more than its single part [113]. Some solutions to this problem are the reduction of training parameters or the calculation of corre-lations between parameters in QNN layers which leads to the dimensionality reduction of the model architecture. It is worth noticing that this solution does not eliminate the problem. As presented in [71] using the quantum back-propagation method the quantum neural network achieved equal accuracy as classical neural networks in image compression but in half epochs compared to the classical neural network. In general, the training of quantum computer vision models is similar to classical methods since most of the models have parts in their main process that will be executed in classical computers like the loss function calculation and learnable parameters update. The evaluation is made using common methods such as 10-fold cross-validation.

To close this section quantum machine and deep-learning models are very promising in the significantly increasing computational capabilities unlock-ing new areas of investigation. More attention has been given to generative models, where the quantum computer vision domain can be given new capa-bilities since their quantum version performs better compared to the state of the art with less time and computational complexity. Even though quantum computer vision methods and models are tested and evolved, the evolution of quantum hardware and tools will lead to further improvements in these implementations.

6.3.2   Tools and hardware

In this section, an extensive review will take place around the tools that are accessible for quantum computing with references to models and methods for the quantum computer vision domain. Next, a presentation of the available quantum computers will take place along with some of their characteristics.

6.3.2.1   Quantum computing tools

To start the review of quantum computing tools Table 6.1 presents the most common quantum computing libraries, programming languages, and universal languages. Quantum computing tools appeared before quantum hardware and are the reason behind today's quantum computing success. They were used for the development of classical algorithms or for debugging procedures. In this review, the interest is targeted to the QC tools that have an interaction with a physical quantum computer.

Starting from the bottom of Table 6.1 there are quantum imperative pro-gramming languages also known as procedural languages. These languages are responsible for the compilation, execution, post-processing of quantum soft-ware in a quantum computer and they are built on QRAM model [114] and are implemented in a quantum assembly language. Before analyzing each of the

TABLE 6.1

Quantum computing tools along with their providers.

Company	IBM	Rigetti	DWave	Xanadu	Google	Microsoft
Quantum Circuits	Qiskit Terra	pyquil	qbsolv	Strawberry Fields	Cirq	Q#
Full-stack libraries	Qiskit	Forest	Ocean SDK	Strawberry Fields	Cirq	Quantum Dev Kit
Imperative Language	Open QASM	Quil	QMASM	Blackbird	QASM	Other

imperative languages the three basic phases of executing quantum programs through imperative languages on real quantum hardware are presented. First, the compilation phase takes place on a classical computer where the source code of the quantum program is converted to a combination of quantum-classical expressions. At this phase, there is no connection with a real quantum computer. Next, the circuit generation takes place again on a classical computer where the quantum circuits are generated from the compilation output. In this phase, a connection may occur with the real quantum hardware depending on the declaration of information. Then the execution phase takes place, where the quantum circuits along with every parameter for the runtime control are executed on a low-level quantum controller and the results are streamed back to the high-level controller which is classical.

Imperative languages

Starting from bottom to start in Table 6.1 the first imperative language is OpenQASM from Quantum Assembly Language [115], a gate-based intermediate representation for quantum programs used in IBM quantum computers. OpenQASM allows for abstraction in the form of quantum gates which are form quantum circuits in a hierarchical manner and they provide measurements of single qubits. At the same base is Quil [116] which is used under Rigetti quantum computer. From its different quantum computer architecture, D-Wave has implemented QMASM from a quantum macro assembler [117] and contains some functionalities for low-level D-programming similar to a macro assembler for low-level programming. QMASM can execute quantum programs in physical annealers specifically D-wave 2x quantum annealers. Blackbird is another version [118] of quantum assembly language sharing the same features with OpenQASM under a quantum hardware implementation of photonics.

Full-stack libraries give functionalities and algorithm development capabilities for quantum programs. The below-presented libraries are exploiting the imperative languages presented above. Starting with IBM Qiskit [119] or Quantum Information Science Kit is an open-source programming framework and it is one of the most commercial. Qiskit provides an environment where developers can implement and process OpenQASM programs with python programming language. It is famous for each abstraction capability like the fabrication of quantum circuits that include numerous quantum gates and allows certain unitary transformations. It includes a simulation environment that can be executed on both CPU and GPU. The whole framework of Qiskit can be divided into four software frameworks. Terra which provides the basic data structures of quantum computing, Aer which contains the simulation backends for circuit execution, Aqua which provides generalized and customized quantum algorithms including domains like machine-learning, deep-learning, and Ignis, which is a framework for studying noise in quantum programs. Some of the most famous quantum algorithms are included in Qiskit like Grover's, Shor's algorithms which have been widely used in quantum machines and deep-learning models in their quantum version. Forest SDK, developed by Rigetti [120] uses the imperative language Quil. Forest SDK also contains the pyQuil library based on which developers can implement quantum circuits on Rigetti QPUs. Ocean SDKs is a python-based software framework [121] for programming in D-Wave quantum annealers. It differs from the above-mentioned libraries and frameworks because all the computation expressions are modeled in an optimization scheme in order to be executed on quantum annealers which simulate a quantum phenomenon that will be presented at the end of this section. For solving the computation expressions two binary quadratic models (BQM) are included, quadratic unconstrained binary optimizer (QUBO), and Ising model. Ocean SDK has the most implementations around quantum computer vision applications and methods since it provides a real hardware execution with large capabilities. Strawberry Fields is the quantum SDK [122] of a photonic quantum computer fabricated by Xanadu. Strawberry Field SDK is also different from the other libraries and frameworks because photonic quantum computers represent quantum states in an infinite-dimensional Hilbert space and the quantum operators are continuous. Based on these differences quantum gates and measurements are also different from other implementations. Cirq is the SDK that was created by Google. Cirq is very similar to Qiskit since it provides basically the same features about writing, manipulating, and optimizing quantum circuits. Quantum Development Kit (QDK) was created by Microsoft and it is provided under quantum-focused programming language Q# based on C#. Also contains codes for optimization, simulation, and algorithms. QDK is fabricated with the concept that will be executed in topological qubits that use anyons as a physical unit to store and manipulate quantum information [123]. At this moment there is no existence of quantum computers on topological qubits and so the implementations of the algorithms can only be simulated.

To close the review of quantum computing tools cross-platform frameworks will be presented. Cross-platform frameworks are quantum implementations of libraries that give the potential to create and execute quantum programs in a variety of quantum hardware. In Table 6.2 some of the cross-platform frameworks are presented and analyzed.

ProjectQ is a cross-platform framework [124] unifying quantum program executions to different quantum hardware schemes that appertain to IBM enterprise hardware implementation. XACC stands for extreme-scale accelerator

TABLE 6.2

Quantum cross platform frameworks.

Company	IBM	Rigetti	DWave	Xanadu	Google
Cross-platform Frameworks	–	–	–	–	–
Cross-platform Frameworks	ProjectQ	–	–	–	–
Cross-platform Frameworks	XACC	XACC	XACC	–	–
Cross-platform Frameworks	Pennylane	Pennylane	Pennylane	Pennylane	Pennylane

and follows a coprocessor machine model that is independent of the underlying quantum hardware, thus it gives the capability to execute quantum programs under various types of QPUs including D-Wave, IBM, and Rigetti. It is built based on the C++ programming language and provides a fully programmable API for gate programming and annealing quantum computers. PennyLane is also a unified framework that can execute quantum programs in several quantum computing vendors. The main target of PennyLane is quantum-classical machine-learning algorithms. Also, it can bridge QPUs, GPUs, and CPUs integrating popular machine-learning frameworks like Tensorflow and PyTorch.

All the above-mentioned quantum computing tools are the most common among the various research communities that are interested in the quantum computing domain. As quantum hardware evolves the above-mentioned frameworks and libraries will provide higher-level tools to implement quantum algorithms. It is crystal clear that the programming style of quantum computing is different than classical and it will take time until quantum computing programming tools reach the higher levels of abstraction where classical computing programming tools already have.

6.3.2.2 Quantum hardware

The basic challenge of engineers and scientists around quantum computers is related to controlling, measuring, conserving, and isolating quantum qubits and their interactions with the uncontrolled environment. Before the presentation of various quantum hardware architectures, it is important to present some basic differences between classical and quantum computers in the hardware level along with some basic characteristics of quantum computer hardware from quantum physics. A fundamental difference between classical and quantum computers is the memory management. Classical computers can write and read data from memory maintaining or processing them and writing them again. This flexibility does not exist in quantum computers and is extremely difficult to maintain quantum information for a long time and by definition quantum operations overwrite the quantum data in memory according to the non-cloning theory. So at this moment, quantum states can be characterized as fragile and naturally decaying. To point out the different scales of quantum and classical computers it is calculated that 8GB RAM with millisecond access times would need quadrillions of qubits. It is clear that quantum computer hardware is less effective in comparison to classical computers. Another characteristic of quantum computers is the presence of noise because quantum states are in atomic dimensions where everything is

more fragile and can cause energy shiftiness. Thus noise is observed in the
measurements of the quantum states and the next step is to equip quantum
computers with error correction algorithms in order to fix partially corrupted
quantum states. The problem with error correction algorithms is that they
cannot be the same as in classical computers since it is very hard to measure
a quantum state. As a result, quantum error correction algorithms cannot be
applied directly to corrupted quantum states. In addition, error correction
algorithms need physical resources in order to be applied, leading to performance loss. Some basic meanings of quantum physics that concern quantum
computing are:

- Coherence time is one of the most important metrics in quantum computing since it counts for how long a quantum state that is represented by
  qubits can remain stable. The bigger the coherence time the more operations can be done by quantum algorithms and lower execution times for
  error correction algorithms since the quantum state will remain stable.

- Gate Latency is the time that describes how long a quantum gate will take
  to perform an operation. The lower the latency the bigger the coherence
  time of the quantum state.

- Gate Fidelity metrics show how stable a quantum gate will perform without the presence of any errors. It is proof of why coherence time and gate
  latency are not linearly related. While lower gate latency means higher
  coherence time but less gate fidelity, which means higher decoherence opportunities. For the record quantum operations have lower fidelity than
  classical operations.

- Mobility is how many qubits can be physically moved. There are some
  implementations like superconductors where qubits are not moving and
  others like photons that never stop moving. Mobility is a characteristic of
  quantum devices and shows how information is transferred and how qubits
  are entangling, which synthesizes the noise appearance in the quantum
  system.

With the above-mentioned metrics, the performance of a quantum computer can be extracted. In Table 6.3 the various kinds of quantum computers
architectures are presented along with some of their characteristics.

From Table 6.3 it is obvious that trapped ions are a quantum computer
hardware implementation that has achieved by far larger coherence time compared to others. The implementation of the phenomenon is conducted with
the placement of atomic ions within nanometer space with electromagnetic
fields. The next famous implementation is superconductors which are materials that can conduct electricity from one atom to another without resistance.
There are three different categories charge, flux and phase. They have a short
decoherence time and require very low temperature for maintenance. Photonics implementation is based on the polarization of photons which are free
of decoherence and they have very promising characteristics with the main

TABLE 6.3

Quantum computers architectures.

Qubit Technology	Advantage	Disadvantage	Coherence	Gate fidelity	Gate operation time
Trapped ions qubits	higher coherence time	absolute zero cooling	50s	99.9%	$3\text{-}50\mu s$
Superconducting qubits	Low nose	absolute zero cooling	$50\text{-}200\mu s$	99.4%	10-50ns
Silicon qubits	lowest disorder, higher energy	Multi-qubit fabrication	1-10s	90%	1ns
Photonic qubits	easy manipulation, room temperature	Nanofabrication of diamonds	$3\text{-}150\mu s$	98%	$1\text{-}10\mu s$

challenge, the fabrication and interaction between two photons. Finally, silicon quantum computers have not been tested yet but they possess some very strong characteristics including the cheap materials for production and lower cooling requirements than superconductors.

6.4 Conclusions

Based on this study, the quantum supremacy in theory and performance is blurred over the computer vision domain. There are several reasons that support the above statement. To begin with, quantum hardware computers have several limitations that don't let quantum computer vision be fully formatted. Some of these limitations are the small number of physical qubits and the huge computation resources that are demanded by algorithms and problem solutions. From the quantum computer vision domain, there is an absence of discussion around quantum models deployment and real-time inference which is a vital step in order to increase the capability of applications and methods to implement for quantum computers. Another problem is the presence of noise in quantum hardware which creates difficulties for quantum computer vision to rise. It is well known that the computer vision domain is formed mainly from deep-learning and machine-learning where methods and models have been adopted to analyze images. The power to detect and analyze image patterns is based on nonlinear operations in order to find nonlinear relationships. This comes into contradiction with quantum mechanics and quantum computing methods which use completely linear operations. The latest trend to overcome this limitation is the use of quantum-classical operations to approach nonlinear operations like VQA. The problem with this solution is that increases the circuit depth which denotes negative events like computation time, increased noise, and quantum state collapse. Based on this fact the superiority of quantum computing over machine-learning and deep-learning in general, needs further verification. Computer vision is supported by quantum machine-learning and deep-learning domains and so the limitation of these two domains are transferred to quantum computer vision evolution. Another fact is that the deep-learning domain mainly supports computer vision methods and applications since images have denoted the computational resources problem. Despite the nonlinear problem, quantum deep-learning models are characterized as presented in the previous sections by complex architectures and various operations with their own conditions, which are difficult to implement under quantum computing principles like VGG16 due to huge quantum computational resources requirements. In general quantum computer vision methods and applications that process and analyze images have not met a clear advantage in the whole domain and it justifies the evolution path of the quantum computer vision domain.

In order to make a complete transition to the Quantum Computing domain several scientific breakthroughs should take place related to the evolution of Quantum Computers. Most of the work in the literature leverages simulation tools for the experiments, which is acceptable, but not enough to validate the experiments in real hardware due to the current status of quantum computers. One of the biggest challenges is related to coherence time in quantum computing, which does not exceed the 1-minute barrier. This short period of time is not enough for all the necessary operations which are required for the execution of quantum algorithms in real hardware that depends on the classical-quantum interaction model along with the operation itself. The further evolution of quantum hardware will open the path for more robust and capable quantum computation capabilities. Image analysis is one of those domains, which in their classical forms require lots of computational resources, so investigating quantum solutions is the next step to create faster algorithms and systems. Based on quantum software simulations there is a plethora of proposed methodologies for image analysis and processing, although the validation of those methods on real hardware is a great challenge.

Authors' biographies

Konstantinos Tziridis received the Diploma [Computer Engineering], the M.Sc [MPHIL in Advanced Technologies in Information and Computers] Degrees from the Department of Computer Science (CS), International Hellenic University (IHU), Greece, in 2018 and 2021 respectively. He is currently a Ph.D. student in the field of Quantum Image Analysis since 2022. His research interests are mainly in computer vision, machine learning (ML) and deep learning (DL). As a researcher he is a member of MLV (Machine Learning & Vision) Research Group of the Department of CS (IHU). He has (co)authored 10 publications in indexed journals and international conferences.

Theofanis Kalampokas received the Diploma [Computer Engineering], the M.Sc [MPHIL in Advanced Technologies in Information and Computers] Degrees from the Department of Computer Science (CS), International Hellenic University (IHU), Greece, in 2018 and 2021 respectively. He is currently a Ph.D. student in the field of Quantum Machine Learning since 2022. His research interests are mainly in machine learning (ML), deep learning (DL), computer vision and pattern recognition. As a researcher he is a member of MLV (Machine Learning & Vision) Research Group of the Department of CS (IHU). He has (co)authored 15 publications in indexed journals and international conferences.

George A. Papakostas has received a Diploma in Electrical and Computer Engineering in 1999 and the M.Sc. and Ph.D. degrees in Electrical and Computer Engineering in 2002 and 2007, respectively, from the Democritus University of Thrace (DUTH), Greece. Dr. Papakostas serves as a Tenured Full Professor in the Department of Computer Science, International Hellenic University, Greece. Papakostas has 10 years of experience in large-scale systems design, as a senior software engineer and technical manager in INTRACOM TELECOM S.A. and INTALOT S.A. companies. Currently, he is the Director of the "Machine Learning & Vision Research Group". He has (co)authored more than 190 publications in indexed journals, international conferences and book chapters, 1 book (in greek), 2 edited books and 5 journal special issues. His publications have more than 3000 citations with h-index 31 (GoogleScholar). His research interests include machine/deep learning, computer/machine vision, pattern recognition, computational intelligence. Papakostas served as a reviewer in numerous journals and conferences and he is a member of the IAENG, MIR Labs, EUCogIII and the Technical Chamber of Greece (TEE).

Bibliography

[1] Paul Benioff. The computer as a physical system: A microscopic quantum mechanical hamiltonian model of computers as represented by turing machines. *Journal of Statistical Physics*, 22, 1980.

[2] Roman S. Ingarden. Quantum information theory. *Reports on Mathematical Physics*, 10, 1976.

[3] Yu. I. Manin. Computable and non-computable, sovetskoe radio. page 128, 1980.

[4] Richard P. Feynman. Simulating physics with computers. *International Journal of Theoretical Physics*, 21, 1982.

[5] D. C. Tsui, H. L. Stormer, and A. C. Gossard. Two-dimensional magnetotransport in the extreme quantum limit. *Physical Review Letters*, 48, 1982.

[6] Charles H. Bennett, Gilles Brassard, Claude Crépeau, Richard Jozsa, Asher Peres, and William K. Wootters. Teleporting an unknown quantum state via dual classical and einstein-podolsky-rosen channels. *Physical Review Letters*, 70, 1993.

[7] Peter W. Shor. Algorithms for quantum computation: Discrete logarithms and factoring. 1994.

[8] Lov K. Grover. A fast quantum mechanical algorithm for database search. volume Part F129452, 1996.

[9] Alexander Yu Vlasov. Quantum computations and images recognition, 1997.

[10] Ralf Schützhold. Pattern recognition on a quantum computer. *Physical Review A - Atomic, Molecular, and Optical Physics*, 67, 2003.

[11] G. Beach, C. Lomont, and C. Cohen. Quantum image processing (quip). volume 2003-January, 2004.

[12] Phuc Q. Le, Abdullahi M. Iliyasu, Fangyan Dong, and Kaoru Hirota. A flexible representation and invertible transformations for images on quantum computers. *Studies in Computational Intelligence*, 372, 2011.

[13] Bo Sun, Phuc Q. Le, Abdullah M. Iliyasu, Fei Yan, J. Adrian Garcia, Fangyan Dong, and Kaoru Hirota. A multi-channel representation for images on quantum computers using the rgbα color space. 2011.

[14] Yi Zhang, Kai Lu, Yinghui Gao, and Mo Wang. Neqr: A novel enhanced quantum representation of digital images. *Quantum Information Processing*, 12, 2013.

[15] Simona Caraiman and Vasile I. Manta. Quantum image filtering in the frequency domain. *Advances in Electrical and Computer Engineering*, 13, 2013.

[16] Yi Zhang, Kai Lu, Yinghui Gao, and Kai Xu. A novel quantum representation for log-polar images. *Quantum Information Processing*, 12, 2013.

[17] Hai Sheng Li, Zhu Qingxin, Song Lan, Chen Yi Shen, Rigui Zhou, and Jia Mo. Image storage, retrieval, compression and segmentation in a quantum system. *Quantum Information Processing*, 12, 2013.

[18] Hai Sheng Li, Qingxin Zhu, Ri Gui Zhou, Ming Cui Li, Lan Song, and Hou Ian. Multidimensional color image storage, retrieval, and compression based on quantum amplitudes and phases. *Information Sciences*, 273, 2014.

[19] Bo Sun, Abdullah M. Iliyasu, Fei Yan, Fangyan Dong, and Kaoru Hirota. An rgb multi-channel representation for images on quantum computers. *Journal of Advanced Computational Intelligence and Intelligent Informatics*, 17, 2013.

[20] Suzhen Yuan, Xia Mao, Yuli Xue, Lijiang Chen, Qingxu Xiong, and Angelo Compare. Sqr: A simple quantum representation of infrared images. *Quantum Information Processing*, 13, 2014.

[21] Nan Jiang, Jian Wang, and Yue Mu. Quantum image scaling up based on nearest-neighbor interpolation with integer scaling ratio. *Quantum Information Processing*, 14, 2015.

[22] Jianzhi Sang, Shen Wang, and Qiong Li. A novel quantum representation of color digital images. *Quantum Information Processing*, 16, 2017.

[23] Mona Abdolmaleky, Mosayeb Naseri, Josep Batle, Ahmed Farouk, and Li Hua Gong. Red-green-blue multi-channel quantum representation of digital images. *Optik*, 128, 2017.

[24] Nanrun Zhou, Xingyu Yan, Haoran Liang, Xiangyang Tao, and Guangyong Li. Multi-image encryption scheme based on quantum 3d arnold transform and scaled zhongtang chaotic system. *Quantum Information Processing*, 17, 2018.

[25] Engin Şahin and İhsan Yilmaz. Qrmw: Quantum representation of multi wavelength images. *Turkish Journal of Electrical Engineering and Computer Sciences*, 26:768–779, 1 2018.

[26] Hai Sheng Li, Xiao Chen, Shuxiang Song, Zhixian Liao, and Jianying Fang. A block-based quantum image scrambling for gneqr. *IEEE Access*, 7, 2019.

[27] Kai Liu, Yi Zhang, Kai Lu, Xiaoping Wang, and Xin Wang. An optimized quantum representation for color digital images. *International Journal of Theoretical Physics*, 57, 2018.

[28] Hai Sheng Li, Xiao Chen, Haiying Xia, Yan Liang, and Zuoshan Zhou. A quantum image representation based on bitplanes. *IEEE Access*, 6, 2018.

[29] Guanlei Xu, Xiaogang Xu, Xun Wang, and Xiaotong Wang. Order-encoded quantum image model and parallel histogram specification. *Quantum Information Processing*, 18, 2019.

[30] Rabia Amin Khan. An improved flexible representation of quantum images. *Quantum Information Processing*, 18, 2019.

[31] Ling Wang, Qiwen Ran, Jing Ma, Siyuan Yu, and Liying Tan. Qrci: A new quantum representation model of color digital images. *Optics Communications*, 438, 2019.

[32] Xingbin Liu, Di Xiao, Wei Huang, and Cong Liu. Quantum block image encryption based on arnold transform and sine chaotification model. *IEEE Access*, 7, 2019.

[33] Bing Wang, Meng qi Hao, Pan chi Li, and Zong bao Liu. Quantum representation of indexed images and its applications. *International Journal of Theoretical Physics*, 59, 2020.

[34] Ling Wang, Qiwen Ran, and Jing Ma. Double quantum color images encryption scheme based on dqrci. *Multimedia Tools and Applications*, 79, 2020.

[35] Fei Yan, Nianqiao Li, and Kaoru Hirota. Qhsl: A quantum hue, saturation, and lightness color model. *Information Sciences*, 577, 2021.

[36] Jian Wang, Nan Jiang, and Luo Wang. Quantum image translation. *Quantum Information Processing*, 14, 2015.

[37] Raphael Smith, Adrian Basarab, Bertr Georgeot, and Denis Kouame. Adaptive transform via quantum signal processing: Application to signal and image denoising. 2018.

[38] Jianzhi Sang, Shen Wang, and Xiamu Niu. Quantum realization of the nearest-neighbor interpolation method for frqi and neqr. *Quantum Information Processing*, 15, 2016.

[39] Ri Gui Zhou, Zhi bo Chang, Ping Fan, Wei Li, and Tian tian Huan. Quantum image morphology processing based on quantum set operation. *International Journal of Theoretical Physics*, 54, 2015.

[40] Suzhen Yuan, Xia Mao, Tian Li, Yuli Xue, Lijiang Chen, and Qingxu Xiong. Quantum morphology operations based on quantum representation model. *Quantum Information Processing*, 14, 2015.

[41] Ping Fan, Ri Gui Zhou, Wen Wen Hu, and Naihuan Jing. Quantum circuit realization of morphological gradient for quantum grayscale image. *International Journal of Theoretical Physics*, 58, 2019.

[42] Suzhen Yuan, Xia Mao, Lijiang Chen, and Yuli Xue. Quantum digital image processing algorithms based onquantum measurement. *Optik*, 124, 2013.

[43] Rashmi Dubey, Rajesh Prasad Singh, Sarika Jain, and Rakesh Singh Jadon. Quantum methodology for edge detection: A compelling approach to enhance edge detection in digital image processing. 2014.

[44] Yi Zhang, Kai Lu, and Ying Hui Gao. Qsobel: A novel quantum image edge extraction algorithm. *Science China Information Sciences*, 58, 2015.

[45] Abdelilah El Amraoui, Lhoussaine Masmoudi, Hamid Ez-Zahraouy, and Youssef El Amraoui. Quantum edge detection based on shannon entropy for medical images. volume 0, 2016.

[46] Xi Wei Yao, Hengyan Wang, and Zeyang Liao. Quantum image processing and its application to edge detection: Theory and experiment. *Physical Review X*, 7, 2017.

[47] S. Djemame, M. Batouche, H. Oulhadj, and P. Siarry. Solving reverse emergence with quantum pso application to image processing. *Soft Computing*, 23, 2019.

[48] Ping Fan, Ri Gui Zhou, Wenwen Hu, and Naihuan Jing. Quantum image edge extraction based on classical sobel operator for neqr. *Quantum Information Processing*, 18, 2019.

[49] Ri Gui Zhou, Han Yu, Yu Cheng, and Feng Xin Li. Quantum image edge extraction based on improved prewitt operator. *Quantum Information Processing*, 18, 2019.

[50] Rajib Chetia, S. M.B. Boruah, S. Roy, and P. P Sahu. Quantum image edge detection based on four directional sobel operator. volume 11941 LNCS, 2019.

[51] Panchi Li, Tong Shi, Aiping Lu, and Bing Wang. Quantum implementation of classical marr–hildreth edge detection. *Quantum Information Processing*, 19, 2020.

[52] Dongxia Chang, Yao Zhao, and Changwen Zheng. A real-valued quantum genetic niching clustering algorithm and its application to color image segmentation. 2011.

[53] Fang Liu, Yingying Liu, and Hongxia Hao. Unsupervised sar image segmentation based on quantum-inspired evolutionary gaussian mixture model. 2009.

[54] Yangyang Li, Hongzhu Shi, Licheng Jiao, and Ruochen Liu. Quantum evolutionary clustering algorithm based on watershed applied to sar image segmentation. *Neurocomputing*, 87, 2012.

[55] Chih Cheng Hung, Ellis Casper, Bor Chen Kuo, Wenping Liu, Xiaoyi Yu, Edward Jung, and Ming Yang. A quantum-modeled fuzzy c-means clustering algorithm for remotely sensed multi-band image segmentation. 2013.

[56] Sandip Dey, Siddhartha Bhattacharyya, and Ujjwal Maulik. Quantum inspired meta-heuristic algorithms for multi-level thresholding for true colour images. 2013.

[57] Simona Caraiman and Vasile I. Manta. Image segmentation on a quantum computer. *Quantum Information Processing*, 14, 2015.

[58] Xiangluo Wang, Chunlei Yang, Guo-Sen Xie, and Zhonghua Liu. Image thresholding segmentation on quantum state space. *Entropy*, 20, 2018.

[59] Fahad Parvez Mahdi and Syoji Kobashi. Quantum particle swarm optimization for multilevel thresholding-based image segmentation on dental x-ray images. 2018.

[60] Fengcai Huo, Xueting Sun, and Weijian Ren. Multilevel image threshold segmentation using an improved bloch quantum artificial bee colony algorithm. *Multimedia Tools and Applications*, 79, 2020.

[61] Inès Hilali-Jaghdam, Anis Ben Ishak, S. Abdel-Khalek, and Amani Jamal. Quantum and classical genetic algorithms for multilevel segmentation of medical images: A comparative study. *Computer Communications*, 162, 2020.

[62] Mohamed Abd Elaziz, Davood Mohammadi, Diego Oliva, and Khodakaram Salimifard. Quantum marine predators algorithm for addressing multilevel image segmentation. *Applied Soft Computing*, 110, 2021.

[63] R. Radha and R. Gopalakrishnan. A medical analytical system using intelligent fuzzy level set brain image segmentation based on improved quantum particle swarm optimization. *Microprocessors and Microsystems*, 79, 2020.

[64] Ri Gui Zhou and Ya Juan Sun. Quantum multidimensional color images similarity comparison. *Quantum Information Processing*, 14, 2015.

[65] Abdullah M. Iliyasu, Fei Yan, and Kaoru Hirota. Metric for estimating congruity between quantum images. *Entropy*, 18, 2016.

[66] Ri Gui Zhou, Xing Ao Liu, Changming Zhu, Lai Wei, Xiafen Zhang, and Hou Ian. Similarity analysis between quantum images. *Quantum Information Processing*, 17, 2018.

[67] Xing Ao Liu, Ri Gui Zhou, Ahmed El-Rafei, Feng Xin Li, and Rui Qing Xu. Similarity assessment of quantum images. *Quantum Information Processing*, 18, 2019.

[68] Huifang Li and Mo Li. A new method of image compression based on quantum neural network. volume 1, 2010.

[69] Qi gao Feng and Hao yu Zhou. Research of image compression based on quantum bp network. *TELKOMNIKA Indonesian Journal of Electrical Engineering*, 12, 2014.

[70] Qing Lei, Baoju Zhang, Wei Wang, Jiasong Mu, and Xiaorong Wu. A reconstructed algorithm based on qpso in compressed sensing. 2013.

[71] Songlin Du, Yaping Yan, and Yide Ma. Quantum-accelerated fractal image compression: An interdisciplinary approach. *IEEE Signal Processing Letters*, 22, 2015.

[72] Nan Jiang, Xiaowei Lu, Hao Hu, Yijie Dang, and Yongquan Cai. A novel quantum image compression method based on jpeg. *International Journal of Theoretical Physics*, 57, 2018.

[73] Yu Guang Yang, Juan Xia, Xin Jia, and Hua Zhang. Novel image encryption/decryption based on quantum fourier transform and double phase encoding. *Quantum Information Processing*, 12, 2013.

[74] Yu Guang Yang, Xin Jia, Si Jia Sun, and Qing Xiang Pan. Quantum cryptographic algorithm for color images using quantum fourier transform and double random-phase encoding. *Information Sciences*, 277, 2014.

[75] Li Li, Bassem Abd-El-Atty, Ahmed A.Abd El-Latif, and Ahmed Ghoneim. Quantum color image encryption based on multiple discrete chaotic systems. 2017.

[76] Ahmed A. Abd El-Latif, Bassem Abd-El-Atty, and Muhammad Talha. Robust encryption of quantum medical images. *IEEE Access*, 6, 2017.

[77] Siddhartha Bhattacharyya, Pankaj Pal, and Sandip Bhowmick. Binary image denoising using a quantum multilayer self organizing neural network. *Applied Soft Computing*, 24, 2014.

[78] Mario Mastriani. Quantum boolean image denoising. *Quantum Information Processing*, 14, 2015.

[79] Panchi Li, Xiande Liu, and Hong Xiao. Quantum image median filtering in the spatial domain. *Quantum Information Processing*, 17, 2018.

[80] Xian Hua Song, Shen Wang, Shuai Liu, Ahmed A. Abd El-Latif, and Xia Mu Niu. A dynamic watermarking scheme for quantum images using quantum wavelet transform. *Quantum Information Processing*, 12, 2013.

[81] Yu Guang Yang, Xin Jia, Peng Xu, and Ju Tian. Analysis and improvement of the watermark strategy for quantum images based on quantum fourier transform, 2013.

[82] Abdullah M. Iliyasu. Towards realising secure and efficient image and video processing applications on quantum computers, 2013.

[83] Mona M. Soliman, Aboul Ella Hassanien, and Hoda M. Onsi. An adaptive watermarking approach based on weighted quantum particle swarm optimization. *Neural Computing and Applications*, 27, 2016.

[84] Chao Yang Pang, Ri Gui Zhou, Ben Qiong Hu, Wen Wen Hu, and Ahmed El-Rafei. Signal and image compression using quantum discrete cosine transform. *Information Sciences*, 473, 2019.

[85] Shen Wang, Xianhua Song, and Xiamu Niu. Quantum cosine transform based watermarking scheme for quantum images. *Chinese Journal of Electronics*, 24, 2015.

[86] Y. LeCun, B. Boser, J. S. Denker, D. Henderson, R. E. Howard, W. Hubbard, and L. D. Jackel. Backpropagation applied to handwritten zip code recognition. *Neural Computation*, 1, 1989.

[87] Esma Aïmeur, Gilles Brassard, and Sébastien Gambs. Machine learning in a quantum world. volume 4013 LNAI, 2006.

[88] Maria Schuld and Francesco Petruccione. Supervised learning with quantum computers. 2018.

[89] Hsin Yuan Huang, Michael Broughton, Masoud Mohseni, Ryan Babbush, Sergio Boixo, Hartmut Neven, and Jarrod R. McClean. Power of data in quantum machine learning. *Nature Communications*, 12, 2021.

[90] Seth Lloyd, Masoud Mohseni, and Patrick Rebentrost. Quantum algorithms for supervised and unsupervised machine learning. 7 2013.

[91] Yijie Dang, Nan Jiang, Hao Hu, Zhuoxiao Ji, and Wenyin Zhang. Image classification based on quantum k-nearest-neighbor algorithm. *Quantum Information Processing*, 17, 2018.

[92] Nan Run Zhou, Xiu Xun Liu, Yu Ling Chen, and Ni Suo Du. Quantum k-nearest-neighbor image classification algorithm based on k-l transform. *International Journal of Theoretical Physics*, 60, 2021.

[93] Corinna Cortes and Vladimir Vapnik. Support-vector networks. *Machine Learning*, 20, 1995.

[94] Patrick Rebentrost, Masoud Mohseni, and Seth Lloyd. Quantum support vector machine for big data classification. *Physical Review Letters*, 113, 2014.

[95] Jae-Eun Park, Brian Quanz, Steve Wood, Heather Higgins, and Ray Harishankar. Practical application improvement to quantum svm: theory to practice. 12 2020.

[96] Sergey Gushanskiy and Viktor Potapov. Investigation of quantum algorithms for face detection and recognition using a quantum neural network. 2021.

[97] Anurag Rana, Pankaj Vaidya, and Gaurav Gupta. A comparative study of quantum support vector machine algorithm for handwritten recognition with support vector machine algorithm. *Materials Today: Proceedings*, 2021.

[98] Warren S. McCulloch and Walter Pitts. A logical calculus of the ideas immanent in nervous activity. *The Bulletin of Mathematical Biophysics*, 5, 1943.

[99] F Rosenblatt. The perceptron - a perceiving and recognizing automaton, 1 1957.

[100] Edward Farhi and Hartmut Neven. Classification with quantum neural networks on near term processors. 2 2018.

[101] E. Torrontegui and J. J. Garcia-Ripoll. Unitary quantum perceptron as efficient universal approximator. *EPL*, 125, 2019.

[102] Adam Paetznick and Krysta M. Svore. Repeat-until-success: Nondeterministic decomposition of single-qubit unitaries. *Quantum Information and Computation*, 14:1277–1301, 11 2013.

[103] Kinshuk Sengupta and Praveen Ranjan Srivastava. Quantum algorithm for quicker clinical prognostic analysis: an application and experimental study using ct scan images of covid-19 patients. *BMC Medical Informatics and Decision Making*, 21, 2021.

[104] Iris Cong, Soonwon Choi, and Mikhail D. Lukin. Quantum convolutional neural networks. *Nature Physics*, 15, 2019.

[105] Seth Lloyd and Christian Weedbrook. Quantum generative adversarial learning. *Physical Review Letters*, 121, 2018.

[106] He Liang Huang, Yuxuan Du, Ming Gong, and Youwei Zhao. Experimental quantum generative adversarial networks for image generation. *Physical Review Applied*, 16, 2021.

[107] Sirui Lu, Lu Ming Duan, and Dong Ling Deng. Quantum adversarial machine learning. *Physical Review Research*, 2, 2020.

[108] Eric R. Anschuetz and Cristian Zanoci. Near-term quantum-classical associative adversarial networks. *Physical Review A*, 100, 2019.

[109] Mohammad H. Amin, Evgeny Andriyash, Jason Rolfe, Bohdan Kulchytskyy, and Roger Melko. Quantum boltzmann machine. *Physical Review X*, 8, 2018.

[110] Jennifer Sleeman, John Dorband, and Milton Halem. A hybrid quantum enabled rbm advantage: Convolutional autoencoders for quantum image compression and generative learning. page 9, 1 2020.

[111] Vivek Dixit, Raja Selvarajan, Muhammad A. Alam, Travis S. Humble, and Sabre Kais. Training restricted boltzmann machines with a d-wave quantum annealer. *Frontiers in Physics*, 9, 2021.

[112] Amir Khoshaman, Walter Vinci, Brandon Denis, Evgeny Andriyash, Hossein Sadeghi, and Mohammad H. Amin. Quantum variational autoencoder. *Quantum Science and Technology*, 4, 2018.

[113] Jarrod R. McClean, Sergio Boixo, Vadim N. Smelyanskiy, Ryan Babbush, and Hartmut Neven. Barren plateaus in quantum neural network training landscapes. *Nature Communications*, 9, 2018.

[114] E. Knill. Conventions for quantum pseudocode. 6, 1996.

[115] Andrew W. Cross, Lev S. Bishop, John A. Smolin, and Jay M. Gambetta. Open quantum assembly language. 7, 2017.

[116] Robert S. Smith, Michael J. Curtis, and William J. Zeng. A practical quantum instruction set architecture. 8, 2016.

[117] Scott Pakin. A quantum macro assembler. 2016.

[118] Nathan Killoran, Josh Izaac, Nicolás Quesada, Ville Bergholm, Matthew Amy, and Christian Weedbrook. A software platform for photonic quantum computing. *Quantum*, 3, 2019.

[119] Gadi Aleksandrowicz, Thomas Alexander, and Panagiotis Barkoutsos. Qiskit: An open-source framework for quantum computing. 1, 2019.

[120] Peter J. Karalekas, Nikolas A. Tezak, Eric C. Peterson, Colm A. Ryan, Marcus P. Da Silva, and Robert S. Smith. A quantum-classical cloud platform optimized for variational hybrid algorithms. *Quantum Science and Technology*, 5, 2020.

[121] D-wave ocean software documentation — ocean documentation 4.4.0 documentation.

[122] Strawberry fields - software framework for photonic quantum computing.

[123] Franklin de Lima Marquezino, Renato Portugal, and Carlile Lavor. A primer on quantum computing. 2019.

[124] Damian S. Steiger, Thomas Häner, and Matthias Troyer. Projectq: An open source software framework for quantum computing. *Quantum*, 2, 2018.

7

Visual Analytics for Automated Behavior Understanding in Learning Environments: A Review of Opportunities, Emerging Methods, and Challenges

Zafi Sherhan Syed

Mehran University of Engineering and Technology, Pakistan

Faisal Karim Shaikh

Mehran University of Engineering and Technology, Pakistan

Muhammad Shehram Shah Syed

RMIT University, Australia

Abbas Syed

University of Louisville, Louisville, KY 40202, USA

CONTENTS

7.1	Introduction	221
7.2	Human Behavior Understanding in Learning Environments	223
7.3	Methods for Dataset Collection and Annotation	227
7.4	Methods for Feature Engineering and Machine-Learning	231
7.5	Challenges, Limitations, and Future Directions	237
	7.5.1 Data collection	237
	7.5.2 Data annotation	238
	7.5.3 Feature engineering and machine-learning	238
	7.5.4 Multimodal sensing systems for classroom behavior understanding	239

7.1 Introduction

The field of automated human behavior understanding, also known as affective computing [1], social signal processing [2], and behavioral signal processing [3] is a multidisciplinary field of science that involves psychology,

DOI: 10.1201/9781003053262-7

computer science, and engineering. The goal is to create automated systems for detecting, inferring, and understanding human behavioral characteristics in order to enable social intelligence for machines. The past decade has witnessed unprecedented growth in the field of automated human behavior understanding. Several AI-based systems have been proposed that leverage visual sensing methods to measure aspects of human behavior which include emotion recognition [4, 5], screening of mental health [6–8], and recognition of gait [9, 10], amongst others.

An emerging application of automated human behavior understanding is its use in an academic setting, such as a classroom. Here, it is important to recognize student behavior since it is representative of their progress in learning. For example, the action of hand-raise can be an indicator of student participation [11], whereas a closed upper-body pose may indicate that a student is not receptive to learning [12]. Student sitting location in the classroom has been shown to impact their participation [13]. The students who consistently sit at the back of the class offer minimal engagement [14]. Facial expressions and facial action units are important behavioral social signals which enable the identification of students who are confused or surprised [15].

Teachers' behavioral characteristics are essential because they can show the quality of instruction they provide to students. For example, Hesler [16] reported that teachers who stood behind the table while delivering lectures were deemed to be less inclusive as compared to those who walked around the classroom. Teachers who maintain mutual gaze (in terms of head-pose) are known to have a higher rapport with students, who also find such teachers more approachable [12]. It has been argued that student behavior and their synchrony with the teacher is indicative of a students' interest in learning [17,18].

While appealing and promising, building systems that can assess human behavior in an 'in the wild' setting is a challenging process. To begin, data collection and annotation are non-trivial and time-consuming. Ethical considerations, which are necessary when dealing with human subjects, add to the complexity of the data collection process. Other important factors include the selection of relevant feature engineering and machine-learning methods.

The rest of the chapter is organized as follows. Section 7.2 discusses potential opportunities of visual sensing in learning environments for understanding student-teacher behavior. Section 7.3 includes a survey of strategies used for dataset collection and annotation. Section 7.4 presents a review of computational methods used for the recognition of behavioral cues. To conclude the chapter, Section 7.5 discusses various challenges, limitations, and future directions for human behavior understanding in learning environments.

7.2 Human Behavior Understanding in Learning Environments

The classroom is a dynamic learning environment that is characterized by multimodal dyadic communication between students and the teacher. In a traditional classroom, a teacher is expected to evaluate students' affect and adapt the teaching methodology accordingly. However, it is not always possible for a teacher to track the classroom behavior of all students. It can be argued that such a situation is well suited to the application of automated human behavior understanding. The aim here is to measure, quantify, and record student behavior and provide it as feedback to the teacher so that the teacher can take necessary affirmative action. Figure 7.1 illustrates the conceptual block diagram for a visual sensing system for a learning environment. This system consists of cameras that focus on students, teachers, or both as subjects, and a signal processing unit that computes high-level behavioral attributes from video recordings on a frame-level basis.

The first task for the system is to locate each subject in the video frame. A popular choice for this task is to use the multi-task cascaded convolutional neural network (MTCNN) [19] algorithm. Once the subject is located within the video frame, it is possible to compute many behavioral attributes depending on whether appearance-based image features or body keypoints are used. For example, using appearance features, one can infer the underlying emotions using facial expression recognition [20]. These, in turn, can be used to measure a student's response to teaching methodology. It is customary practice to categorize facial expressions in terms of six basic emotions described by Paul Ekman [21], which include *anger, happiness, sadness, surprise, contempt*, and *disgust*. A *neutral* emotional state is often added to Ekman's basic emotions for the sake of completeness.

In addition to appearance-based features, one can also use keypoints for the face, arms, and torso to determine behavioral attributes. The keypoints from the face region can provide information regarding head-pose, eye-gaze, and facial landmarks which can be used to detect blinking, smiles as well as a myriad of subtle changes in facial expressions that do not fall under the categories of Ekman's basic emotions. The keypoints from the torso can identify whether the subject is standing, sitting, or it is leaning backward or forwards. Finally, the keypoints from the arm can be used to detect actions such as hand-raise, hands-crossed, or whether the subject is using a mobile phone or playing with their hair. In theory, a classroom behavioral understanding (CBU) system can recognize these and more behavioral attributes, but its performance is often limited due to practical constraints. Limitations in terms of video/image quality, processing power, occlusions in the recording environment that hide parts of the human body, and in-the-wild subject movements. It is, therefore, necessary to identify standard practices as well as those that are emerging in the field to build systems with improved performance.

Given the data-driven nature of the task at hand, one can categorize methods for CBU in terms of (1) data collection and annotation and (2) feature engineering and machine-learning. Next, we provide a general overview of systems that introduced new aspects of data-driven knowledge in the context of CBU.

In [22], Ahuja et al. proposed the EduSense sensing system for classroom environments which integrates open-source software tools to produce theoretically motivated audio and visual attributes for recognizing classroom activities. These activities include *sitting* or *standing*, *hand raise*, three categories of upper body pose (i.e., *arms at rest*, *arms crossed*, or *hands on face*), *smile detection*, *mouth opening detection*, *head-orientation* (as a proxy for class gaze), *classroom topology* (i.e., student layout in the classroom), *synthetic accelerometry* to track the motion of body across frames, *student originated speech* or *instructor originated speech*, and *speech act delimitation*. They propose a toolkit as an off-the-shelf solution for classroom behavior monitoring. Meanwhile, Goldberg et al. [23] investigated the efficacy of computer vision based methods to measure cognitive engagement, student involvement, and situational interest in classrooms and validated their approach with a self-reported score for engagement from students. It was reported that eye gaze, head-pose, and facial expressions (computed using OpenFace tool [24]) at individual student level and synchrony at group level can be used to estimate of students' self-assessed level of the overall engagement. However, the authors acknowledge that limitation in terms of dataset size reduced the statistical power of their correlation analysis but also point out that limited data is an intrinsic shortcoming of the field.

Araya et al. [25] investigated the efficacy of a data-driven approach to determine the values of yaw and pitch angles for head position and body orientation. They argue that head-pose (which they refer to as gaze, although the term gaze typically refers to eye-gaze) and orientation of the body is an essential indicator of attention in a classroom environment. They aimed to demonstrate that videos recorded through smartphone cameras carry adequate quality such these can be used to estimate the direction of the head-pose and body orientation of students and teachers.

In [26], a two-step method is proposed to train a deep neural network that can classify between engaged and disengaged subjects based on their facial images. It was argued that their approach is an effective solution to the issue of data scarcity in the field of CBU. In the first step, they pre-trained a deep neural network to recognize facial expressions. The motivation for the pre-training step is that a considerable amount of data is available for the task of facial expression recognition. Once this model had achieved adequate performance, they fine-tuned it for a new dataset which they collected for the engagement recognition task. It was reported that the fine-tuned model outperformed those which are trained directly on the engagement recognition dataset. It was also reported that fine-tuned model outperformed a machine-learning model that uses handcrafted appearance features (histogram-of-oriented gradients [27]) and support vector machine as the classifier.

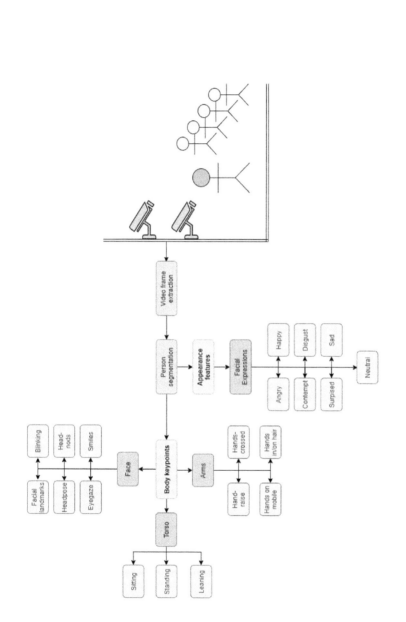

FIGURE 7.1

Conception of classroom behavior understanding system that can recognize various student/teacher activities based on facial appearance features and keypoints from the face, arms, and torso.

Vanneste et al. [28] conducted a study to develop methods based on visual computing techniques to measure student engagement. They considered two cases for student engagement, i.e., engagement of every student (individual-level) and engagement at classroom-level. For individual students, they wanted to determine the accuracy with which actions such as *note-taking* and *hand-raising* can be detected. Meanwhile, at the classroom level, they wanted to detect indicators such as the *synchronicity of behavioral actions, reaction times*, and *shared eye-gaze interactions*. In the end, the correlation between engagement scores predicted through a machine-learning pipeline and the students' self-reported engagement scores was computed. The authors reported that their study was not able to prove the existence of a strong link between computer-vision-based models and students' self-reported engagement, at least based on correlation analysis. It was surmised that this may be because their system relied on visual behavioral indicators alone whereas emotional and cognitive components of engagement may have been necessary to measure student engagement. Although these experiments did not prove their hypothesis, this study provides a detailed report on negative results which can be used to build better systems for classroom behavior understanding.

In [29], Jiang et al. proposed a method to measure student concentration based on information provided by head-pose and sitting position in the classroom. They curated a dataset of 25,625 face images from a real-world classroom environment and trained a convolutional neural network (CNN) based computer vision model to recognize between three head-pose states, i.e., *gazing state (student faces the teacher)*, *non-gazing state*, and *unrecognizable*. It was reported that the model achieved an accuracy of 94.6% for the classification task. It can be argued that their approach only offers a rudimentary take on quantifying student concentration. Similarly, Anh et al. [30] build an automated system that used computer vision techniques to provide a summary of student behavior in classrooms. Their system could identify when a student does not pay attention and reports them to the instructor. Here, head-pose information was used to quantify whether the student looked at the board, laptop, or any other place in the classroom with an accuracy of 93.33%. The major limitation of their work is that the system would only work for a limited selection of head-poses and can be influenced by the angle with which the video is recorded through the camera.

Renawi et al. [31] proposed a rule-based system that can identify when a student's head-pose is oriented towards the center-front of the classroom. Their solution was based on a webcam, standard computer, and computer vision algorithms. The Dlib libary [32] was used to compute 2D facial landmarks, which were subsequently used to estimate head-pose. They appreciated that raw values of head-pose angles cannot be used since these will be different for students sitting at far left and far right of the classroom even if they were looking towards the whiteboard at center-front. The proposed solution to this problem was to calibrate their model with data recorded at various points in the classroom by considering pitch and yaw angles as well as face depth and its location in the classroom. They empirically determined pitch and yaw angle values that corresponded to subjects looking at the whiteboard and used this information for rule-based classification. If head-pose angles were outside the limits, a subject was assumed to be non-attentive and their overall attention score was decreased.

In [33], Zhang et al. proposed a computer vision based approach to recognize three types of body postures, i.e., *reading, hunching*, and *looking (in different directions)* in classroom environment. They used the YOLOv3 algorithm [34] for object recognition and fine-tuned it to detect human subjects. Next, the squeeze-and-excitation network (SENET) [35] was used along with high-resolution networks (HRNET) [36] to estimate body keypoints. These keypoints were then used to train machine-learning models that can classify between the three categories of the body pose. A labeled dataset of 14,470 student body postures was created from 2894 images collected as part of their experiments. The authors suggested that their proposed method is better than the one based on OpenPose since it can estimate the correct locations of body keypoints for hunching posture whereas OpenPose could not. It was reported that their proposed approach achieved an overall accuracy of 90.1% compared to 64.2% to a classifier based on OpenPose and support vector machine classifier.

7.3 Methods for Dataset Collection and Annotation

In this section, we shall provide a review of data collection and annotation strategies used for human behavior understanding in classrooms. The goal is to identify the merits and limitations of these approaches and identify emerging trends. Table 7.1 provides a summary of tasks related to CBU from a selection of published research literature. Most studies [22,33,37–42], have focused on proposing methods to recognize student activities that are based on body posture and head-pose of pupils. Some studies have leveraged activity recognition to measure latent behavioral states such as attention [25,30,31], engagement [26,28], and concentration [43] in classrooms. The table also provides information about the hardware aspect of visual sensing. Here, it is clear that diverse camera setups were used that include specialized mounting rigs for cameras [22] and special types of cameras such as a wide-angle camera [39], Kinect V2 [37], camcorders and handycams [38], smartphone-based cameras [25], webcams [31,40], and dome-shaped surveillance cameras [33]. It should be noted that most studies did not provide detailed descriptions of the visual sensing system such as the camera resolution, frame rate, the height at which the camera is fixed, or its location in the room. These are important parameters that can influence the performance of the visual sensing system.

In terms of data collection, we found that all studies had tested the performance of their proposed methods on a dataset with spontaneous activities. In some studies, authors created a special dataset consisting of scripted (acted) behavioral attributes which were then used as training data for

TABLE 7.1

Summary of study objectives, behavioral attributes, recording setup, dataset type, and annotation methods for a selection of systems for classroom behavior understanding.

Ref.	Objective	Behavioral attributes	Recording Setup	Dataset type	Annotation method
[22]	Activity recognition	10 types of body posture patterns, 3 types of acoustic event, classroom topology	Multi-viewpoint capture rig with 12 Lorex LNB8950AB cameras with 4K resolution (3840x2160 pixels) recorded at 3 fps	Scripted (source) Spontaneous (target)	Scripted data collection with manual annotation
[25]	Attention recognition	Pitch and yaw values for head- and body-pose within a predecided range	Smartphone cameras, recordings took place from the front and back of the classroom	Spontaneous	Manual annotation
[26]	Engagement recognition	4 types of facial expressions	Videos recorded at 20 fps	Spontaneous	Manual annotation
[28]	Engagement recognition	8 types of body posture patterns	Camera mounted at the front of the class	Spontaneous	Manual annotation
[31]	Attention recognition	Pitch and yaw values for head-pose within predecided range	Webcam	Spontaneous	Manual annotation
[33]	Activity recognition	3 types of body posture patterns	Dahua network dome surveillance cameras	Spontaneous	Manual annotation
[37]	Activity recognition	5 types of body posture patterns	Kinect V2 and an RGB camera with a resolution of 1920×1080 pixels	Scripted (source) Spontaneous (target)	Scripted data collection with manual annotation
[38]	Activity recognition	4 types of body posture patterns and 12 types of facial expressions	Two handy-cams (Sony HDR-TD10 and Sony HDR-PJ600VE) and a Nikon 3300 DSLR	Scripted (source) Spontaneous (target)	Scripted data collection with manual annotation
[39]	Activity recognition	13 types of posture patterns	Wide-angle camera (Marshall Electronics CV505)	Scripted (source) Spontaneous (target)	Scripted data collection with manual annotation
[40]	Activity recognition	3 types of head-pose patterns	Webcam	Spontaneous	Manual annotation
[42]	Activity recognition	4 types of body posture patterns	Two cameras; one at the front, another in the corner	Scripted (source) Spontaneous (target)	Scripted data collection with manual annotation
[43]	Concentration recognition	2 types of facial expressions	Not specified	Scripted	Scripted data collection with manual annotation

machine-learning models. The main motivation behind this effort was to simplify the process of data annotation – especially since manual annotation was almost always the preferred method. It is reminded that labeling a small set of actions, which is the case in scripted data, is easier than labeling for a large set of actions which is typically the case for data with spontaneous activities.

In [37], a method is proposed to quantify student behavior in terms of hand gestures, facial expressions, head pose in a classroom environment. They used two types of cameras to record visual data, the first of which was a traditional camera that captured 2D RGB images and the second was a Kinect that captured RGBD data. Their motivation for using the Kinect was that it can generate 3D skeleton points through its software development kit without the need for a third-party algorithm. However, the authors noted that the Kinect was limited in terms of range (5 meters) and the number of subjects (6 subjects) it could track accurately. This is a major limitation since classrooms are typically larger and contain many more students. As a solution, the authors used the model from [44] to compute 2D body-pose keypoints from the RGB images. The work is important since it highlights the limitation of cameras with depth-sensing even though these may initially appear ideal for near real-time human pose estimation.

Nezami et al. [26] used two datasets in order to train a deep neural network for engagement recognition. The first dataset is the facial expression recognition 2013 (FER-2013) [45] which consists of 35,887 images labeled for Ekman's basic emotions [21] of happiness, anger, sadness, surprise, fear, disgust, and neutral. In addition to this publicly available dataset, the authors have collected a new dataset which they refer to as *Engagement Recognition* dataset. The raw version of this dataset was collected in form of video recordings of 20 students who were learning scientific knowledge and research skills using an e-learning platform. To generate gold-truth labels, bespoke annotation software was used to annotate selected but representative video frames. Each of these frames was annotated by at least 6 student members of the Psychology department. These annotators were asked to label on behavioral and emotional dimensions of engagement. Labels for behavioral dimension include *on-task*, *off-task*, and *can't decide*, whereas labels for Emotional dimension include *satisfied*, *confused*, *bored*, or *can't decide*.

In [22], Ahuja et al. built the EduSense system, a custom multi-viewpoint capture rig that consisted of three heavy-duty stage tripods and 12 Lorex LNE8950AB cameras as part of their recording setup. Multiple tripods and cameras ensured that the classroom environment was recorded with a variety of viewpoints. All camera streams were recorded at 4K resolution at 3 FPS and saved to disk for subsequent processing at a later time, thus, processing was performed offline. They evaluated the performance of EduSense in five exemplary classrooms with 5 instructors and 25 student participants. Participants were given an orientation on the scripted actions that they would need to perform as part of the evaluation experiment. The scripted dataset was used to develop a machine-learning model that could recognize various behavioral attributes and the efficacy of the trained model was then tested on a small dataset curated from video recordings in a real-world setting.

Ashwin et al. [46] introduced a new database that can be used to build affect recognition systems for e-learning and classroom environments. The dataset consists of images with acted emotions and videos with spontaneous body postures. It was suggested that this dataset can be used to trained robust machine-learning algorithms for affect recognition since it provides training examples of images with varying degrees of occlusion, background clutter, changes in pose, illumination, and camera angles, as well as different ethnicities of subjects, etc. They used two handy-cams (Sony HDR-TD10 and Sony HDR-PJ600VE) and a Nikon 3300 DSLR to record pictures and videos of subjects. For acted emotions, the dataset includes 700 single-person frames and 1450 multi-person frames that were labeled for Ekman's basic emotions, whereas 50 videos were recorded from 350 students amounting to more than 20 hours (72,000 frames) that were labeled for *hand-raise* and body postures for *normal sitting, half bent*, and *completely leaning*. Initially, images were annotated manually by more than 30 faculty members but once sufficient frames were annotated a semi-automated annotation process was used to speed up the process of data annotation.

In [39], Chen et al. proposed a pipeline to simplify the difficult task of data annotation and model building for novel recording environments. Their approach relies on recording a dataset of scripted posture patterns from the same classroom (as the real-world spontaneous setting) before collecting the main dataset for CBU. They argue that since the scripted dataset can be collected and annotated in a relatively short amount of time, a posture recognition system for a classroom environment can be trained with relative ease. Visual data was collected using a single wide-angle camera (Marshall Electronics CV505) so that all subjects could fit within the camera frame. The camera exclusively focused on students and the instructor was not recorded. Before recording took place, informed consent was obtained from all participants of the study. To begin, the authors collected a dataset of scripted body postures in which subjects were asked to play out a set of pre-decided 13 body posture patterns when prompted. These actions include *checking phone, looking at computer, looking down, looking up, looking at front-left, looking at front-right, looking left, looking right, performing question and answers, talking with neighbors, raising left hand, raising right hand*, and *writing*. This dataset consisted of 7-minute videos collected from 11 subjects. Then, a set of features was computed based on the key points supplied by the OpenPose tool, and then a machine-learning model was trained to recognize the scripted body postures. The experiments were conducted to test their hypothesis using a non-scripted dataset that was separately recorded. This dataset consisted of video recordings in a classroom environment of 22 students from 22 sessions of a 75-minute class throughout the semester.

Araya et al. [25] used a smartphone camera to record videos of subjects in a classroom environment. Their work is a digression from the traditional approach of using a high-quality camera setup for visual sensing. They recorded data from 4 first-grade classes with each session of 60 minutes in duration. From these videos, they obtained a sample of 1991 frames for head-pose and body orientation. Each of these frames was annotated by four elementary school teachers for yaw and pitch angles using a web-based annotation tool.

In [28], Vanneste collected visual data for their experiments from a hybrid virtual classroom environment where students were asked to attend certain lectures in person and other lectures online. A front-mounted camera in the classroom was used to make video recordings for the duration of the lecture. Vanneste et al. collected data from two groups of participants. The first group included 14 students (4 female and 10 male) from secondary school. These students were recorded during six lectures that last around 70 minutes each. During each lecture, students were prompted to self-report their level of engagement on a Likert scale with scores between 0 to 2. A score of 0 meant *totally disengaged* and a score of 2 meant *totally engaged*. In total, the authors were able to collect 580 self-reported data points for engagement from the six lectures. The second cohort of participants included 51 university students (38 female and 13 male) who attended six lectures (face-to-face) with each lecture lasting 1 hour and 20 minutes. The data from the first group of students was used to analyze the engagement of students on an individual level. Here, each frame was manually annotated for 8 actions which the authors considered to be relevant for student engagement. These actions include *hand-raising, note-taking, hand on face, working with laptop, looking back, fiddling with hair, playing with cellphone*, and *crossing arms*. The second dataset was used for classroom-level engagement.

Delgado et al. [40] introduced the *Student Engagement Dataset*. The dataset was collected using a front-facing camera from a laptop which recorded visual feedback of student gestures as they were solving maths problems whilst using an e-learning platform. This dataset consists of 400 videos (18,721 video frames after data cleaning) that were recorded from 19 subjects. These images were labeled by crowd-workers from the Amazon Mechanical Turk platform for three labels, that identified whether subjects were *looking at the paper, looking at the screen*, or *wandering*. These three outcomes were then used as gold-truth labels to represent student engagement. The authors reported that their initial dataset was highly imbalanced. For example, frames with the *looking at the screen* class occurred 14 times more than samples from the *wander* class and 3 times more than the *paper* class. Their solution to deal with this class imbalance and create a stratified dataset was to remove frames that looked similar within each class. Interestingly, the stratified dataset was much smaller with only 1973 frames compared to 18,721 frames compared to its raw version.

7.4 Methods for Feature Engineering and Machine-Learning

Table 7.2 provides a summary of objectives for machine-learning tasks, feature engineering mechanisms, and classification/regression methods used within a

TABLE 7.2

Summary of machine-learning objectives and methods for feature engineering and classification/regression for a selection of systems for classroom behavior understanding.

Ref.	ML Objectives	Feature Engineering Methods	ML Methods
[22]	Head-pose, body-pose, and audio classification	Geometric features with OpenPose [24] keypoints	MLP and SVM classifiers
[25]	Head and body-pose regression	Geometric features with OpenPose keypoints	Decision Trees and Random Forest regressors
[26]	Facial affect recognition	Fine-tuned CNN	DNN classifier
[28]	Body-pose classification	Fine-tuned CNN, Clustering for keypoints from OpenPose	i3D model [47] for classification Multi-level regression
[29]	Head-pose classification	Fine-tuned CNN	DNN classifier
[30]	Head-pose classification	Feature embeddings from Arcface [48], head-pose estimation using Hopenet [49], and imbalanced learning using SMOTE [50]	Weighted KNN, SVM, Decision tree, Gradient boosting, and Random forest classifiers
[31]	Head-pose estimation	Facial landmarks from Dlib library [32]	Rule-based classification
[33]	Head-pose classification	Head-pose estimation using HRNET [51] with SENET [35] and OpenPose	SVM classifier
[37]	Head-pose and facial expressions classification	Facial expressions based on [52] and head-pose using [53]	Rule-based classification
[38]	Body-pose classification	Appearance based Visual features, finetuned Inceptionv3 model [54]	SVM and DNN classifiers
[39]	Body-pose classification	Geometric features based on keypoints from OpenPose	Random Forest classifier
[40]	Head-pose classification	Head-pose based on FSA-Net model [55], finetuned VGG16 [56], finetuned MobileNet [57], and finetuned Xception [58]	Rule-based classification, Logistic Regression, and DNN classifiers
[41]	Posture recognition	Keypoint extraction using OpenPose with fine-tuned CNN	DNN classifier
[42]	Body-pose classification	Geometric features based on keypoints from OpenPose	DNN classifier
[43]	Facial affect recognition	Pre-trained VGG-16 and VGG-19 [56], and Fine-tuned VGG-16	DNN classifier

selection of systems proposed for classroom behavior understanding. Given that the lower torso region is often occluded by tables in the classroom, it makes sense that most systems focused on social signals exhibited through facial expressions, head pose, and upper body posture. Some studies, such as [28, 38, 39, 41, 42] and [29–31, 33, 40] used information provided by either body posture or head pose, whereas only [22, 25, 37] used information from both types of social signals.

Amongst feature engineering methods, one can see two distinct strategies. The first strategy is based on computing appearance features [38] from video frames using either handcrafted features or Deep Neural Network (DNN) based on convolutional kernels. The second strategy leverages pre-trained models for human body keypoint estimation followed by computation of geometric features. The OpenPose algorithm was by far the most popular tool for estimation face and body keypoints, although Zhang et al. [33] proposed their method and demonstrated that it performed better than OpenPose. Amongst machine-learning methods, the support vector machine (SVM) classifier and a DNN with fully connected layers were the popular choices. It is interesting to note that rule-based classifiers, built using if-else statements, were used in at least three studies [31, 37, 40]. This shows that some systems were designed as bespoke solutions for a particular classroom setting and may not necessarily work when the system is deployed in a classroom with a different structure. Next, we provide a detailed review of key contributions in terms of feature engineering and machine-learning methods used in visual sensing systems for CBU.

Ku et al. [37] used the Kinect SDK to compute 3D body keypoints for small classrooms but observed that Kinect did not accurately estimate keypoints for large classrooms. Therefore, the authors have used a model from [44] when the recording took place in a large classroom setting. Regardless, they used rule-based classification in order to recognize actions such as *handraise, look left, look right, facing front,* and *facing back* gestures. They also integrated facial expression and head-pose into their system. Here, the method from [52] was used for facial expression recognition, and the method from [53] was used to estimate the head-pose. Moreover, a deep neural network based on the GoogLetNet architecture [59] was trained to detect hand gestures representing numbers from 0 to 5. The authors did not discuss their motivation for recognizing these six gestures.

In [26], Nezami et al. trained 3 deep-learning-based models along with a traditional machine-learning model that used HOG features with an SVM classifier as part of their work on engagement recognition. The FER-2013 dataset was partitioned into training, validation, and test partitions with 3224, 715, and 688 examples, respectively, in each partition. It was ensured that these partitions were created in a subject-independent manner to avoid bias. The *Engagement Recognition* dataset (see Section 7.3) was also partitioned into 3 parts with 3224, 715, and 668 examples in training, validation, and test partitions. Due to the imbalanced nature of datasets, the authors computed avg-f1score and Area Under Curve (AUC) metrics alongside classification accuracy. It was reported that the HOG-SVM model achieved an avg-f1score of 67.38% whereas the best-performing CNN model achieved 73.90% – which is an improvement of approximately 6%. Thus the deep-learning-based model was proposed as the solution for engagement recognition.

Ahuja et al.'s [22] multi-module EduSense system used the OpenPose tool to compute body keypoints for each subject in the video recordings. For modules that recognized *sitting* or *standing, hand raise, upper body pose, smiles,* and *mouth opening detection,* they computed direction vectors and distance between all pairs of keypoints. The resultant feature vectors were normalized to alleviate the scaling effect that occurs when certain subjects are nearer to the camera as compared to others who may be further away from it. Finally, the normalized feature vectors were used to train a Multilayer Perceptron (MLP) and SVM classifiers to detect various types of student activities. For acoustic sensing, the Ubicoustics model from Laput et al. [60] was used to predict whether short-time audio frame consisted of speech or silence. Authors reported that certain modules of the EduSense system were able to achieve "reasonable accuracy", such as *student segmentation* with accuracy $\geq 95\%$, *upper body pose* ˜80%, *gaze estimation* ≤ 15, and *speech detection* $\geq 80\%$, whereas other modules needed further improvement. They noted that limited camera resolution and occlusion were the main causes of limitations of their system. The authors recommended that the EduSense platform be deployed in a classroom with a maximum length of no more than 8 meters and with high-resolution cameras placed sufficiently high in order to get a good view of the classroom.

In [46], Ashwin et al. reported the performance of baseline machine-learning models for their newly introduced dataset to recognize affect and body postures. The authors have experimented with a variety of visual features including local binary patterns [61], Gabor filters [62], local Gabor binary patterns [63], and Pyramid histogram of oriented gradients [64] along with Principal Component Analysis (PCA) for dimensionality reduction, and SVM for classification. The authors reported that although these methods work well for frontal face data, the models failed to provide adequate performance for non-ideal cases. In fact, an overall accuracy of less than 50% was achieved for their dataset. Ashwin et al. also experimented with the Inceptionv3 model [54] which outperformed the traditional machine-learning pipeline by achieving an overall classification accuracy of 72% for their dataset.

Araya et al. [25] used a brute-force approach to estimate head-pose and body orientation for a dataset collected through smartphone cameras. They first computed face and body keypoints using the OpenPose tool and then computed a large set of geometric features (47,900) to describe body posture. Later, two machine-learning algorithms for regression, namely decision trees and random forests, were used to estimate the value of pitch and yaw angles for head-pose and body orientation. The dataset was partitioned into training and test partitions, with 1333 frames used for training the regressor and 658 for testing its performance. Given that the objective was to estimate angles (non-discrete labels), the authors chose root-mean-square error (RMSE) between gold-truth labels and those predicted using regression as the performance metric. The authors report that their best model produced an RMSE that is 11% lower than the RMSE between human annotators, which they argue highlights the effectiveness of their proposed solution. Given that their test data is much smaller (658 samples) than the number of features (47,900), we believe that their model is prone to over-fitting to the small-sized dataset. A more robust testing setup could have been developed through the use of repeated K-fold cross-validation.

In [39], Chen et al. used the OpenPose tool to compute upper body keypoints that included head, neck, and arms. Then, a set of 24 attributes were computed based on the geometric relationship between keypoints. Although the authors did not provide an exhaustive list, examples of such features include *neck to nose distance, left shoulder to nose distance, left hand to nose distance* etc. Then, a random forest classifier was trained using leave-one-subject-out cross-validation to identify each of the 13 student activities in a one-class-versus-rest setting with area-under-curve chosen as the metric to quantify the classification performance. The authors reported that certain postural patterns, such as *look-right, look-left,* and *raise-right-hand* had better classification performance with AUC scores of 0.88, 0.84, and 0.84, respectively, whereas other posture patterns had relatively poor performance such as *check-phone* and *look-down* with AUC scores of 0.51 and 0.53, respectively. We find it particularly interesting that the AUC for *look-down* stands at 0.53 although the *look-up* action has a score of 0.80, even though both actions involve a change in head-pose due to movement in opposite directions. We believe that this may be because other actions such as *check-phone* also involve a similar change in head-pose as the *look-down* action. Perhaps the one-versus-rest classification setup used by the authors is not well suited to the task at hand and multilabel classification, at least for similar actions, would have been more appropriate.

As part of their efforts to assess engagement at the individual student level, Vanneste et al. [28] trained a deep neural network that takes a video of about 2 seconds in duration for recognition of 8 types of student actions. Here, they leveraged the concept of transfer learning to use the i3D model [47] as the image feature extractor and supplied the resultant DNN embeddings to a classifier based on a fully connected neural network with 3 layers. For the regression task, they trained a multi-level regressor to predict self-reported engagement scores. The authors reported that the classifier predicted certain classes more accurately than others. For example, *taking notes* and *raising hands* had recall of 69% and 67%, whereas, *fiddling with hair, working on laptop,* and *playing with cellphone* had recall of 17%, 26%, and 22%. Vanneste et al. provided multiple reasons to explain the poor performance which include (a) relatively small number of training and test examples in the dataset, (b) similarity between actions and only subtle differences between them, and (c) occlusion due to which certain body parts were not visible in video clips. Furthermore, they reported that multi-level regression showed no correlation between self-reported engagement scores and hand-raising.

For collective assessment of engagement, Vanneste et al. started with the OpenPose tool to compute keypoints for the body-pose of each student in every video frame. Since videos were recorded in a classroom environment, some keypoints related to legs were not visible for all subjects, therefore, only upper body keypoints including head, arms, and torso were considered. Next, they undertook collective analysis using a method that we find particularly interesting. After normalizing keypoints, they use k-means clustering (separately) for keypoints related to head, arms, and torso to group students who had a similar body-pose. This procedure was applied for all frames in the video. Then, they determined the time it took for students to change cluster (i.e. to change their body pose) after an educational event. Finally, they computed variance in cluster changing time (VCCT), which was shown by Raca et al. [65] to be an indicator of student engagement. It was hypothesized that if the classroom is engaged then the VCCT is lower such that more students change their body posture simultaneously (or at least at a short temporal distance). On the other hand, if more students are disengaged, their VCCT will be larger as more students respond slowly to changes in an educational event. Based on their experiments, the authors reported that clusters from head-pose were more informative than clusters from arms and torso. They did not even consider clusters from the latter for further analysis since they appeared to contain random movements. It should be mentioned that the authors only found weak evidence of association to class engagement for head-pose clusters. It was also reported that student reaction times did not relate to the self-reported engagement scores.

Delgado et al. [40] used two types of classification regimes to recognize whether the students were engaged or disengaged (wandering) in a classroom environment based on facial information. In both regimes, 80% of the dataset was used for training the machine-learning model whereas the remaining 20% was used for testing it. The first classification regimes used handcrafted features, more specifically the yaw and pitch angles computed using the FSA-Net model [55], along with multiple non-deep-learning-based classifiers. Their first attempt within this regime used rule-based classification where the rules were implemented using if-else statements. This was based on the authors' observation the values for yaw and pitch were close to 0 when the subjects looked at the computer screen (i.e., when they were engaged). In addition to the rule-based classifier, the authors also trained a Logistic Regression classifier trained on yaw and pitch angles. The second classification regime involved fine-tuning three DNN models, i.e., VGG16 [56], MobileNet [57], and Xception [58] that were previously trained on the Imagenet dataset. It should be noted that Imagenet is a collection of 1,281,167 images for training and 50,000 images for validation, and it is assumed that the three models have already learned useful feature representations that enable them to behave as image feature extractors. They used data augmentation for generating new data synthetically to increase the number of training examples in their dataset. The authors reported that fine-tuned MobileNet model yielded the best classification accuracy (94%) for the test partition, followed by fine-tuned Xception

(88%), and fine-tuned VGG16 (85%). Unsurprisingly, models based on non-deep-learning regimes, i.e., rule-based classification, and Logistic Regression classifier achieved much poorer performance with an accuracy of 60% and 55%. These results highlight the efficacy of deep neural networks for engagement classification.

7.5 Challenges, Limitations, and Future Directions

In this section, we shall summarize the current challenges for classroom behavior understanding as identified in the surveyed research literature and discuss the emerging solutions that have been proposed to alleviate some of the limitations caused by the challenging task at hand. This discussion has been organized from the point of view of data collection, data annotation, feature engineering, and machine-learning.

7.5.1 Data collection

The process of data collection is one of the most challenging aspects of human behavior research. While ethical constraints on data collection are unquestionably necessary, they make the process of recruiting volunteers challenging. This is especially true when biometric data such as a person's face, body, or voice must be captured. Even if a subject agrees to donate their data, this does not indicate that the data can be freely shared with the research community. These constraints do not bode well for data-driven systems that rely on the collecting and processing of massive amounts of data for their success. It is worth noting that these limits are not exclusive to CBU and apply to other types of human behavior research as well, such as tasks involving the detection of depression, bipolar illness, autism, or Parkinson's disease [66–68]. A potential solution for the issue of privacy in a CBU setup was proposed by Ahuja et al. [22], in which the video will be deleted immediately without human intervention after computing keypoints from video frames.

Another challenge in data collection for visual sensing is the existence of clutter in the recording environment that causes occlusions to various parts of the body. This is particularly true for a classroom-based learning environment where tables typically occlude the lower torso of subjects. Furthermore, depending on the camera location, some students may occlude others who sit behind them. These issues were also highlighted by [22, 28, 38, 39]. In fact, Vanneste et al. [28] list occlusions as one of the reasons why their visual sensing system failed to achieve desired performance. A solution to this problem was proposed by Ajuha et al. [22] which used a multi-viewpoint rig to mount 12 high-quality cameras that recorded at 4K resolution. They argued that recording subjects from multiple viewpoints would ensure that subjects are not occluded. A major drawback of their solution is the cost associated with it.

Based on our survey, we cannot recommend the use of a particular recording setup. As summarized in Table 7.2, most studies have not provided sufficient information to enable a comparison of the design choices. We also found that whereas some studies had used high-quality cameras as part of their setup, others had used webcams, smartphone cameras, and security cameras as part of their setup. Some studies simply did not provide any information regarding the setup.

7.5.2 Data annotation

Dataset annotation is yet another challenging task that must be performed to generate either ground-truth or gold-standard labels for new datasets. Ground-truth labels can be assigned to data for which labels can be provided with absolute certainties, such as posture patterns for *sitting* or *standing*. Gold-standard labels, generated by consensus from multiple annotators, are assigned to data for which labels are based on opinions. Examples include behavioral states such as *less engaged* or *more engaged*. The problem here is that gold-standard labels may not necessarily be the ground-truth labels since they are based on opinions. This means that results of classification performance of CBU systems based on gold-truth labels should also be interpreted with care. Labels can also be generated through self-report questionnaires such as those used by Vanneste et al. [28] to quantify student engagement, however, such a labeling strategy is also prone to subjective bias [69].

From our survey, we found that almost all studies relied on a manual annotation to generate labels for various types of behavioral attributes. It should be noted that recruiting annotators is a challenging task and reasonable compensation needs to be made for their service. Moreover, the labels generated through manual annotation fall under the category of gold-standard labels. To this end, we found the approach used by Ashwin et al. [38] to be promising. They used the manual annotation method to label a subset of their corpus and then leveraged a semi-automated method to generate labels for the remainder of their data.

We also noticed a trend where researchers created a special dataset with scripted behavioral attributes and used it for the data annotation process. The motivation behind such an approach is to simplify the task of labeling each video frame. This is because it is easier and less time-consuming to annotate a finite set of behavioral attributes (as is the case in a scripted dataset) than a large number of possible actions that is typically the case for spontaneous datasets.

7.5.3 Feature engineering and machine-learning

The feature engineering and machine-learning parts of the visual sensing system are perhaps the least challenging aspect of the task at hand. This is due to the rapid growth in the field of AI where off-the-shelf tools are available for use in the form of software APIs. Based on our survey, we found that most studies used body-keypoints generated from the OpenPose tool. It should be kept in mind that the performance of OpenPose is adversely affected by occlusions, camera resolution, and the number of subjects in the video frame. Moreover, a computational machine-with significant processing power is required to run the OpenPose algorithm [24].

Amongst other trends, we note that deep neural networks performed much better than handcrafted visual features. This is expected given that deep-learning methods have essentially replaced handcrafted machine-learning models in the field of computer vision [70]. However, such models also have their share of limitations. Computer vision algorithms based on DNN have indeed achieved near-human-level accuracy for many classification tasks, but they have high computational complexity and as a result, require long processing times. Such systems may not offer the ideal solution if the goal is to achieve near real-time video analytics. One can argue that the performance of systems may need to be traded-off in favor of reduced computational complexity.

Another important aspect to consider is whether such systems should be built using psychology-inspired interpretable features or 'black box' machine-learning systems [71]. Whereas the latter may achieve better performance against a machine-learning metric, it lacks the interpretability and understanding which is required for projects on human behavior understanding. In our survey (refer Table 7.2), we found that some studies had used rule-based classifiers as well as decision trees – both of which offer a degree of interpretability. The choice between the two types of classification methods shall eventually depend on the goals of the research study.

7.5.4 Multimodal sensing systems for classroom behavior understanding

Although this chapter has focused solely on visual analytics for learning environments, it is generally recognized that combining data from several modalities, such as voice, physiological signals, and body movements, can aid in the understanding of human behavior [72]. As a result, we shall now argue for the use of multimodal sensing in a learning environment by citing the success of various modalities in disciplines of human behavior understanding.

Computer audition of voice, in terms of linguistic and paralinguistic analysis, has the potential to provide useful insights into cognitive and emotional aspects of human behavior [67, 73]. Such methods have already been shown to be useful for recognition of depression [74, 75], bipolar disorder [8], autism spectrum disorder [66], and Alzheimer's dementia [76]. We note that there has recently been an increased interest in using speech analytics for learning environments [77, 78] but there exist many unexplored research avenues. Computational paralinguistics, for example, can be used to develop an automated tool that detects public speaking anxiety. Such information can then be used to take affirmative and help students in dealing with their fears. Similarly, topic-modeling from natural language processing can be used alongside speech transcription tools to analyze the content of taught courses in an automated manner, whereas sentiment analysis can be used to measure student feedback from free-text forms without the need for human intervention. Thus, automated systems can be built that can discern subtle behavioral straits through acoustic sensing.

The utilization of physiological signals provides another method for assessing human emotions and moods. Changes in heart rate, blood pressure, electrodermal activity, electroencephalogram, and temperature are all controlled by the central nervous system [79, 80], and hence are influenced by human behavior. With the advancements in Internet of Things (IoT) [81], Health-IoT [82], and Wireless Body Area Networks [83, 84] the acquisition of physiological signals is relatively easy. Physiological signals have been used to evaluate cognitive load [85–87], stress, and anxiety [88, 89] in the past. Some studies have recently exploited the use of physiological signals in the context of academic learning [90–92], however, these have primarily focused on student engagement. One may imagine research opportunities to investigate how varying levels of concentration, classroom attention, or stress and anxiety during tests affect physiological signals and whether these changes can be quantified.

Recent years have witnessed great progress in the field of human activity recognition through wearable devices that incorporate inertial measurement unit (IMU) sensors [93] or smartphones [94] and using various AI algorithms to identify the activities [95, 96]. Devices such as the Empatica E4 band and Fitbit have been used to record body movement information in learning environments [92, 97], which is an encouraging change from visual sensing. We believe that this emerging trend has a lot of potential. This is because human activity can be recorded without recording any biometric information associated with a subject which alleviates privacy-related constraints that exist for visual sensing.

Authors' biographies

Zafi Sherhan Syed received his BEng.in Telecommunication Engineering from Mehran University, Pakistan, in 2008, MSc. in Communications, Control, and Digital Signal Processing from the University of Strathclyde, UK, and Ph.D. in Computer Science and Informatics from Cardiff University in 2019. He currently works as an Assistant Professor in the Department of Telecommunication Engineering at Mehran University, Pakistan. Dr. Syed's research interests include human behavior understanding, in particular the use of linguistic and paralinguistic qualities of speech for the development of automated screening methods. Dr. Syed is an active member of the IEEE Communications Society and has served multiple terms as Secretary of the IEEE Communications Society's Karachi Chapter.

Faisal K. Shaikh is working as a Professor at the Department of Telecommunication Engg., Mehran University of Engineering & Technology

(MUET), Jamshoro. He received Ph.D. in Computer Science from the Technische Universität Darmstadt, Germany. His research areas include the Internet of Things (IoT), Wireless Sensor Networks, Vehicular Adhoc Networks, Smart Homes and Cities, Body Area Networks, and Underwater Sensor Networks. He is the founder of the IoT Research Laboratory at MUET. He has published more than 125 refereed journals, conferences, and book chapters. He is a senior member of IEEE and a life member of PEC.

Muhammad Shehram Shah Syed received his bachelor's degree in Software Engineering from Mehran University, Pakistan in 2009, and his master's in Advanced Computer Science from the University of Strathclyde, UK. He is currently pursuing his doctoral studies at RMIT University, Australia. His research interests include affective computing and machine-learning.

Abbas Syed is an Assistant Professor in the Department of Electronics Engineering, Mehran University of Engineering and Technology, Pakistan. He received his BEng. in Electronic Engineering from Mehran University of Engineering and Technology, Pakistan as a silver medalist and an MSc in Electrical and Electronic Engineering from the University of Strathclyde, the UK with Distinction. He has been teaching subjects related to Digital Instrumentation, Embedded System Design in Undergraduate as well as Postgraduate programs at Mehran UET. His research interests are Embedded Systems and Machine-learning for health applications. He has publications in the mentioned areas in various conferences and journals. Currently, he is pursuing a Ph.D. in Computer Science and Engineering from the University of Louisville, USA.

Bibliography

[1] Rosalind W Picard. Affective Computing for HCI. In *ACM International on Human-Computer Interaction: Ergonomics and User Interfaces*, pages 829–833, 1999.

[2] Alessandro Vinciarelli, Maja Pantic, and Herve Bourlard. Social Signal Processing: Survey of an emerging domain. *Image and Vision Computing*, 27(12):1743–1759, 2009.

[3] Shrikanth Narayanan and Panayiotis G Georgiou. Behavioral Signal Processing: Deriving Human Behavioral Informatics From Speech and Language. *Proceedings of the IEEE*, 101(5):1203–1233, 2013.

[4] Shan Li and Weihong Deng. Deep facial expression recognition: A survey. *IEEE Transactions on Affective Computing*, pages 1–25, 2020.

[5] Xianye Ben, Yi Ren, Junping Zhang, Su-Jing Wang, Kidiyo Kpalma, Weixiao Meng, and Yong-Jin Liu. Video-based facial micro-expression analysis: A survey of datasets, features and algorithms. *IEEE Transactions on Pattern Analysis and Machine Intelligence*, 2021.

[6] Zafi Sherhan Shah, Kirill Sidorov, and David Marshall. Psychomotor Cues for Depression Screening. In *IEEE International Conference on Digital Signal Processing*, pages 1–5, 2017.

[7] Anastasia Pampouchidou, Panagiotis G Simos, Kostas Marias, Fabrice Meriaudeau, Fan Yang, Matthew Pediaditis, and Manolis Tsiknakis. Automatic assessment of depression based on visual cues: A systematic review. *IEEE Transactions on Affective Computing*, 10(4):445–470, 2017.

[8] Zafi Sherhan Syed, Kirill Sidorov, and David Marshall. Automated Screening for Bipolar Disorder from Audio/Visual Modalities. In *ACM International Workshop on Audio/Visual Emotion Challenge*, pages 39–45, 2018.

[9] Jasvinder Pal Singh, Sanjeev Jain, Sakshi Arora, and Uday Pratap Singh. Vision-based gait recognition: A survey. *IEEE Access*, 6:70497–70527, 2018.

[10] Navleen Kour, Sakshi Arora, and Others. Computer-vision based diagnosis of Parkinson's disease via gait: A survey. *IEEE Access*, 7:156620–156645, 2019.

[11] Kelly A. Rocca. Student participation in the college classroom: An extended multidisciplinary literature review. *Communication Education*, 59(2):185–213, 2010.

[12] Virginia P Richmond. Teacher nonverbal immediacy. *Communication for Teachers*, 65:82, 2002.

[13] Tory Parker, Olivia Hoopes, and Dennis Eggett. The Effect of Seat Location and Movement or Permanence on Student-Initiated Participation. *College Teaching*, 59(2):79–84, 2011.

[14] Nan Gao, Mohammad Saiedur Rahaman, Wei Shao, Kaixin Ji, and Flora D. Salim. Individual and Group-wise Classroom Seating Experience: Effects on Student Engagement in Different Courses. *arXiv:2112.12342*, 2021.

[15] Joseph F. Grafsgaard, Joseph B. Wiggins, Kristy Elizabeth Boyer, Eric N. Wiebe, and James C. Lester. Automatically recognizing facial expression: Predicting engagement and frustration. In *International Conference on Educational Data Mining*, pages 1–8, 2013.

[16] Marjorie Walsh Hesler. *An Investigation of Instructor Use of Space*. Purdue University, 1972.

[17] Zipora Shechtman and Judy Leichtentritt. Affective teaching: A method to enhance classroom management. *European Journal of Teacher Education*, 27(3):323–333, 2004.

[18] Wenhai Zhang and Jiamei Lu. The practice of affective teaching: A view from brain science. *International Journal of Psychological Studies*, 1(1):35, 2009.

[19] Bin Jiang, Qiang Ren, Fei Dai, Jian Xiong, Jie Yang, and Guan Gui. Multi-task cascaded convolutional neural networks for real-time dynamic face recognition method. In *International Conference in Communications, Signal Processing, and Systems*, pages 59–66. Springer, 2018.

[20] Ying-Li Tian, Takeo Kanade, JeffreyF. Cohn, and Jeffrey F. Tian, Ying-Li; Kanade, Takeo ; Cohn. Facial Expression Analysis. In *Handbook of Face Recognition*, pages 247–275. Springer New York, 2005.

[21] Paul Ekman. An argument for basic emotions. *Cognition & Emotion*, 6(3):169–200, 1992.

[22] Karan Ahuja, Dohyun Kim, Franceska Xhakaj, Virag Varga, Anne Xie, Stanley Zhang, Jay Eric Townsend, Chris Harrison, Amy Ogan, and Yuvraj Agarwal. EduSense: Practical Classroom Sensing at Scale. *ACM on Interactive, Mobile, Wearable and Ubiquitous Technologies*, 3(3):1–26, 2019.

[23] Patricia Goldberg, Ömer Sümer, Kathleen Stürmer, Wolfgang Wagner, Richard Göllner, Peter Gerjets, Enkelejda Kasneci, and Ulrich Trautwein. Attentive or Not? Toward a Machine Learning Approach to Assessing Students' Visible Engagement in Classroom Instruction. *Educational Psychology Review*, 33:27–49, 2021.

[24] Zhe Cao, Gines Hidalgo, Tomas Simon, Shih-En Wei, and Yaser Sheikh. OpenPose: Realtime multi-person 2D pose estimation using Part Affinity Fields. arXiv 2018. *arXiv:1812.08008*, 2018.

[25] Roberto Araya and Jorge Sossa-Rivera. Automatic Detection of Gaze and Body Orientation in Elementary School Classrooms. *Frontiers in Robotics and AI*, pages 1–11, 2021.

[26] Omid Mohamad Nezami, Mark Dras, Len Hamey, Deborah Richards, Stephen Wan, and Cécile Paris. Automatic Recognition of Student Engagement Using Deep Learning and Facial Expression. In *Lecture Notes in Computer Science (including subseries Lecture Notes in Artificial Intelligence and Lecture Notes in Bioinformatics)*, volume 11908 LNAI, pages 273–289, 2020.

[27] Navneet Dalal and Bill Triggs. Histograms of Oriented Gradients for Human Detection. In *IEEE Computer Society Conference on Computer Vision and Pattern Recognition*, pages 886–893, 2005.

[28] Pieter Vanneste, José Oramas, Thomas Verelst, Tinne Tuytelaars, Annelies Raes, Fien Depaepe, and Wim Van den Noortgate. Computer vision and human behaviour, emotion and cognition detection: A use case on student engagement. *Mathematics*, 9(3):1–20, 2021.

[29] Bo Jiang, Wei Xu, Chunlin Guo, Wenqi Liu, and Wenqing Cheng. A classroom concentration model based on computer vision. In *ACM Turing Celebration Conference-China*, pages 1–6, 2019.

[30] Bui Ngoc Anh, Ngo Tung Son, Phan Truong Lam, Phuong Le Chi, Nguyen Huu Tuan, Nguyen Cong Dat, Nguyen Huu Trung, Muhammad Umar Aftab, Tran Van Dinh, and Others. A computer-vision based application for student behavior monitoring in classroom. *Applied Sciences*, 9(22):4729, 2019.

[31] Abdulrahman Renawi, Fady Alnajjar, Medha Parambil, Zouheir Trabelsi, Munkhjargal Gochoo, Sumaya Khalid, and Omar Mubin. A simplified real-time camera-based attention assessment system for classrooms: pilot study. *Education and Information Technologies*, pages 1–18, 2021.

[32] Christos Sagonas, Epameinondas Antonakos, Georgios Tzimiropoulos, Stefanos Zafeiriou, and Maja Pantic. 300 Faces In-The-Wild Challenge: database and results. *Image and Vision Computing*, 47:3–18, 2016.

[33] Yiwen Zhang, Tao Zhu, Huansheng Ning, and Zhenyu Liu. Classroom student posture recognition based on an improved high-resolution network. *EURASIP Journal on Wireless Communications and Networking*, 2021(1):1–15, 2021.

[34] Joseph Redmon and Ali Farhadi. Yolov3: An incremental improvement. *arXiv:1804.02767*, 2018.

[35] Jie Hu, Li Shen, and Gang Sun. Squeeze-and-excitation networks. In *IEEE Conference on Computer Vision and Pattern Recognition*, pages 7132–7141, 2018.

[36] Ke Sun, Bin Xiao, Dong Liu, and Jingdong Wang. Deep high-resolution representation learning for human pose estimation. In *IEEE/CVF Conference on Computer Vision and Pattern Recognition*, pages 5693–5703, 2019.

[37] Yu Te Ku, Han Yen Yu, and Yi Chi Chou. A Classroom Atmosphere Management System for Analyzing Human Behaviors in Class Activities. In *International Conference on Artificial Intelligence in Information and Communication*, 2019.

[38] T. S. Ashwin and Ram Mohana Reddy Guddeti. Affective database for e-learning and classroom environments using Indian students' faces, hand gestures and body postures. *Future Generation Computer Systems*, 108:334–348, 2020.

[39] Lujie Karen Chen and David Gerritsen. Building Interpretable Descriptors for Student Posture Analysis in a Physical Classroom. In *International Conference on Artificial Intelligence in Education*, 2021.

[40] Kevin Delgado, Juan Manuel Origgi, Tania Hasanpoor, Hao Yu, Danielle Allessio, Ivon Arroyo, William Lee, Margrit Betke, Beverly Woolf, and Sarah Adel Bargal. Student Engagement Dataset. In *IEEE/CVF International Conference on Computer Vision Workshops*, pages 3621–3629, 2021.

[41] Kehan Chen. Sitting posture recognition based on OpenPose. In *IOP Conference Series: Materials Science and Engineering*, volume 677, page 32057. IOP Publishing, 2019.

[42] Feng-Cheng Lin, Huu-Huy Ngo, Chyi-Ren Dow, Ka-Hou Lam, and Hung Linh Le. Student Behavior Recognition System for the Classroom Environment Based on Skeleton Pose Estimation and Person Detection. *Sensors*, 21(16):5314, 2021.

[43] Taoufik Ben Abdallah, Islam Elleuch, and Radhouane Guermazi. Student Behavior Recognition in Classroom using Deep Transfer Learning with VGG-16. *Procedia Computer Science*, 192:951–960, 2021.

[44] Zhe Cao, Tomas Simon, Shih En Wei, and Yaser Sheikh. Realtime multi-person 2D pose estimation using part affinity fields. In *IEEE Computer Vision and Pattern Recognition (CVPR)*, pages 7291–7299, 2017.

[45] Ian J Goodfellow, Dumitru Erhan, Pierre Luc Carrier, Aaron Courville, Mehdi Mirza, Ben Hamner, Will Cukierski, Yichuan Tang, David Thaler, Dong-Hyun Lee, and Others. Challenges in representation learning: A report on three machine learning contests. In *International Conference on Neural Information Processing*, pages 117–124. Springer, 2013.

[46] T. S. Ashwin and Ram Mohana Reddy Guddeti. Impact of inquiry interventions on students in e-learning and classroom environments using affective computing framework. *User Modeling and User-Adapted Interaction*, 30:759–801, 2020.

[47] Joao Carreira and Andrew Zisserman. Quo vadis, action recognition? a new model and the kinetics dataset. In *IEEE Conference on Computer Vision and Pattern Recognition*, pages 6299–6308, 2017.

[48] Jiankang Deng, Jia Guo, Niannan Xue, and Stefanos Zafeiriou. ArcFace: Additive angular margin loss for deep face recognition. In *IEEE Computer Society Conference on Computer Vision and Pattern Recognition*, volume 2019-June, pages 4685–4694, 2019.

[49] Nataniel Ruiz, Eunji Chong, and James M. Rehg. Fine-grained head pose estimation without keypoints. In *IEEE Computer Society Conference on Computer Vision and Pattern Recognition Workshops*, volume 2018-June, pages 2155–2164, 2018.

[50] Nitesh V. Chawla, Kevin W. Bowyer, Lawrence O. Hall, and W. Philip Kegelmeyer. SMOTE: Synthetic minority over-sampling technique. *Journal of Artificial Intelligence Research*, 16:321–357, 2002.

[51] Chi Sun, Xipeng Qiu, Yige Xu, and Xuanjing Huang. How to Fine-Tune BERT for Text Classification? In *Lecture Notes in Computer Science (including subseries Lecture Notes in Artificial Intelligence and Lecture Notes in Bioinformatics)*, 2019.

[52] Gil Levi and Tal Hassner. Emotion recognition in the wild via convolutional neural networks and mapped binary patterns. In *ACM on International Conference on Multimodal Interaction*, pages 503–510, 2015.

[53] Philipp Werner, Frerk Saxen, and Ayoub Al-Hamadi. Landmark based head pose estimation benchmark and method. In *IEEE International Conference on Image Processing*, pages 3909–3913, 2017.

[54] Christian Szegedy, Vincent Vanhoucke, Sergey Ioffe, and Jon Shlens. Rethinking the Inception Architecture for Computer Vision. In *IEEE Computer Vision and Pattern Recognition (CVPR)*, pages 2818–2826, 2016.

[55] Tsun-Yi Yang, Yi-Ting Chen, Yen-Yu Lin, and Yung-Yu Chuang. Fsanet: Learning fine-grained structure aggregation for head pose estimation from a single image. In *IEEE/CVF Conference on Computer Vision and Pattern Recognition*, pages 1087–1096, 2019.

[56] Karen Simonyan and Andrew Zisserman. Very Deep Convolutional Networks for Large-Scale Image Recognition. In *International Conference on Learning Representations*, pages 1–14, 2015.

[57] Hartwig Adam Andrew G. Howard, Menglong Zhu, Bo Chen, Dmitry Kalenichenko, Weijun Wang, Tobias Weyand, Marco Andreetto. MobileNets: Efficient Convolutional Neural Networks for Mobile Vision Applications. *arXiv:1704.04861*, pages 1–9, 2017.

[58] François Chollet. Xception: Deep learning with depthwise separable convolutions. In *IEEE Conference on Computer Vision and Pattern Recognition*, volume 2017-Janua, pages 1800–1807, 2017.

[59] Christian Szegedy, Wei Liu, Yangqing Jia, Pierre Sermanet, Scott Reed, Dragomir Anguelov, Dumitru Erhan, Vincent Vanhoucke, and Andrew Rabinovich. Going deeper with convolutions. In *IEEE Computer Vision and Pattern Recognition*, pages 1–9, Boston, MA, USA, 2015.

[60] Gierad Laput, Karan Ahuja, Mayank Goel, and Chris Harrison. Ubicoustics: Plug-and-play acoustic activity recognition. In *ACM Symposium on User Interface Software and Technology*, pages 213–224, 2018.

[61] Timo Ojala, Matti Pietikainen, and Topi Maenpaa. Multiresolution grayscale and rotation invariant texture classification with local binary patterns. *IEEE Transactions on Pattern Analysis and Machine Intelligence*, 24(7):971–987, 2002.

[62] Anil K Jain and Farshid Farrokhnia. Unsupervised texture segmentation using Gabor filters. *Pattern Recognition*, 24(12):1167–1186, 1991.

[63] Wenchao Zhang, Shiguang Shan, Wen Gao, Xilin Chen, and Hongming Zhang. Local Gabor binary pattern histogram sequence (LGBPHS): a novel non-statistical model for face representation and recognition. In *IEEE International Conference on Computer Vision*, volume 1, pages 786–791. IEEE, 2005.

[64] Yu Zhang, Jack Jiang, and Douglas A. Rahn. Studying vocal fold vibrations in Parkinson's disease with a nonlinear model. *Chaos*, 15(3):1–11, 2005.

[65] Mirko Raca, Roland Tormey, and Pierre Dillenbourg. Sleepers' lag-study on motion and attention. In *International Conference on Learning Analytics and Knowledge*, pages 36–43, 2014.

[66] Riccardo Fusaroli, Anna Lambrechts, Dan Bang, Dermot M. Bowler, and Sebastian B. Gaigg. Is voice a marker for Autism spectrum disorder? A systematic review and meta-analysis. *Autism Research*, pages 1–50, 2016.

[67] Bjorn Schuller, Felix Weninger, Yue Zhang, Fabien Ringeval, Anton Batliner, Stefan Steidl, Florian Eyben, Erik Marchi, Alessandro Vinciarelli, Klaus Scherer, Mohamed Chetouani, and Marcello Mortillaro. Affective and behavioural computing: Lessons learnt from the First Computational Paralinguistics Challenge. *Computer Speech and Language*, 1(1):1–25, 2018.

[68] Yiling Li, Yi Lin, Hongwei Ding, and Chunbo Li. Speech databases for mental disorders: A systematic review. *General Psychiatry*, 32(3), 2019.

[69] Nan Gao, Mohammad Saiedur Rahaman, Wei Shao, and Flora D Salim. Investigating the Reliability of Self-report Survey in the Wild: The Quest for Ground Truth. *arXiv preprint arXiv:2107.00389*, 2021.

[70] Athanasios Voulodimos, Nikolaos Doulamis, Anastasios Doulamis, and Eftychios Protopapadakis. Deep learning for computer vision: A brief review. *Computational Intelligence and Neuroscience*, 2018, 2018.

[71] Kshitij Sharma, Zacharoula Papamitsiou, and Michail Giannakos. Building pipelines for educational data using AI and multimodal analytics: A "grey-box" approach. *British Journal of Educational Technology*, 50(6):3004–3031, 2019.

[72] Xavier Alameda-Pineda, Elisa Ricci, and Nicu Sebe. Multimodal behavior analysis in the wild: An introduction. In *Multimodal Behavior Analysis in the Wild*, pages 1–8. Elsevier, 2019.

[73] Zafi Sherhan Syed, Julien Schroeter, Kirill Sidorov, and David Marshall. Computational Paralinguistics: Automatic Assessment of Emotions, Mood, and Behavioural State from Acoustics of Speech. In *INTERSPEECH*, pages 511–515, 2018.

[74] Nicholas Cummins, Stefan Scherer, Jarek Krajewski, Sebastian Schnieder, Julien Epps, and Thomas F. Quatieri. A review of depression and suicide risk assessment using speech analysis. *Speech Communication*, 71:10–49, 2015.

[75] Zafi Sherhan Syed, Kirill Sidorov, and David Marshall. Depression Severity Prediction Based on Biomarkers of Psychomotor Retardation. In *ACM International Workshop on Audio/Visual Emotion Challenge (AVEC)*, pages 37–43, 2017.

[76] Zafi Sherhan Syed, Muhammad Shehram Shah Syed, Margaret Lech, and Elena Pirogova. Tackling the ADRESSO Challenge 2021: The MUET-RMIT System for Alzheimer's Dementia Recognition from Spontaneous Speech. In *INTERSPEECH*, pages 3815–3819, 2021.

[77] Elizabeth Dyer, Cynthia D'Angelo, Nigel Bosch, Stina Krist, and Joshua Rosenberg. Analyzing Learning with Speech Analytics and Computer Vision Methods: Technologies, Principles, and Ethics. In *International Conference of the Learning Sciences: The Interdisciplinarity of the Learning Sciences*, pages 2651–2653. International Society of the Learning Sciences (ISLS), 2020.

[78] Pipit Rahayu, Yenni Rozimela, and Jufrizal Jufrizal. Students' Public Speaking Assessment for Persuasive Speech. In *67th TEFLIN International Virtual Conference & the 9th ICOELT 2021*, pages 255–261. Atlantis Press, 2022.

[79] Jonghwa Kim and Elisabeth André. Emotion recognition based on physiological changes in music listening. *IEEE Transactions on Pattern Analysis and Machine Intelligence*, 30(12):2067–2083, 2008.

[80] Dale Purves, George J Augustine, David Fitzpatrick, Lawrence C Katz, Anthony-Samuel LaMantia, James O McNamara, and S Mark Williams. Physiological changes associated with emotion. In *Neuroscience*. Sunderland (MA): Sinauer Associates, 2nd edition, 2001.

[81] Faisal Karim Shaikh, Sherali Zeadally, and Ernesto Exposito. Enabling technologies for green internet of things. *IEEE Systems Journal*, 11(2):983–994, 2015.

[82] Zartasha Baloch, Faisal Karim Shaikh, and Mukhtiar Ali Unar. A context-aware data fusion approach for health-iot. *International Journal of Information Technology*, 10(3):241–245, 2018.

[83] Anum Talpur, Faisal Karim Shaikh, Natasha Baloch, Emad Felemban, Abdelmajid Khelil, and Muhammad Mahtab Alam. Validation of wired and wireless interconnected body sensor networks. *Sensors*, 19(17):3697, 2019.

[84] Tabassum Waheed, Faisal Karim, Sayeed Ghani, et al. Qos enhancement of aodv routing for mbans. *Wireless Personal Communications*, 116(2):1379–1406, 2021.

[85] Jianlong Zhou, Kun Yu, Fang Chen, Yang Wang, and Syed Z Arshad. Multimodal behavioral and physiological signals as indicators of cognitive load. In *The Handbook of Multimodal-Multisensor Interfaces: Signal Processing, Architectures, and Detection of Emotion and Cognition-Volume 2*, pages 287–329. 2018.

[86] Roger D Dias, Marco A Zenati, Ronald Stevens, Jennifer M Gabany, and Steven J Yule. Physiological synchronization and entropy as measures of team cognitive load. *Journal of Biomedical Informatics*, 96:103250, 2019.

[87] Rebecca L Charles and Jim Nixon. Measuring mental workload using physiological measures: A systematic review. *Applied Ergonomics*, 74:221–232, 2019.

[88] Philip Schmidt, Attila Reiss, Robert Duerichen, and Kristof Van Laerhoven. Introducing WeSAD, a multimodal dataset for wearable stress and affect detection. In *International Conference on Multimodal Interaction*, pages 400–408, 2018.

[89] Akshi Kumar, Kapil Sharma, and Aditi Sharma. Hierarchical deep neural network for mental stress state detection using IoT based biomarkers. *Pattern Recognition Letters*, 145:81–87, 2021.

[90] Elena Di Lascio, Shkurta Gashi, and Silvia Santini. Unobtrusive Assessment of Students' Emotional Engagement during Lectures Using Electrodermal Activity Sensors. *ACM Conference on Interactive, Mobile, Wearable and Ubiquitous Technologies*, 2(3):1–21, 2018.

[91] Shkurta Gashi, Elena Di Lascio, and Silvia Santini. Using unobtrusive wearable sensors to measure the physiological synchrony between presenters and audience members. *Proceedings of the ACM on Interactive, Mobile, Wearable and Ubiquitous Technologies*, 3(1):1–19, 2019.

[92] Nan Gao, Wei Shao, Mohammad Saiedur Rahaman, and Flora D. Salim. N-Gage: Predicting in-class Emotional, Behavioural and Cognitive Engagement in the Wild. *Proceedings of the ACM on Interactive, Mobile, Wearable and Ubiquitous Technologies*, 4(3):1–26, 2020.

[93] E Ramanujam, Thinagaran Perumal, and S Padmavathi. Human activity recognition with smartphone and wearable sensors using deep learning techniques: A review. *IEEE Sensors Journal*, 21(12):13029–13040, 2021.

[94] Mehak Fatima Qureshi, Rizwan Ahmed Kango, Nafeesa Zaki, and Faisal Karim Shaikh. Activity monitoring of the potential covid'19 individuals in quarantine facility. In *IECON 2021–47th Annual Conference of the IEEE Industrial Electronics Society*, pages 1–6. IEEE, 2021.

[95] Zartasha Baloch, Faisal Karim Shaikh, and Mukhtiar Ali Unar. Deep architectures for human activity recognition using sensors. *3C Tecnologia*, 2019.

[96] Mohsin Bilal, Faisal K Shaikh, Muhammad Arif, and Mudasser F Wyne. A revised framework of machine learning application for optimal activity recognition. *Cluster Computing*, 22(3):7257–7273, 2019.

[97] Michail N Giannakos, Kshitij Sharma, Sofia Papavlasopoulou, Ilias O Pappas, and Vassilis Kostakos. Fitbit for learning: Towards capturing the learning experience using wearable sensing. *International Journal of Human-Computer Studies*, 136:102384, 2020.

8

Noise-Estimation-Based Fuzzy C-Means Clustering for Image Segmentation

Cong Wang

School of Artificial Intelligence, OPtics and ElectroNics (iOPEN), Northwestern Polytechnical University, Xi'an, China and also with the Research & Development Institute of Northwestern Polytechnical University in Shenzhen, Shenzhen, China

MengChu Zhou

Institute of Systems Engineering, Macau University of Science and Technology, Macau, China and also with the Helen and John C. Hartmann Department of Electrical and Computer Engineering, New Jersey Institute of Technology, Newark, NJ, USA

Witold Pedrycz

Department of Electrical and Computer Engineering, University of Alberta, Edmonton, AB, Canada, the School of Electro-Mechanical Engineering, Xidian University, Xi'an, China, and also with the Faculty of Engineering, King Abdulaziz University, Jeddah, Saudi Arabia

Zhiwu Li

School of Electro-Mechanical Engineering, Xidian University, Xi'an, China, and also with the Institute of Systems Engineering, Macau University of Science and Technology, Macau, China

CONTENTS

8.1	Introduction	....	252
8.2	Literature Review	....	256
	8.2.1	FCM with spatial information	256
	8.2.2	FCM with kernel distance	256
	8.2.3	Comprehensive FCM	257
8.3	Conventional Fuzzy C-Means (FCM)		257
8.4	Residual-Driven Fuzzy C-Means Framework		258
8.5	RFCM with Weighted ℓ_2-Norm Regularization		261
	8.5.1	Mixed noise model	261
	8.5.2	Analysis of mixed noise distribution	262
	8.5.3	Segmentation model	263

DOI: 10.1201/9781003053262-8

8.5.4   Minimization algorithm 266
8.5.5   Convergence and robustness analysis 269
8.6   Comparison to Deviation-Sparse Fuzzy *C*-Means 271
8.6.1   Deviation-sparse FCM (DSFCM) 271
8.6.2   Proof of conclusion 1 272
8.6.3   Proof of conclusion 2 276
8.7   Experimental Studies 279
8.7.0.1   Evaluation indicators 279
8.7.1   Dataset descriptions 279
8.7.2   Parameter settings 280
8.7.3   Comparison with FCM-related methods 281
8.7.3.1   Results on synthetic images 281
8.7.3.2   Results on medical images 283
8.7.3.3   Results on real-world images 284
8.7.3.4   Performance improvement 288
8.7.3.5   Overhead analysis 289
8.7.4   Comparison with non-FCM methods 290
8.8   Conclusions and Future Work 292

8.1 Introduction

As an important approach to data analysis and processing, fuzzy clustering has been widely applied to a number of visible domains such as pattern recognition [1,2], data mining [3], granular computing [4], and image processing [5]. One of the most popular fuzzy clustering methods is a Fuzzy *C*-Means (FCM) algorithm [6–9]. It plays a significant role in image segmentation; yet it only works well for noise-free images. In real-world applications, images are often contaminated by different types of noises, especially mixed or unknown noises, produced in the process of image acquisition and transmission. Therefore, to make FCM robust to noise, FCM is refined resulting in many modified versions in two main means, i.e., introducing spatial information into its objective function [10–15] and substituting its Euclidean distance with a kernel distance (function) [16–23]. Even though such versions improve its robustness to some extent, they often fail to account for high computing overhead of clustering. To balance the effectiveness and efficiency of clustering, researchers have recently attempted to develop FCM with the aid of mathematical technologies such as Kullback-Leibler divergence [24,25], sparse regularization [26,27], morphological reconstruction [28–30] and gray level histograms [31,32], as well as pre-processing and post-processing steps like image pixel filtering [33], membership filtering [31], and label filtering [27,33,34]. In brief, such existing studies make evident efforts to improve FCM's robustness to noise by the aid of noise removal in each iteration or before and after clustering. Such noise removal shows an impact on the final attribution of a partition matrix of FCM. However, in most cases, noise removal causes the loss of effective image information. Furthermore, all the algorithms mentioned above directly use observed image data rather than their ideal values (noise-free image data) estimated from them in clustering. As a result, their segmentation results are not fully reliable and can be further improved.

Generally speaking, noise can be modeled as the residual between an observed image and its ideal value (noise-free image). Intuitively, using a noise-free image instead of its observed values (noisy image) as clustering data can greatly enhance FCM's segmentation performance. Superior to the direct use of noise removal, integrating noise estimation into FCM is an effective way to improve FCM's robustness, thus resulting in noise-estimation-based FCM (NFCM), where a regularization term is considered as a part of FCM. In this sense, NFCM can not only effectively enhance FCM's segmentation performance, but also lead to FCM's theoretical development.

Accurate noise estimation is a very challenging task. To the best of our knowledge, there is only one attempt [35] to improve FCM by revealing the sparsity of the residual. To be specific, in [35], noise sparsity in an observed image is revealed under an asumption that most of image pixels have small or near zero noise/outliers. In the sequel, ℓ_1-norm regularization can be used to characterize the sparsity of the residual, thus forming deviation-sparse FCM (DSFCM). When spatial information is used, it upgrades to its augmented version, named as DSFCM_N. From the optimal solution to DSFCM, we see that a soft thresholding operation is used to realize noise estimation. In essence, DSFCM works well only when impulse noise is involved. Except impulse noise, the distribution of most types of noise does not exhibit such sparsity. It makes more sense that noise or outliers are sparse in some transformed domain. Such transforms can be Fourier transform, orthonormal wavelets [36], translation-invariant wavelets [37], framelets [38–41], etc. Therefore, due to the inappropriate assumption, DSFCM cannot realize accurate noise estimation and recover the ideal values from observed image data precisely when coping with many other noise types than impulse noise.

Motivated by [35], to address a wide range of noise types, we elaborate a residual-driven FCM (RFCM) framework as shown in Figure 8.1, which furthers FCM's performance. It is modeled by introducing a regularization term on residual as a part of the objective function of FCM. This term makes residual accurately estimated. It is determined by a noise distribution, e.g., an ℓ_2-norm regularization term corresponds to Gaussian noise and an ℓ_1-norm one suits impulse noise. Thus, RFCM achieves more accurate residual estimation than DSFCM and DSFCM_N that were introduced [35] since it is modeled by an analysis of noise distribution characteristic replacing noise sparsity. Their technical cores are illustrated in Figure 8.2.

In real-world applications, since images are often corrupted by mixed or unknown noises, a specific noise distribution is difficult to be obtained. To deal with this issue, by analyzing the distribution of a wide range of mixed noises, especially a mixture of Poisson, Gaussian and impulse noises, we present a

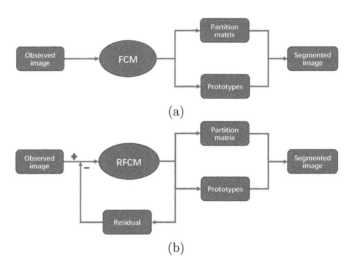

FIGURE 8.1
A comparison between the frameworks of FCM and RFCM. (a) FCM and (b)
RFCM.

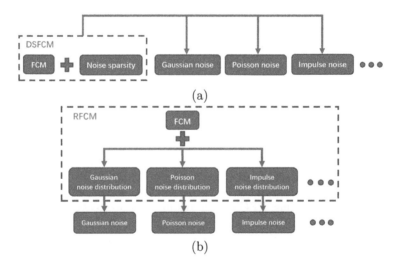

FIGURE 8.2
Comparison of technical cores of DSFCM and RFCM. (a) DSFCM and (b)
RFCM.

weighted ℓ_2-norm regularization term in which each residual is assigned a weight, thus resulting in an augmented version namely WRFCM for image segmentation with mixed or unknown noises. To obtain better noise suppression, we also consider spatial information of image pixels in WRFCM since it is naturally encountered in image segmentation. In addition, we design a two-step iterative algorithm to minimize the objective function of WRFCM. The first step is to employ the Lagrangian multiplier method to optimize the partition matrix, prototypes and residual when fixing the assigned weights. The second step is to update the weights by using the calculated residual. Finally, based on the optimal partition matrix and prototypes, a segmented image is obtained.

This work makes threefold contributions to advance FCM for image segmentation:

1) For the first time, an RFCM framework is proposed for image segmentation by introducing a regularization term derived from a noise distribution into FCM. It relies on accurate residual estimation to greatly improve FCM's performance, which is absent from existing FCMs.

2) Built on the RFCM framework, WRFCM is presented by weighting mixed noise distribution and incorporating spatial information. The use of spatial information makes resulting residual estimation more reliable. It is regarded as a universal RFCM algorithm for coping with mixed or unknown noises.

3) It makes a thorough analysis of the performance of RFCM and DSFCM, thus resulting in two conclusions: 1. RFCM realizes more accurate noise estimation than DSFCM; and 2. DSFCM is a particular case of RFCM. Specifically, when only impulse noise is involved in an image, DSFCM is the same as RFCM.

The originality of this work comes with a realization of accurate residual estimation from observed images, which benefits FCM's performance enhancement. In essence, the proposed algorithm is an unsupervised method. Compared with commonly used supervised methods such as convolutional neural networks (CNNs) [42–46] and dictionary learning [47,48], it realizes the residual estimation precisely by virtue of a regularization term rather than using any image samples to train a residual estimation model. Hence, it needs low computing overhead and can be experimentally executed by using a low-end CPU rather than a high-end GPU, which means that its practicality is high. In addition, it is free of the aid of mathematical techniques and achieves the superior performance over some recently proposed comprehensive FCMs. Therefore, we conclude that WRFCM is a fast and robust FCM algorithm. Finally, its minimization problem involves an ℓ_2 vector norm only. Thus it can be easily solved by using a well-known Lagrangian multiplier method.

Section 8.2 reviews the state of the art relevant to this work. Section 8.3 briefly introduces conventional FCM. Section 8.4 formulates an RFCM framework. Section 8.5 details WRFCM. Section 8.6 makes a comparison between

RFCM and DSFCM. Section 8.7 reports experimental results. Conclusions and some open issues are given in Section 8.8.

8.2   Literature Review

In 1984, Bezdek et al. [8] first proposed FCM. So far, it has evolved into the most popular fuzzy clustering algorithm. However, it cannot work well for segmenting observed (noisy) images. It has been improved by mostly considering spatial information [10–15], kernel distances (functions) [16–23], and various mathematical techniques [24–34]. In this chapter, we mainly focus on the improvement of FCM with regard to its robustness to noise for image segmentation. Therefore, we introduce related work about it in this section.

8.2.1   FCM with spatial information

Over the past two decades, using spatial information to improve FCM's robustness achieved remarkable successes, thus resulting in many improved versions [10–15]. For instance, Ahmed et al. [10] introduce a neighbor term into the objective function of FCM so as to improve its robustness by leaps and bounds, thus yielding FCM_S where S refers to "spatial information". To further improve it, Chen and Zhang [11] integrate mean and median filters into a neighbor term, thus resulting in two FCM_S variants labeled as FCM_S1 and FCM_S2. However, their computing overhead is very high. To lower it, Szilagyi et al. [12] propose an enhanced FCM (EnFCM) where a weighted sum image is generated by the observed pixels and their neighborhoods. Based on it, Cai et al. [13] substitute image pixels by gray level histograms, which gives rise to fast generalized FCM (FGFCM). Although it has a high computational efficiency, more parameters are required and tuned. Krinidis et al. [14] come up with a fuzzy local information C-means algorithm (FLICM) for simplifying the parameter setting in FGFCM. Nevertheless, FLICM considers only non-robust Euclidean distance that is not applicable to arbitrary spatial information.

8.2.2   FCM with kernel distance

To address the serious shortcoming of FLICM [14], kernel distances (functions) are used to replace Euclidean distance in FCM. They realize the transformation from an original data space to a new one. As a result, a collection of kernel-based FCMs have been put forward [16–23]. For example, Gong et al. [16] propose an improved version of FLICM, namely KWFLICM, which augments a tradeoff weighted fuzzy factor and a kernel metric into FCM. Even though it is generally robust to extensive noise, it is more time-consuming than most of existing FCMs. Zhao et al. [21] take a neighborhood weighted distance into account, thus presenting a novel algorithm called NWFCM.

Although it runs faster than KWFLICM, its segmentation performance is worse. Moreover, it exhibits lower computational efficiency than other FCMs. More recently, Wang et al. [23] consider tight wavelet frames as a kernel function so as to present wavelet frame-based FCM (WFCM), which takes full advantage of the feature extraction capacity of tight wavelet frames. In spite of its rarely low computational cost, its segmentation effects can be further improved by using various mathematical techniques.

8.2.3 Comprehensive FCM

To keep a sound trade-off between performance and speed of clustering, comprehensive FCMs involving various mathematical techniques has been put forward [24–34]. For instance, Gharieb et al. [24] present an FCM framework based on Kullback-Leibler (KL) divergence. It uses KL divergence to optimize the membership similarity between a pixel and its neighbors. Yet it has slow clustering speed. Gu et al. [26] report a fuzzy double C-Means algorithm (FDCM) through the utility of sparse representation, which addresses two datasets simultaneously, i.e., a basic feature set associated with an observed image and a feature set learned from a spare self-representation model. Overall, FDCM is robust and applicable to a wide range of image segmentation problems. However, its computational efficiency is not satisfactory. Lei et al. [31] present a fast and robust FCM algorithm (FRFCM) by using gray level histograms and morphological gray reconstruction. In spite of its fast clustering, its performance is sometimes unstable since morphological gray reconstruction may cause the loss of useful image features. More recently, Lei et al. [32] propose an automatic fuzzy clustering framework (AFCF) by incorporating threefold techniques, i.e., superpixel algorithms, density peak clustering, and prior entropy. It overcomes two difficulties in existing algorithms [23, 26, 31]. One is to select the number of clusters automatically. The other one is to employ superpixel algorithms and the prior entropy to improve image segmentation performance. However, AFCF's results are unstable.

In this work, the proposed algorithm differs from all algorithms mentioned above in the sense that we take a wide range of noise distribution characteristics used for noise estimation as a starting point and directly minimize the objective function of RFCM formulated by using multiple regularization without dictionary learning and CNNs, and archives outstanding performance in image segmentation tasks.

8.3 Conventional Fuzzy C-Means (FCM)

For a domain $\Omega = \{1, 2, \cdots, K\}$, an observed image defined on Ω is denoted as $\boldsymbol{X} = \{\boldsymbol{x}_i \in \mathbb{R}^L : i \in \Omega\}$, where $\boldsymbol{x}_i$ has L channels, i.e., $\boldsymbol{x}_i = \{x_{il} : l = 1, 2, \cdots, L\}$. To be specific, $\boldsymbol{X}$ is a gray image for $L = 1$ while it is a

Red-Green-Blue color image for $L = 3$. Since there exists a noise perturbation in the process of image acquisition and transmission, $\boldsymbol{X}$ can be modeled as:

$$\boldsymbol{X} = \widetilde{\boldsymbol{X}} + \boldsymbol{R} \tag{8.1}$$

where $\widetilde{\boldsymbol{X}} = \{\tilde{\boldsymbol{x}}_i \in \mathbb{R}^L : i \in \Omega\}$ denotes an unknown noise-free image and $\boldsymbol{R} = \{\boldsymbol{r}_i \in \mathbb{R}^L : i \in \Omega\}$ is a noise perturbation, e.g., Gaussian noise, impulse noise, and mixed Gaussian and impulse noise. Formally speaking, $\widetilde{\boldsymbol{X}}$ is an ideal value of $\boldsymbol{X}$ and $\boldsymbol{R}$ is viewed as a residual (difference) between $\boldsymbol{X}$ and $\widetilde{\boldsymbol{X}}$.

FCM is often used to divide an observed image into several nonoverlapping clusters. Given is an observed image $\boldsymbol{X}$, FCM is used to segment it by minimizing the objective function:

$$J^{\mathrm{FCM}}(\boldsymbol{U}, \boldsymbol{V}) = \sum_{i=1}^{K} \sum_{j=1}^{c} u_{ij}^m \|\boldsymbol{x}_i - \boldsymbol{v}_j\|^2 \tag{8.2}$$

where $\boldsymbol{U} = [u_{ij}]_{K \times c}$ is a partition matrix with constraints $\sum_{j=1}^{c} u_{ij} = 1$ for $\forall i$ and $0 < \sum_{i=1}^{K} u_{ij} < K$ for $\forall j$, $\boldsymbol{V} = \{\boldsymbol{v}_j \in \mathbb{R}^L : j \in \{1, 2, \cdots, c\}\}$ denotes a prototype set, $\|\cdot\|$ represents a Euclidean distance, and m is a fuzzification exponent. m assumes values greater than 1.

An alternating iteration scheme [8] is used to minimize Equation 8.1. Each iteration is realized as follows:

$$u_{ij}^{(t+1)} = \frac{(\|\boldsymbol{x}_i - \boldsymbol{v}_j^{(t)}\|^2)^{-1/(m-1)}}{\sum_{q=1}^{c} (\|\boldsymbol{x}_i - \boldsymbol{v}_q^{(t)}\|^2)^{-1/(m-1)}}$$

$$\boldsymbol{v}_j^{(t+1)} = \frac{\sum_{i=1}^{K} \left(u_{ij}^{(t+1)}\right)^m \boldsymbol{x}_i}{\sum_{i=1}^{K} \left(u_{ij}^{(t+1)}\right)^m}$$

Here, $t = 0, 1, 2, \cdots$ is an iterative step and $l = 1, 2, \cdots, L$. By presetting a threshold ε, the procedure stops when $\|\boldsymbol{U}^{(t+1)} - \boldsymbol{U}^{(t)}\| < \varepsilon$.

8.4 Residual-Driven Fuzzy C-Means Framework

Due to the presence of a noise perturbation, the direct use of an observed image gives rise to bad segmentation results. Intuitively, using its noise-free image instead of it as data to be clustered benefits FCM's robustness improvement.

Therefore, an ideal variant of FCM's objective function Equation 8.2 for image segmentation is expressed:

$$J^{\text{FCM}}(\boldsymbol{U}, \boldsymbol{V}) = \sum_{i=1}^{K} \sum_{j=1}^{c} u_{ij}^{m} \|\tilde{\boldsymbol{x}}_i - \boldsymbol{v}_j\|^2 \qquad (8.3)$$

By Equation 8.1, we can transform Equation 8.3 into:

$$J^{\text{FCM}}(\boldsymbol{U}, \boldsymbol{V}, \boldsymbol{R}) = \sum_{i=1}^{K} \sum_{j=1}^{c} u_{ij}^{m} \|\boldsymbol{x}_i - \boldsymbol{r}_i - \boldsymbol{v}_j\|^2 \qquad (8.4)$$

To make Equation 8.4 executed, a main task is to realize accurate estimation of $\boldsymbol{R}$. Therefore, it is reasonable to impose a regularization term $\Gamma(\boldsymbol{R})$ to Equation 8.4, which characterizes some properties of $\boldsymbol{R}$. As a result, residual-driven FCM (RFCM) is proposed. The augmented objective function is defined:

$$J^{\text{RFCM}}(\boldsymbol{U}, \boldsymbol{V}, \boldsymbol{R}) = \sum_{i=1}^{K} \sum_{j=1}^{c} u_{ij}^{m} \|\boldsymbol{x}_i - \boldsymbol{r}_i - \boldsymbol{v}_j\|^2 + \boldsymbol{\beta} \cdot \Gamma(\boldsymbol{R})$$

$$= \sum_{i=1}^{K} \sum_{j=1}^{c} u_{ij}^{m} \|\boldsymbol{x}_i - \boldsymbol{r}_i - \boldsymbol{v}_j\|^2 + \sum_{l=1}^{L} \beta_l \Gamma(\boldsymbol{R}_l) \qquad (8.5)$$

In Equation 8.5, $\boldsymbol{R}$ is rewritten as $\{\boldsymbol{R}_l \in \mathbb{R}^K : l = 1, 2, \cdots, L\}$ with $\boldsymbol{R}_l = (r_{1l}, r_{2l}, \cdots, r_{Kl})^{\text{T}}$ and $\boldsymbol{\beta} = (\beta_1, \cdots, \beta_l, \cdots \beta_L)$ is a positive parameter vector that is used to strike a balance between $\Gamma(\boldsymbol{R}) = (\Gamma(\boldsymbol{R}_1), \cdots, \Gamma(\boldsymbol{R}_l), \cdots, \Gamma(\boldsymbol{R}_L))^{\text{T}}$ and J^{FCM} in Equation 8.4. The regularization term $\Gamma(\boldsymbol{R})$ guarantees that the solution accords with the degradation process of minimizing Equation 8.5. As Equation 8.5 indicates, the main task involved in RFCM is to determine the expression of $\Gamma(\boldsymbol{R}_l)$ based on an analysis of the property of l-th residual channel $\boldsymbol{R}_l$.

By Equation 8.1, a noise-free image $\boldsymbol{X}$ and the residual $\boldsymbol{R}$ can be converted to each other. Motivated by a variational model for image restoration [49], a regularization term $\Gamma(\boldsymbol{R}_l)$ on $\boldsymbol{R}_l$ can be equivalently converted to a fidelity term $\Psi(\widetilde{\boldsymbol{X}_l})$ on $\widetilde{\boldsymbol{X}_l}$, i.e.,

$$\Gamma(\boldsymbol{R}_l) \Leftrightarrow \Psi(\widetilde{\boldsymbol{X}_l}) \text{ for } l = 1, 2, \cdots, L \qquad (8.6)$$

The fidelity term $\Psi(\widetilde{\boldsymbol{X}_l})$ usually penalizes the discrepancy between $\widetilde{\boldsymbol{X}_l}$ and $\boldsymbol{X}_l$. Its formulation leans on an assumed noise distribution. Since the residual $\boldsymbol{R}_l$ stands for such a discrepancy, it is easy to understand that the regularization term $\Gamma(\boldsymbol{R}_l)$ is associated with the assumed noise distribution. From Equation 8.6, we see that once $\Psi(\widetilde{\boldsymbol{X}_l})$ is given, the corresponding regularization model $\Gamma(\boldsymbol{R}_l)$ is determined. Therefore, before formulating $\Gamma(\boldsymbol{R}_l)$, a key step is to derive $\Psi(\widetilde{\boldsymbol{X}_l})$ by using a maximum a posteriori (MAP) method and a likelihood function. The details are formulated as follows.

The MAP estimate of $\widetilde{\boldsymbol{X}}_l$ can be found by maximizing the conditional posterior probability $\mathbf{P}(\widetilde{\boldsymbol{X}}_l|\boldsymbol{X}_l)$, i.e., the probability that $\widetilde{\boldsymbol{X}}_l$ occurs when $\boldsymbol{X}_l$ is observed. Based on the Bayes' theorem, we have:

$$\mathbf{P}(\widetilde{\boldsymbol{X}}_l|\boldsymbol{X}_l) = \frac{\mathbf{P}(\widetilde{\boldsymbol{X}}_l) \cdot \mathbf{P}(\boldsymbol{X}_l|\widetilde{\boldsymbol{X}}_l)}{\mathbf{P}(\boldsymbol{X}_l)} \tag{8.7}$$

By taking the negative logarithm of Equation 8.7 and ignoring $\mathbf{P}(\widetilde{\boldsymbol{X}}_l)$ unassociated with $\boldsymbol{R}$ and a constant $\mathbf{P}(\boldsymbol{X}_l)$, the estimate is acquired by solving the following optimization problem:

$$\max_{\widetilde{\boldsymbol{X}}_l} \log \mathbf{P}(\boldsymbol{X}_l|\widetilde{\boldsymbol{X}}_l) \Leftrightarrow \min_{\widetilde{\boldsymbol{X}}_l} - \log \mathbf{P}(\boldsymbol{X}_l|\widetilde{\boldsymbol{X}}_l) \tag{8.8}$$

The likelihood $\mathbf{P}(\boldsymbol{X}_l|\widetilde{\boldsymbol{X}}_l)$ is used to yield a fidelity term on $\widetilde{\boldsymbol{X}}_l$, which measures the discrepancy between $\widetilde{\boldsymbol{X}}_l$ and $\boldsymbol{X}_l$. Its choice depends on specific noise characteristics.

To make the determination process of $\mathbf{P}(\boldsymbol{X}_l|\widetilde{\boldsymbol{X}}_l)$ easily understood, we take the presence of additive white Gaussian noise as a case study. In the sequel, we derive the expression form of $\Psi(\widetilde{\boldsymbol{X}}_l)$ and $\Gamma(\boldsymbol{R}_l)$. Then, RFCM for Gaussian noise is presented. The derivation procedure is detailed as follows.

Example 8.4.1. Consider an observed image $\boldsymbol{X}$ and the presence of additive white Gaussian noise. For $l = 1, 2, \cdots, L$, $\boldsymbol{R}_l$ obeys an independent normal distribution $\mathbf{N}$ of zero mean and variance σ_l^2, i.e.,

$$\boldsymbol{R}_l \sim \mathbf{N}(0, \sigma_l^2) \tag{8.9}$$

Thus, we arrive at the likelihood:

$$\mathbf{P}(\boldsymbol{X}_l|\widetilde{\boldsymbol{X}}_l) = \mathbf{P}(\boldsymbol{R}_l|\widetilde{\boldsymbol{X}}_l) \simeq \exp\left(-\frac{1}{2\sigma_l^2}(\boldsymbol{X}_l - \widetilde{\boldsymbol{X}}_l)^T(\boldsymbol{X}_l - \widetilde{\boldsymbol{X}}_l)\right) \tag{8.10}$$

Using maximum likelihood, we take the negative logarithm of the normal distribution Equation 8.10, i.e.,

$$-\log \mathbf{P}(\boldsymbol{R}_l|\widetilde{\boldsymbol{X}}_l) \propto \frac{1}{2\sigma_l^2}(\boldsymbol{X}_l - \widetilde{\boldsymbol{X}}_l)^T(\boldsymbol{X}_l - \widetilde{\boldsymbol{X}}_l) \propto \frac{1}{2\sigma_l^2}\|\boldsymbol{X}_l - \widetilde{\boldsymbol{X}}_l\|_{\ell_2}^2 \tag{8.11}$$

By ignoring the coefficient in Equation 8.11, we obtain a fidelity term on $\widetilde{\boldsymbol{X}}_l$:

$$\Psi(\widetilde{\boldsymbol{X}}_l) = \|\boldsymbol{X}_l - \widetilde{\boldsymbol{X}}_l\|_{\ell_2}^2 = \sum_{i=1}^{K} |x_{il} - \tilde{x}_{il}|^2$$

By Equation 8.1 and Equation 8.6, the regularization term on $\boldsymbol{R}_l$ is expressed as

$$\Gamma(\boldsymbol{R}_l) = \|\boldsymbol{R}_l\|_{\ell_2}^2 = \sum_{i=1}^{K} |r_{il}|^2 \tag{8.12}$$

TABLE 8.1

Data regularization models.

Data Regularization Functions	Noise Types
$\Gamma(\boldsymbol{R}_l) = \|\boldsymbol{R}_l\|_{\ell_2}^2$	Gaussian noise [49, 50]
$\Gamma(\boldsymbol{R}_l) = \|\boldsymbol{R}_l\|_{\ell_0}$	impulse noise [51, 52]
$\Gamma(\boldsymbol{R}_l) = \|\boldsymbol{R}_l\|_{\ell_\infty}$	uniform noise [53, 54]
$\Gamma(\boldsymbol{R}_l) = \langle -\boldsymbol{R}_l \circ \log(\boldsymbol{X}_l - \boldsymbol{R}_l), 1 \rangle$	Poisson noise [55, 56]
$\Gamma(\boldsymbol{R}_l) = \langle \log(\boldsymbol{X}_l - \boldsymbol{R}_l) + \boldsymbol{X}_l \circ \frac{1}{\boldsymbol{X}_l - \boldsymbol{R}_l}, 1 \rangle$	Gamma noise [57, 58]
$\Gamma(\boldsymbol{R}_l) = \langle \log(\boldsymbol{X}_l - \boldsymbol{R}_l) + \boldsymbol{X}_l \circ \boldsymbol{X}_l \circ \frac{1}{2\boldsymbol{X}_l - 2\boldsymbol{R}_l}, 1 \rangle$	Rayleigh noise [59, 60]

By substituting Equation 8.12 into Equation 8.5, RFCM's objective function associated with Gaussian noise is formulated:

$$J^{\mathrm{RFCM}}(\boldsymbol{U}, \boldsymbol{V}, \boldsymbol{R}) = \sum_{i=1}^{K} \sum_{j=1}^{c} u_{ij}^m \|\boldsymbol{x}_i - \boldsymbol{r}_i - \boldsymbol{v}_j\|^2 + \sum_{l=1}^{L} \beta_l \sum_{i=1}^{K} |r_{il}|^2 \qquad (8.13)$$

where β_l is associated with σ_l.

By following the above case, we derive the regularization terms associated with typical noise types, as summarized in Table 8.1. Note that $\circ$ performs element-by-element multiplication. Obviously, RFCM can realize accurate noise estimations in presence of different types of noise.

8.5 RFCM with Weighted ℓ_2-Norm Regularization

8.5.1 Mixed noise model

The accurate estimation of $\boldsymbol{R}$ can make $\widetilde{\boldsymbol{X}}$ instead of $\boldsymbol{X}$ participate in clustering so as to improve FCM's robustness. Hence, it is a necessary step to formulate a noise model before constructing an FCM model. In image processing, the models of single noise such as Gaussian, Poisson, and impulse noise are widely used. In this work, in order to construct robust FCM, we consider mixed or unknown noise since it is often encountered in real-world applications. Its specific model is unfortunately hard to be formulated. Therefore, a common solution is to assume the type of mixed noise in advance. In universal image processing, two kinds of mixed noises are the most common, namely mixed Poisson-Gaussian noise and mixed Gaussian and impulse noise. Beyond them, we focus on a mixture of a wide range of noises, i.e., a mixture of Poisson, Gaussian, and impulse noise. We investigate an FCM-related model based on the analysis of the mixed noise model and extend it to image segmentation with mixed or unknown noises.

Formally speaking, a noise-free image $\widetilde{\boldsymbol{X}}$ is defined in a domain $\Omega = \{1, 2, \cdots, K\}$. It is first corrupted by Poisson noise, thus resulting in $\overline{\boldsymbol{X}} = \{\bar{\boldsymbol{x}}_1, \bar{\boldsymbol{x}}_2, \cdots, \bar{\boldsymbol{x}}_K\}$ that obeys a Poisson distribution, or, $\overline{\boldsymbol{X}} \sim \mathbf{P}(\widetilde{\boldsymbol{X}})$. Then additive zero-mean white Gaussian noise $\boldsymbol{R}' = \{\boldsymbol{r}'_1, \boldsymbol{r}'_2, \cdots, \boldsymbol{r}'_K\}$ with standard deviation σ is added. Finally, impulse noise $\boldsymbol{R}'' = \{\boldsymbol{r}''_j, \boldsymbol{r}''_2, \cdots, \boldsymbol{r}''_K\}$ with a given probability $\zeta \in (0, 1)$ is imposed. Hence, for $i \in \Omega$, an arbitrary element in observed image $\boldsymbol{X}$ is expressed as

$$\boldsymbol{x}_i = \begin{cases} \bar{\boldsymbol{x}}_i + \boldsymbol{r}'_i & i \in \Omega_1 \\ \boldsymbol{r}''_i & i \in \Omega_2 := \Omega \backslash \Omega_1 \end{cases} \tag{8.14}$$

where Ω_1 is called an observable region including mixed Poisson-Gaussian noise and Ω_2 denotes the region consisting of the missing information of $\overline{\boldsymbol{X}} + \boldsymbol{R}'$ and is assumed to be unknown with each element being drawn from the whole region Ω by Bernoulli trial with ζ. In image segmentation, mixed noise model Equation 8.14 is for the first time presented.

8.5.2 Analysis of mixed noise distribution

For a common single noise, i.e., Gaussian, Poisson, and impulse noise, the regularization terms in Table 8.1 lead to an MAP solution to such noise estimations. In real-world applications, images are generally contaminated by mixed or unknown noises rather than a single noise. The regularization terms for single noise estimation become inapplicable since the distribution of mixed or unknown noise is difficult to be modeled mathematically. Therefore, one of the main purposes of this work is to design a universal regularization term for mixed or unknown noise estimation.

To reveal the essence of mixed noise distributions, we here consider generic and representative mixed noise, i.e., a mixture of Poisson, Gaussian, and impulse noise. Let us take an example to exhibit its distribution. Here, we impose Gaussian noise ($\sigma = 10$) and a mixture of Poisson, Gaussian ($\sigma = 10$) and random-valued impulse noise ($\zeta = 20\%$) on image "Lena" with size 512×512, respectively. We show original and observed images in Figure 8.3.

As Figure 8.3(b) shows, Gaussian noise is overall orderly. As a common sense, Poisson distribution is a Gaussian-like one under the condition of enough samples. Therefore, due to impulse noise, mixed noise is disorderly and unsystematic as shown in Figure 8.3(c). In Figure 8.4, we portray the distributions of Gaussian and mixed noise, respectively.

Figure 8.4(a) shows noise distribution in a linear domain. To illustrate a heavy tail intuitively, we present it in a logarithmic domain as shown in Figure 8.4(b). Clearly, Poisson noise leads to a Gaussian-like distribution. Nevertheless, impulse noise gives rise to a more irregular distribution with a heavy tail. Therefore, neither ℓ_1 norm nor ℓ_2 norm can precisely characterize the residual $\boldsymbol{R}$ in the sense of the MAP estimation.

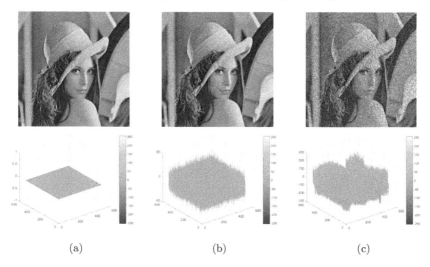

(a)   (b)   (c)

FIGURE 8.3
Noise-free image and two observed ones corrupted by Gaussian and mixed noise, respectively. The first row: (a) noise-free image; (b) observed image with Gaussian noise; and (c) observed image with mixed noise. The second row portrays noise included in three images.

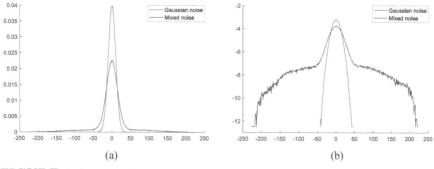

(a)   (b)

FIGURE 8.4
Distributions of Gaussian and mixed noise in different domains. (a) linear domain and (b) logarithmic domain.

8.5.3 Segmentation model

Intuitively, if the regularization term can be modified so as to make mixed noise distribution more Gaussian-like, we can still use ℓ_2 norm to characterize residual $\boldsymbol{R}$. It means that mixed noise can be more accurately estimated. Therefore, we adopt robust estimation techniques [61,62] to weaken the heavy tail, which makes mixed noise distribution more regular. In the sequel, we assign a proper weight w_{il} to each residual r_{il}, which forms a weighted residual $w_{il}r_{il}$ that almost obeys a Gaussian distribution. Given Figure 8.5, we use an example for showing the effect of weighting.

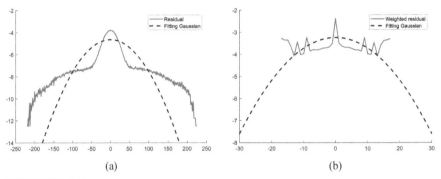

(a)   (b)

FIGURE 8.5
Distributions of residual r_{il} and weighted residual $w_{il}r_{il}$, as well as the fitting Gaussian function in the logarithmic domain.

Figure 8.5(a) shows the distribution of r_{il} and the fitting Gaussian function based on the variance of r_{il}. Figure 8.5(b) gives the distribution of $w_{il}r_{il}$ and the fitting Gaussian function based on the variance of $w_{il}r_{il}$. Clearly, the distribution of $w_{il}r_{il}$ in Figure 8.5(b) is more Gaussian-like than that in Figure 8.5(a), which means that ℓ_2-norm regularization can work on weighted residual $w_{il}r_{il}$ for an MAP-like solution of $\boldsymbol{R}$.

By analyzing Figure 8.5, for $l = 1, 2, \cdots, L$, we propose a weighted ℓ_2-norm regularization term for mixed or unknown noise estimation:

$$\Gamma(\boldsymbol{R}_l) = \|\boldsymbol{W}_l \circ \boldsymbol{R}_l\|_{\ell_2}^2 = \sum_{j=1}^{K} |w_{il}r_{il}|^2 \qquad (8.15)$$

where $\circ$ performs element-by-element multiplication of $\boldsymbol{R}_l = (r_{1l}, r_{2l}, \cdots, r_{Kl})^{\mathrm{T}}$ and $\boldsymbol{W}_l = (w_{1l}, w_{2l}, \cdots, w_{Kl})^{\mathrm{T}}$. For $l = 1, 2, \cdots, L$, $\boldsymbol{W}_l$ makes up a weight matrix $\boldsymbol{W} = [w_{il}]_{K \times L}$. Each element w_{il} is assigned to location (i, l). Since it is inversely proportional to residual r_{il}, it can be automatically determined. In this work, we adopt the following expression:

$$w_{il} = e^{-\xi r_{il}^2} \qquad (8.16)$$

where ξ is a positive parameter, which aims to control the decreasing rate of w_{il}.

By substituting Equation 8.15 into Equation 8.5, we present RFCM with weighted ℓ_2-norm regularization (WRFCM) for image segmentation:

$$
\begin{aligned}
J^{\mathrm{WRFCM}}(\boldsymbol{U}, \boldsymbol{V}, \boldsymbol{R}, \boldsymbol{W}) &= \sum_{i=1}^{K}\sum_{j=1}^{c} u_{ij}^{m} \|\boldsymbol{x}_i - \boldsymbol{r}_i - \boldsymbol{v}_j\|^2 + \sum_{l=1}^{L} \beta_l \|\boldsymbol{W}_l \circ \boldsymbol{R}_l\|_{\ell_2}^2 \\
&= \sum_{i=1}^{K}\sum_{j=1}^{c} u_{ij}^{m} \|\boldsymbol{x}_i - \boldsymbol{r}_i - \boldsymbol{v}_j\|^2 + \sum_{l=1}^{L} \beta_l \sum_{i=1}^{K} |w_{il} r_{il}|^2
\end{aligned}
\tag{8.17}
$$

In addition, the use of spatial information is beneficial to improve FCM's robustness. If the distance between an image pixel and its neighbors is small, there exists a large possibility that they belong to the same cluster. To further improve segmentation performance, we introduce spatial information into the objective function of FCM. Prior to performing the modified objective function, we need to complete an analysis of spatial information. To make spatial information easily understood, refer to Figure 8.6.

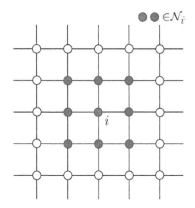

FIGURE 8.6
Illustration of spatial information of pixel i.

In a concise way, a pixel or feature is sometimes loosely represented by its corresponding index while this is not ambiguous. In Figure 8.6, we show an arbitrary pixel i and its spatial information with a local window. $\mathcal{N}_i$ stands for a local window centralized in i including j. We let $|\mathcal{N}_i|$ be the cardinality of $\mathcal{N}_i$, which represents the size of a local window $\mathcal{N}_i$. In Figure 8.6, we have $|\mathcal{N}_i| = 3 \times 3$, which represents that $\mathcal{N}_i$ contains pixel i and its eight neighbors.

According to Equation 8.17, u_{ij} depends on the distance $\|\boldsymbol{x}_i - \boldsymbol{r}_i - \boldsymbol{v}_j\|$. Thus, it is optimized by considering spatial information of $\boldsymbol{x}_i$ and $\boldsymbol{r}_i$. Motivated by [14], we express the spatial information of $\boldsymbol{x}_i$ and $\boldsymbol{r}_i$ as

$$
\sum_{n \in \mathcal{N}_i} \frac{\|\boldsymbol{x}_n - \boldsymbol{r}_n - \boldsymbol{v}_j\|^2}{1 + d_{ni}} \qquad \sum_{n \in \mathcal{N}_i} \frac{|w_{nl} r_{nl}|^2}{1 + d_{ni}}
\tag{8.18}
$$

where n is a neighbor pixel of i, d_{ni} represents the Euclidean distance between n and i, and the factor $1/(d_{ni}+1)$ reflects the spatial structure information. As a result, by substituting Equation 8.18 into Equation 8.17, the modified objective function can be defined as

$$J^{\text{WRFCM}}(\boldsymbol{U},\boldsymbol{V},\boldsymbol{R},\boldsymbol{W}) = \sum_{i=1}^{K}\sum_{j=1}^{c} u_{ij}^{m} \sum_{n\in\mathcal{N}_i} \frac{\|\boldsymbol{x}_n - \boldsymbol{r}_n - \boldsymbol{v}_j\|^2}{1+d_{ni}}$$
$$+ \sum_{l=1}^{L}\beta_l \sum_{i=1}^{K}\sum_{n\in\mathcal{N}_i} \frac{|w_{nl}r_{nl}|^2}{1+d_{ni}} \tag{8.19}$$

Its minimization is completed subject to

$$\sum_{j=1}^{c} u_{ij} = 1, \quad \forall i \in \{1,2,\cdots,K\}$$

In Equation 8.19, an image pixel is sometimes loosely represented by its corresponding index even though this is not ambiguous. Thus, n is a neighbor pixel of i and d_{ni} represents the Euclidean distance between n and i. $\mathcal{N}_i$ stands for a local window centralized at i including i and its size is denoted as $|\mathcal{N}_i|$.

8.5.4 Minimization algorithm

Minimizing Equation 8.19 involves four unknowns, i.e., $\boldsymbol{U}$, $\boldsymbol{V}$, $\boldsymbol{R}$ and $\boldsymbol{W}$. According to Equation 8.16, $\boldsymbol{W}$ is automatically determined by $\boldsymbol{R}$. Hence, we can design a two-step iterative algorithm to minimize Equation 8.19, which fixes $\boldsymbol{W}$ first to solve $\boldsymbol{U}$, $\boldsymbol{V}$ and $\boldsymbol{R}$, then uses $\boldsymbol{R}$ to update $\boldsymbol{W}$. The main task in each iteration is to solve the minimization problem in terms of $\boldsymbol{U}$, $\boldsymbol{V}$ and $\boldsymbol{R}$ when fixing $\boldsymbol{W}$. Assume that $\boldsymbol{W}$ is given. We can apply a Lagrangian multiplier method to minimize Equation 8.19. The Lagrangian function is expressed as

$$\mathcal{L}_\Lambda(\boldsymbol{U},\boldsymbol{V},\boldsymbol{R};\boldsymbol{W}) = \sum_{i=1}^{K}\sum_{j=1}^{c} u_{ij}^{m} \left(\sum_{n\in\mathcal{N}_i} \frac{\|\boldsymbol{x}_n - \boldsymbol{r}_n - \boldsymbol{v}_j\|^2}{1+d_{ni}}\right)$$
$$+ \sum_{l=1}^{L}\beta_l \sum_{i=1}^{K}\sum_{n\in\mathcal{N}_i} \frac{|w_{nl}r_{nl}|^2}{1+d_{ni}}$$
$$+ \sum_{i=1}^{K}\lambda_i \left(\sum_{j=1}^{c} u_{ij} - 1\right) \tag{8.20}$$

where $\Lambda = \{\lambda_i : i=1,2,\cdots,K\}$ is a set of Lagrangian multipliers. The two-step iterative algorithm for minimizing Equation 8.19 is realized in Algorithm 1.

The minimization problem Equation 8.22 can be divided into the following three subproblems:

$$\begin{cases} \boldsymbol{U}^{(t+1)} = \arg\min_{\boldsymbol{U}}\mathcal{L}_\Lambda(\boldsymbol{U},\boldsymbol{V}^{(t)},\boldsymbol{R}^{(t)};\boldsymbol{W}^{(t)}) \\ \boldsymbol{V}^{(t+1)} = \arg\min_{\boldsymbol{V}}\mathcal{L}_\Lambda(\boldsymbol{U}^{(t+1)},\boldsymbol{V},\boldsymbol{R}^{(t)};\boldsymbol{W}^{(t)}) \\ \boldsymbol{R}^{(t+1)} = \arg\min_{\boldsymbol{R}}\mathcal{L}_\Lambda(\boldsymbol{U}^{(t+1)},\boldsymbol{V}^{(t+1)},\boldsymbol{R};\boldsymbol{W}^{(t)}) \end{cases} \tag{8.21}$$

Algorithm 1 Two-step iterative algorithm

Given a threshold ε, input $\boldsymbol{W}^{(0)}$. For $t = 0, 1, \cdots$, iterate:
  Step 1: Find minimizers $\boldsymbol{U}^{(t+1)}$, $\boldsymbol{V}^{(t+1)}$, and $\boldsymbol{R}^{(t+1)}$:

$$\left(\boldsymbol{U}^{(t+1)}, \boldsymbol{V}^{(t+1)}, \boldsymbol{R}^{(t+1)}\right) = \arg\min_{U,V,R} \mathcal{L}_\Lambda(\boldsymbol{U}, \boldsymbol{V}, \boldsymbol{R}; \boldsymbol{W}^{(t)}) \qquad (8.22)$$

  Step 2: Update the weight matrix $\boldsymbol{W}^{(t+1)}$
If $\|\boldsymbol{U}^{(t+1)} - \boldsymbol{U}^{(t)}\| < \varepsilon$, stop; else update t such that
$$0 \leq t \uparrow < +\infty$$

Each subproblem in Equation 8.21 has a closed-form solution. We use an alternative optimization scheme similar to the one used in FCM to optimize $\boldsymbol{U}$ and $\boldsymbol{V}$. The following result is needed to obtain the iterative updates of $\boldsymbol{U}$ and $\boldsymbol{V}$.

Theorem 1. *Consider the first two subproblems of Equation 8.21. By applying the Lagrangian multiplier method to solve them, the iterative solutions are presented as*

$$u_{ij}^{(t+1)} = \frac{\left(\sum_{n \in \mathcal{N}_i} \frac{\|\boldsymbol{x}_n - \boldsymbol{r}_n^{(t)} - \boldsymbol{v}_j^{(t)}\|^2}{1 + d_{ni}}\right)^{-1/(m-1)}}{\sum_{q=1}^{c}\left(\sum_{n \in \mathcal{N}_i} \frac{\|\boldsymbol{x}_n - \boldsymbol{r}_n^{(t)} - \boldsymbol{v}_q^{(t)}\|^2}{1 + d_{ni}}\right)^{-1/(m-1)}} \qquad (8.23)$$

$$\boldsymbol{v}_j^{(t+1)} = \frac{\sum_{i=1}^{K}\left(\left(u_{ij}^{(t+1)}\right)^m \sum_{n \in \mathcal{N}_i} \frac{\boldsymbol{x}_n - \boldsymbol{r}_n^{(t)}}{1 + d_{ni}}\right)}{\sum_{i=1}^{K}\left(\left(u_{ij}^{(t+1)}\right)^m \sum_{n \in \mathcal{N}_i} \frac{1}{1 + d_{ni}}\right)} \qquad (8.24)$$

Proof. Consider the first two subproblems of Equation 8.21. The Lagrangian function Equation 8.20 is reformulated as

$$\mathcal{L}_\Lambda(\boldsymbol{U}, \boldsymbol{V}) = \sum_{i=1}^{K}\sum_{j=1}^{c} u_{ij}^m D_{ij} + \sum_{i=1}^{K} \lambda_i \left(\sum_{j=1}^{c} u_{ij} - 1\right) \qquad (8.25)$$

where $D_{ij} = \sum_{n \in \mathcal{N}_i} \frac{\|\boldsymbol{x}_n - \boldsymbol{r}_n - \boldsymbol{v}_j\|^2}{1 + d_{ni}}$.

By fixing $\boldsymbol{V}$, we minimize Equation 8.25 in terms of $\boldsymbol{U}$. By zeroing the gradient of Equation 8.25 in terms of $\boldsymbol{U}$, one has

$$\frac{\partial \mathcal{L}_\Lambda}{\partial u_{ij}} = m D_{ij} u_{ij}^{m-1} + \lambda_i = 0$$

Thus, u_{ij} is expressed as

$$u_{ij} = \left(\frac{-\lambda_i}{m}\right)^{1/(m-1)} D_{ij}^{-1/(m-1)} \qquad (8.26)$$

Due to the constraint $\sum_{j=1}^{c} u_{ij} = 1$, one has

$$
\begin{aligned}
1 = \sum_{q=1}^{c} u_{iq} \;\; &= \sum_{q=1}^{c}\left(\left(\tfrac{-\lambda_i}{m}\right)^{1/(m-1)} D_{iq}^{-1/(m-1)}\right)\\[2mm]
&= \left(\tfrac{-\lambda_i}{m}\right)^{1/(m-1)} \sum_{q=1}^{c} D_{iq}^{-1/(m-1)}
\end{aligned}
$$

In the sequel, one can get

$$
\left(\frac{-\lambda_i}{m}\right)^{1/(m-1)} = 1 \Big/ \sum_{q=1}^{c} D_{iq}^{-1/(m-1)} \tag{8.27}
$$

Substituting Equation 8.27 into Equation 8.26, the optimal u_{ij} is acquired:

$$
u_{ij} = \frac{D_{ij}^{-1/(m-1)}}{\sum_{q=1}^{c} D_{iq}^{-1/(m-1)}}
$$

By fixing $\mathbf{U}$, we minimize Equation 8.25 in terms of $\mathbf{V}$. By zeroing the gradient of Equation 8.25 in terms of $\mathbf{V}$, one has

$$
\frac{\partial \mathcal{L}_\Lambda}{\partial \boldsymbol{v}_j} - -2 \cdot \sum_{i=1}^{K} \left(u_{ij}^m \sum_{n \in \mathcal{N}_i} \frac{(\boldsymbol{x}_n - \boldsymbol{r}_n - \boldsymbol{v}_j)}{1 + d_{ni}} \right) = 0
$$

The intermediate process is presented as

$$
\sum_{i=1}^{K} u_{ij}^m \left(\sum_{n \in \mathcal{N}_i} \frac{(\boldsymbol{x}_n - \boldsymbol{r}_n)}{1 + d_{ni}} \right) = \sum_{i=1}^{K} u_{ij}^m \left(\sum_{n \in \mathcal{N}_i} \frac{\boldsymbol{v}_j}{1 + d_{ni}} \right)
$$

The optimal $\boldsymbol{v}_j$ is computed:

$$
\boldsymbol{v}_j = \frac{\displaystyle\sum_{i=1}^{K} \left(u_{ij}^m \sum_{n \in \mathcal{N}_i} \frac{\boldsymbol{x}_n - \boldsymbol{r}_n}{1 + d_{ni}} \right)}{\displaystyle\sum_{i=1}^{K} \left(u_{ij}^m \sum_{n \in \mathcal{N}_i} \frac{1}{1 + d_{ni}} \right)}
$$

$\square$

In the last subproblem of Equation 8.21, both $\boldsymbol{r}_i$ and $\boldsymbol{r}_n$ appear simultaneously. Since $\boldsymbol{r}_n$ is dependent on $\boldsymbol{r}_i$, it should not be considered as a constant vector. In other words, n is one of neighbors of i while i is one of neighbors of n symmetrically. Thus, $n \in \mathcal{N}_i$ is equivalent to $i \in \mathcal{N}_n$. Then we have:

$$
\sum_{i=1}^{K} u_{ij}^m \left(f(\boldsymbol{r}_i) + \sum_{\substack{n \in \mathcal{N}_i \\ n \neq i}} f(\boldsymbol{r}_n) \right) = \sum_{i=1}^{K} \sum_{n \in \mathcal{N}_i} u_{nj}^m (f(\boldsymbol{r}_i)) \tag{8.28}
$$

where f represents a function in terms of r_i or r_n. By Equation 8.28, we rewrite Equation 8.19 as

$$
J^{\mathrm{WRFCM}}(\boldsymbol{U},\boldsymbol{V},\boldsymbol{R},\boldsymbol{W}) = \sum_{i=1}^{K}\sum_{j=1}^{c}\sum_{n\in\mathcal{N}_i} \frac{u_{nj}^m \|\boldsymbol{x}_i - \boldsymbol{r}_i - \boldsymbol{v}_j\|^2}{1+d_{ni}}
$$
$$
+ \sum_{l=1}^{L}\beta_l \sum_{i=1}^{K}\sum_{n\in\mathcal{N}_i} \frac{|w_{il}r_{il}|^2}{1+d_{ni}} \tag{8.29}
$$

According to the two-step iterative algorithm, we assume that $\boldsymbol{W}$ in Equation 8.29 is fixed in advance. When $\boldsymbol{U}$ and $\boldsymbol{V}$ are updated, the last subproblem of Equation 8.21 is separable and can be decomposed into $K \times L$ subproblems:

$$
r_{il}^{(t+1)} = \arg\min_{r_{il}} \sum_{j=1}^{c}\left(\sum_{n\in\mathcal{N}_i} \frac{\left(u_{nj}^{(t+1)}\right)^m \|x_{il}-r_{il}-v_{jl}^{(t+1)}\|^2}{1+d_{ni}} \right)
$$
$$
+ \sum_{n\in\mathcal{N}_i} \frac{\beta_l |w_{il}^{(t)} r_{il}|^2}{1+d_{ni}} \tag{8.30}
$$

By zeroing the gradient of the energy function in Equation 8.30 in terms of r_{il}, the iterative solution to Equation 8.30 is expressed as

$$
r_{il}^{(t+1)} = \frac{\displaystyle\sum_{j=1}^{c}\sum_{n\in\mathcal{N}_i} \frac{\left(u_{nj}^{(t+1)}\right)^m \left(x_{il}-v_{jl}^{(t+1)}\right)}{1+d_{ni}}}{\displaystyle\sum_{j=1}^{c}\sum_{n\in\mathcal{N}_i}\frac{\left(u_{nj}^{(t+1)}\right)^m}{1+d_{ni}} + \sum_{n\in\mathcal{N}_i}\frac{\beta_l\left(w_{il}^{(t)}\right)^2}{1+d_{ni}}} \tag{8.31}
$$

Example 8.5.1. Considering a noise-free image shown in Figure 8.7(a), we impose a mixture of Poisson, Gaussian, and impulse noise ($\sigma = 30$, $\zeta = 20\%$) on it. We set c to 4. The settings of ξ and $\boldsymbol{\beta}$ are discussed in the later section. As shown in Figure 8.7, the noise estimation of WRFCM in Figure 8.7(f) is close to the true one in Figure 8.7(e). In addition, it has sound performance for noise-suppression and feature-preserving, which can be visually observed from Figure 8.7(c).

Algorithm 1 is terminated when $\|\boldsymbol{U}^{(t+1)} - \boldsymbol{U}^{(t)}\| < \varepsilon$. Based on optimal $\boldsymbol{U}$ and $\boldsymbol{V}$, a segmented image $\widehat{\boldsymbol{X}}$ is obtained. WRFCM for minimizing Equation 8.19 is realized in Algorithm 2.

8.5.5 Convergence and robustness analysis

In WRFCM, we set $\|\boldsymbol{U}^{(t+1)} - \boldsymbol{U}^{(t)}\| < \varepsilon$ as the termination condition. In order to analyze the convergence and robustness of WRFCM, we take Figure 8.7 as a case study. We set $\varepsilon = 1 \times 10^{-8}$. For convergence analysis, we draw the curves of $\theta = \|\boldsymbol{U}^{(t+1)} - \boldsymbol{U}^{(t)}\|$ and J versus iteration step t, respectively. For robustness analysis, we draw the curve of $\tau = \|\boldsymbol{R}^{t+1} - \widehat{\boldsymbol{R}}\|$ versus iteration step t, where $\widehat{\boldsymbol{R}}$ represents the measured residual (noise) reserved in Figure 8.7(b). The results are presented in Figure 8.8.

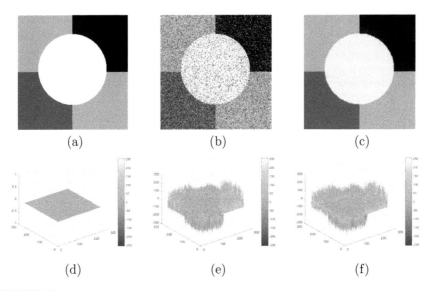

(a)   (b)   (c)

(d)   (e)   (f)

FIGURE 8.7
Noise estimation of WRFCM. (a) noise-free image; (b) observed image; (c) segmented image of WRFCM; (d) noise in the noise-free image; (e) noise in the observed image; (f) noise estimation of WRFCM.

As Figure 8.8(a) indicates, since the prototypes are randomly initialized, the convergence of WRFCM oscillates slightly at the beginning. Nevertheless, it reaches steady state after a few iterations. Even though θ exhibits an oscillating process, the objective function value J keeps decreasing until the iteration stops. Affected by the oscillating process of θ, τ takes on a pattern of increasing and then decreasing and eventually stabilizes. The finding indicates that WRFCM can obtain optimal residual estimation as t increases. To sum up, WRFCM has outstanding convergence and robustness since the weight ℓ_2-norm regularization makes mixed noise distribution estimated accurately

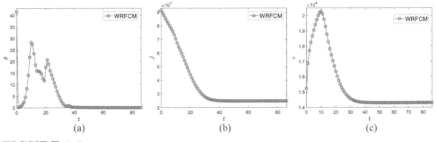

(a)   (b)   (c)

FIGURE 8.8
Convergence and robustness of WRFCM. (a) θ; (b) J; and (c) τ versus t.

Algorithm 2 Residual-driven FCM with weighted ℓ_2-norm regularization (WRFCM)

Input: Observed image $\boldsymbol{X}$, fuzzification exponent m, number of clusters c, and threshold ε.
Output: Segmented image $\widehat{\boldsymbol{X}}$.
  1: Initialize $\boldsymbol{W}^{(0)}$ as a matrix with all elements of 1 and generate randomly prototypes $\boldsymbol{V}^{(0)}$
  2: $t \leftarrow 0$
  3: **repeat**
  4:   Calculate the partition matrix $\boldsymbol{U}^{(t+1)}$ via (8.23)
  5:   Calculate the prototypes $\boldsymbol{V}^{(t+1)}$ via (8.24)
  6:   Calculate the residual $\boldsymbol{R}^{(t+1)}$ via (8.31)
  7:   Update the weight matrix $\boldsymbol{W}^{(t+1)}$ via (8.16)
  8:   $t \leftarrow t + 1$
  9: **until** $\|\boldsymbol{U}^{(t+1)} - \boldsymbol{U}^{(t)}\| < \varepsilon$
 10: **return** $\boldsymbol{U}$, $\boldsymbol{V}$, $\boldsymbol{R}$ and $\boldsymbol{W}$
 11: Generate the segmented image $\widehat{\boldsymbol{X}}$ based on $\boldsymbol{U}$ and $\boldsymbol{V}$

---

so that the residual is gradually separated from observed data as iterations proceed.

8.6 Comparison to Deviation-Sparse Fuzzy C-Means

8.6.1 Deviation-sparse FCM (DSFCM)

In [35], an achievable NFCM version is proposed for image segmentation, which is known as deviation-sparse FCM (DSFCM). It is modeled based on an assumption that most of pixels in an observed image $\boldsymbol{X}$ have small or near zero noise/outliers. As a result, the residual $\boldsymbol{R}$ shows strong sparsity. This study focuses on characterizing noise sparsity by using an ℓ_1-norm regularization term on residual:

$$\Gamma(\boldsymbol{R}_l) = \|\boldsymbol{R}_l\|_{\ell_1} = \sum_{i=1}^{K} |r_{il}| \qquad (8.32)$$

where $\|\cdot\|_{\ell_1}$ denotes an ℓ_1 vector norm and $|\cdot|$ denotes an absolute value operation. By substituting Equation 8.32 into Equation 8.5, DSFCM's objective function is expressed:

$$J^{\text{DSFCM}}(\boldsymbol{U}, \boldsymbol{V}, \boldsymbol{R}) = \sum_{i=1}^{K} \sum_{j=1}^{c} u_{ij}^m \|\boldsymbol{x}_i - \boldsymbol{r}_i - \boldsymbol{v}_j\|^2 + \sum_{l=1}^{L} \beta_l \sum_{i=1}^{K} |r_{il}| \qquad (8.33)$$

For image segmentation, an arbitrary image pixel is close to its neighbors.

Therefore, neighbor or spatial information is usually introduced into FCM, which usually makes a critical impact on the final attribution of a partition matrix U. By combining neighbor information and DSFCM, an enhanced version, labeled as DSFCM_N, is given:

$$
J^{\text{DSFCM- N}}(U, V, R) = \sum_{i=1}^{K} \sum_{j=1}^{c} u_{ij}^{m} \left(\sum_{n \in \mathcal{N}_i} \frac{\|x_n - r_n - v_j\|^2}{1 + d_{ni}} \right)
$$
$$
+ \sum_{l=1}^{L} \beta_l \sum_{i=1}^{K} \sum_{n \in \mathcal{N}_i} \frac{|r_{nl}|}{1 + d_{ni}} \tag{8.34}
$$

In this chapter, we make a thorough analysis of the performance of RFCM and DSFCM, thus resulting in two conclusions:

1. RFCM realizes more accurate noise estimation than DSFCM.

2. DSFCM is a particular case of RFCM. Specifically, when only impulse noise is involved in an image, DSFCM is the same as RFCM.

In this following, we provide the proof of the above conclusions.

8.6.2   Proof of conclusion 1

To convincingly demonstrate that RFCM can realize more accurate noise estimation than DSFCM, we need to determine that the residual results produced by RFCM are closer to true noise than DSFCM's. Here, we take the presence of additive white Gaussian noise as a case study. Formally speaking, the objective functions of DSFCM and RFCM have been previously presented as Equation 8.33 and Equation 8.13. By applying the Lagrangian multiplier method to minimize Equation 8.33 and Equation 8.13, we can acquire the expressions of a partition matrix U, prototypes V and residuals R. RFCM and DSFCM have the same expressions of U and V, which can be found as Eqs. (15) and (16) in [35]. We here focus on their expressions of R, which are, respectively, presented as

$$
r_{il}^{\text{RFCM}} = \frac{\sum\limits_{j=1}^{c} u_{ij}^{m}(x_{il} - v_{jl})}{\sum\limits_{j=1}^{c} u_{ij}^{m} + \beta_l} \tag{8.35}
$$

$$
r_{il}^{\text{DSFCM}} = \frac{\mathcal{S}(\sum\limits_{j=1}^{c} u_{ij}^{m}(x_{il} - v_{jl}), \beta_l/2)}{\sum\limits_{j=1}^{c} u_{ij}^{m}} \tag{8.36}
$$

where $\mathcal{S}$ is a soft-thresholding operator [63–65] defined as

$$
\mathcal{S}(\theta, \delta) = \text{sign}(\theta) \cdot \max\{|\theta| - \delta, 0\}
$$

For convenience, we denote true Gaussian noise located at the ith pixel in the lth channel as $r_{il}^* = x_{il} - \tilde{x}_{il}$. For the lth pixel channel, we measure the discrepancies between true Gaussian noise and residual results of RFCM and DSFCM, respectively. We denote them as $\boldsymbol{\vartheta}_l^{\text{RFCM}} = \{\vartheta_{il}^{\text{RFCM}} : i \in \Omega\}$ and $\boldsymbol{\vartheta}_l^{\text{DSFCM}} = \{\vartheta_{il}^{\text{DSFCM}} : i \in \Omega\}$, respectively:

$$\vartheta_{il}^{\text{RFCM}} = |r_{il}^* - r_{il}^{\text{RFCM}}|, \quad \vartheta_{il}^{\text{DSFCM}} = |r_{il}^* - r_{il}^{\text{DSFCM}}|$$

If $\vartheta_{il}^{\text{DSFCM}} > \vartheta_{il}^{\text{RFCM}}$, r_{il}^{RFCM} is closer to r_{il}^* than r_{il}^{DSFCM}. We denote the discrepancy between $\boldsymbol{\vartheta}_l^{\text{DSFCM}}$ and $\boldsymbol{\vartheta}_l^{\text{RFCM}}$ as $\boldsymbol{\Theta}_l = \{\Theta_{il} : i \in \Omega\}$, which is defined as

$$\Theta_{il} = \vartheta_{il}^{\text{DSFCM}} - \vartheta_{il}^{\text{RFCM}}$$

If the expectation $\mathbf{E}(\boldsymbol{\Theta}_l) > 0$, the average residual results of RFCM are closer to true Gaussian noise than DSFCM's. Then, we can confirm that RFCM realizes more accurate noise estimations than DSFCM.

Since an alternating iteration scheme [8] is used to update u_{ij}, v_{jl} and r_{il}, for all cases $c \geq 2$, it is hard to illustrate $\mathbf{E}(\boldsymbol{\Theta}_l) > 0$. For the sake of a simple comparison, let us consider the case in terms of two clusters, i.e., $c = 2$. It is since in this case two membership degrees of an arbitrary pixel belonging to two clusters are significantly different, i.e., extremely approximate to 0 or 1. To do so, we provide Theorem 2 and its proof.

Theorem 2. *Given an observed image $\mathbf{X}$, assume that it is corrupted with Gaussian noise $\mathbf{R}$ with $\mathbf{R}_l \sim \mathbf{N}(0, \sigma_l^2)$. For $c = 2$, we have that the expectation $\mathbf{E}(\boldsymbol{\Theta}_l) > 0$ if and only if $\beta_l > 0.76\sigma_l$.*

Proof. For $c = 2$, there usually exist two deviations between v_{jl} and $\tilde{x}_{il}$, which are denoted as Δ_{1l} and Δ_{2l}, respectively. Thus, we have:

$$v_{1l} = \tilde{x}_{il} - \Delta_{1l}, \quad v_{2l} = \tilde{x}_{il} - \Delta_{2l}$$

Assume that x_{il} belongs to the first cluster. The known conditions contain: $u_{i1} \approx 1$, $u_{i2} \approx 0$, Δ_{1l} is small positive number or near 0 while Δ_{2l} is much larger than Δ_{1l}, and $\beta_l \gg \Delta_{1l}$. Thus, we have:

$$r_{il}^{\text{RFCM}} \approx \frac{x_{il} - v_{1l}}{1 + \beta_l} = \frac{r_{il}^* + \Delta_{1l}}{1 + \beta_l}$$

$$r_{il}^{\text{DSFCM}} \approx \text{sign}(r_{il}^* + \Delta_{1l}) \cdot \max\{|r_{il}^* + \Delta_{1l}| - \beta_l/2, 0\}$$

Since a Gaussian distribution is symmetric about $r_{il}^* = 0$, the proof process is similar in both cases, i.e., $r_{il}^* < 0$ and $r_{il}^* \geq 0$. Thus, we only focus on $r_{il}^* \geq 0$. Then, we have:

$$\vartheta_{il}^{\text{RFCM}} = \frac{\beta_l r_{il}^*}{1 + \beta_l} - \frac{\Delta_{1l}}{1 + \beta_l} \tag{8.37}$$

$$\vartheta_{il}^{\text{DSFCM}} = \begin{cases} r_{il}^*, & r_{il}^* < \frac{\beta_l}{2} - \Delta_{1l} \\ \frac{\beta_l}{2} - \Delta_{1l}, & r_{il}^* \geq \frac{\beta_l}{2} - \Delta_{1l} \end{cases} \tag{8.38}$$

For $r_{il}^* < \frac{\beta_l}{2} - \Delta_{1l}$, we have:

$$\Theta_{il} = r_{il}^* - \frac{\beta_l r_{il}^*}{1 + \beta_l} + \frac{\Delta_{1l}}{1 + \beta_l} \tag{8.39}$$

For $r_{il}^* \geq \frac{\beta_l}{2} - \Delta_{1l}$, we have:

$$\Theta_{il} = \frac{\beta_l}{2} - \Delta_{1l} - \frac{\beta_l r_{il}^*}{1 + \beta_l} + \frac{\Delta_{1l}}{1 + \beta_l} \tag{8.40}$$

As mentioned above, Δ_{1l} is small or near 0, and $\beta_l \gg \Delta_{1l}$. Let us consider two extreme conditions, i.e., $\Delta_{1l} \to 0$ and $\frac{\beta_l}{1+\beta_l} \to 1$. Thus, for $r_{il}^* < \frac{\beta_l}{2}$, Equation 8.39 is rewritten as

$$\Theta_{il} = 0 \tag{8.41}$$

For $r_{il}^* \geq \frac{\beta_l}{2}$, Equation 8.40 is rewritten as

$$\Theta_{il} = \frac{\beta_l}{2} - r_{il}^* \tag{8.42}$$

Given a random variable $\boldsymbol{\xi}(x)$, its expectation is defined as

$$\mathbf{E}(\boldsymbol{\xi}) = \int_{-\infty}^{+\infty} x\mathbf{P}(x)\mathrm{d}x \tag{8.43}$$

By Equation 8.43, we can acquire the expectation of $\boldsymbol{\Theta}_l$. For $r_{il}^* < \frac{\beta_l}{2}$, we have:

$$\mathbf{E}(\boldsymbol{\Theta}_l) = 0 \tag{8.44}$$

In this case, both algorithms have the same or similar noise estimation. For $r_{il}^* \geq \frac{\beta_l}{2}$, we have:

$$\begin{aligned} \mathbf{E}(\boldsymbol{\Theta}_l) &= \frac{\beta_l}{2} - \int_{\frac{\beta_l}{2}}^{+\infty} \frac{1}{\sqrt{2\pi}\sigma_l} \exp(-\frac{(r_{il}^*)^2}{2\sigma_l^2}) r_{il}^* \mathrm{d}r_{il}^* \\ &= \frac{\beta_l}{2} - \frac{\sigma_l}{\sqrt{2\pi}} \exp(-\frac{\beta_l^2}{8\sigma_l^2}) \end{aligned} \tag{8.45}$$

Since β_l is associated with σ_l, we can define it into

$$\beta_l = 2\tau\sigma_l \tag{8.46}$$

where τ is a positive parameter. Thus, Equation 8.45 is equivalently converted into

$$\begin{aligned} \mathbf{E}(\boldsymbol{\Theta}_l) &= \tau\sigma_l - \frac{\sigma_l}{\sqrt{2\pi}} \exp(-\frac{\tau^2}{2}) \\ &= \sigma_l(\tau - \frac{1}{\sqrt{2\pi}} \exp(-\frac{\tau^2}{2})) \end{aligned} \tag{8.47}$$

In order for $\mathbf{E}(\boldsymbol{\Theta}_l)$ in Equation 8.47 to be greater than 0, we only require $f(\tau) = \tau - \frac{1}{\sqrt{2\pi}} \exp(-\frac{\tau^2}{2}) > 0$. By using a bisection method [66], we have that $f(\tau) > 0$ if and only if $\tau > 0.38$. According to Equation 8.46, we have $\beta_l > 0.76\sigma_l$.

$\square$

Remark 1. By Theorem 2, if and only if $\beta_l > 0.76\sigma_l$, RFCM can realize more accurate noise estimation than DSFCM. We see that $\beta_l > 0.76\sigma_l$ is a prerequisite for RFCM to be superior to DSFCM in theory. Let us illustrate its underlying essence. Assume that true Gaussian noise is covered in a set $\mathbb{S}$. $|\mathbb{S}|$ denotes the cardinality of $\mathbb{S}$, which represents the amount of true Gaussian noise. Here, we have $|\mathbb{S}| = K$. According to the probability density function of a Gaussian distribution, it is clear that most of noise is small or near zero while a small proportion of noise is large. Therefore, we can divide Gaussian noise into two subsets, i.e., $\mathbb{S}_1 = \{r_{il}^* : |r_{il}^*| < \frac{\beta_l}{2}\}$ and $\mathbb{S}_2 = \{r_{il}^* : |r_{il}^*| \geq \frac{\beta_l}{2}\}$. We assume that $\mathbb{S}_1$ collects most of Gaussian noise being small or near zero. The remaining noise is put into $\mathbb{S}_2$. As a result, the following inequality satisfies:

$$\frac{|\mathbb{S}_1|}{|\mathbb{S}|} \in (0.5, 1) \tag{8.48}$$

Consider $\beta_l = 0.76\sigma_l$ and $\mathbb{S}_1 = \{r_{il}^* : |r_{il}^*| < 0.38\sigma_l\}$. By referring to the numerical table of a normal distribution function [67], we obtain:

$$\Phi = \frac{1}{\sqrt{2\pi}} \int_{-\infty}^{0.38\sigma_l} \exp(-\frac{(r_{il}^*)^2}{2}) dr_{il}^* = 0.648$$

In the sequel, we get:

$$\frac{|\mathbb{S}_1|}{|\mathbb{S}|} = \frac{1}{\sqrt{2\pi}} \int_{-0.38\sigma_l}^{0.38\sigma_l} \exp(-\frac{(r_{il}^*)^2}{2}) dr_{il}^*$$
$$= 0.648 - (1 - \Phi)$$
$$= 0.296$$

Thus, when $\beta_l > 0.76\sigma_l$, we have $\frac{|\mathbb{S}_1|}{|\mathbb{S}|} \in (0.296, 1)$. Obviously, Equation 8.48 is always satisfied. Therefore, when Gaussian noise is involved in real-world applications, RFCM can always realize more accurate noise estimation than DSFCM.

By following the above case and combining Table 8.1, we conclude that RFCM realizes accurate noise estimations in presence of different types of noise. It has more expressive forms and is more selective than DSFCM. Therefore, it is concluded that RFCM is more applicable to different types of noise than DSFCM. Next, we experimentally show that RFCM realizes more accurate noise estimation than DSFCM. We cover an example, as portrayed in Figure 8.9.

Example 8.6.1. Consider a noise-free image shown in Figure 8.9(a), it is contaminated by Gaussian noise with level 30, impulse noise with 30%, and a mixture of Poisson, Gaussian (level 30) and impulse noise (level 20%)), respectively. We offer a noise estimation comparison between DSFCM and RFCM. As shown in Figure 8.9, in presence of Gaussian and mixed noise, the noise estimation of DSFCM in Figure 8.9(c) and (i) is not accurate and even far from the true one in Figure 8.9(b) and (h). By contrast, DSFCM works well for impulse noise estimation, as shown in Figure 8.9(f), which is since a low level of impulse noise exhibits sparsity. Superior to it, RFCM achieves better noise estimation results, as shown in Figure 8.9(d), (g) and (j). Moreover, it achieves better results on noise-suppression and feature-preserving than DSFCM does.

8.6.3   Proof of conclusion 2

Impulse noise is a common type of noise in image acquisition and transmission, which mainly results from faulty sensors or analog-to-digital converter errors. Since scratches in photos and video sequences are sparse, they are sometimes regarded as a type of impulse noise. However, they are difficult to be removed, since their intensities are indistinguishable from those of their neighbors and the distribution of damaged image pixels is random. Generally speaking, two types of impulse noise are often encountered, i.e., salt-and-pepper and random-valued impulse noise [51, 52]. For the l-th pixel channel, we denote the minimum and maximum pixel values as $\check{x}$ and $\hat{x}$, respectively. Moreover, the clean and observed pixel values at location i are represented as $\tilde{x}_{il}$ and x_{il}, respectively.

Salt-and-pepper impulse noise: A certain percentage of pixels are altered to be either $\check{x}$ or $\hat{x}$:

$$x_{il} = \begin{cases} \check{x}, & \text{with probability } \zeta/2 \\ \hat{x}, & \text{with probability } \zeta/2 \\ \tilde{x}_{il}, & \text{with probability } 1 - \zeta \end{cases} \tag{8.49}$$

Random-valued impulse noise: A certain percentage of pixels are altered to take on a uniform random number $x'_{il} \in [\check{x}, \hat{x}]$:

$$x_{il} = \begin{cases} x'_{il}, & \text{with probability } \zeta \\ \tilde{x}_{il}, & \text{with probability } 1 - \zeta \end{cases} \tag{8.50}$$

where ζ in Equations 8.49 and 8.50 stands for an impulse noise level, which implies that a percentage ζ of pixels in an observed image are corrupted.

Theorem 3. *Given an observed image $\boldsymbol{X}$, assume that it is polluted with impulse noise only, i.e., satisfying Equation 8.49 or Equation 8.50. Then, when segmenting $\boldsymbol{X}$ in the l-th pixel channel, RFCM has the same regularization term as DSFCM, which is formulated as;*

$$\Gamma(\boldsymbol{R}_l) = \|\boldsymbol{R}_l\|_{\ell_1}$$

In other words, DSFCM is a special case of RFCM if only impulse noise is embedded in an image.

Proof. By solving Equation 8.8, we can acquire the MAP estimate of $\widetilde{\boldsymbol{X}}_l$. From the definition of impulse noise in Equations 8.49 and 8.50, we have:

$$\mathbf{P}(\boldsymbol{X}_l|\widetilde{\boldsymbol{X}}_l) = 1 - \zeta = 1 - \frac{\|\boldsymbol{X}_l - \widetilde{\boldsymbol{X}}_l\|_{\ell_0}}{K} \tag{8.51}$$

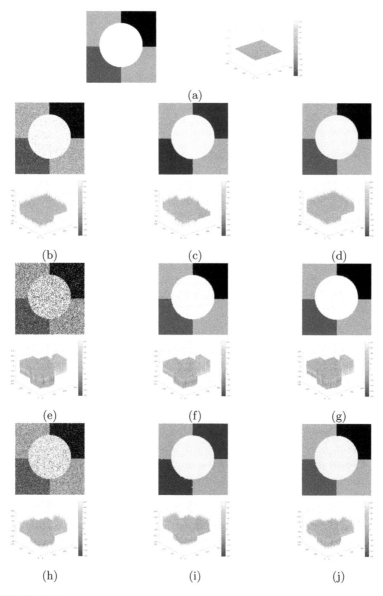

FIGURE 8.9
Noise estimation comparison between DSFCM and RFCM. The row 1 shows a noise-free image. In other rows, from left to right: observed images (from top to bottom: Gaussian noise ($\sigma = 30$), impulse noise ($\zeta = 30\%$), and a mixture of Poisson, Gaussian ($\sigma = 30$) and impulse noise ($\zeta = 20\%$)) and the results of DSFCM and RFCM.

where $\| \cdot \|_{\ell_0}$ denotes an ℓ_0 vector norm, which counts the number of non-zero entries in a vector. By Equation 8.51, we see that the likelihood $\mathbf{P}(\boldsymbol{X}_l | \widetilde{\boldsymbol{X}}_l)$ is negatively correlated with $\| \boldsymbol{X}_l - \widetilde{\boldsymbol{X}}_l \|_{\ell_0}$, i.e.,

$$\mathbf{P}(\boldsymbol{X}_l | \widetilde{\boldsymbol{X}}_l) \propto -\frac{1}{K} \| \boldsymbol{X}_l - \widetilde{\boldsymbol{X}}_l \|_{\ell_0}$$

Since the likelihood $\mathbf{P}(\boldsymbol{X}_l | \widetilde{\boldsymbol{X}}_l)$ is not directly related to the value of variable $\widetilde{\boldsymbol{X}}_l$, we have:

$$-\log \mathbf{P}(\boldsymbol{X}_l | \widetilde{\boldsymbol{X}}_l) \propto \frac{1}{K} \| \boldsymbol{X}_l - \widetilde{\boldsymbol{X}}_l \|_{\ell_0} \tag{8.52}$$

By ignoring the coefficient in Equation 8.52, we obtain a fidelity term on $\widetilde{\boldsymbol{X}}_l$:

$$\Psi(\widetilde{\boldsymbol{X}}_l) = \| \boldsymbol{X}_l - \widetilde{\boldsymbol{X}}_l \|_{\ell_0} \tag{8.53}$$

By Equation 8.1 and Equation 8.6, the regularization term on $\boldsymbol{R}_l$ corresponding to Equation 8.53 is expressed as

$$\Gamma(\boldsymbol{R}_l) = \| \boldsymbol{R}_l \|_{\ell_0} \tag{8.54}$$

In essence, $\Gamma(\boldsymbol{R}_l)$ in Equation 8.54 characterizes the sparsity of impulse noise. By substituting Equation 8.54 into Equation 8.5, RFCM's objective function associated with impulse noise is expressed as

$$J^{\mathrm{RFCM}}(\boldsymbol{U}, \boldsymbol{V}, \boldsymbol{R}) = \sum_{i=1}^{K} \sum_{j=1}^{c} u_{ij}^m \| \boldsymbol{x}_i - \boldsymbol{r}_i - \boldsymbol{v}_j \|^2 + \sum_{l=1}^{L} \beta_l \sum_{i=1}^{K} |r_{il}|_0 \tag{8.55}$$

with

$$|r_{il}|_0 = \begin{cases} 1, & r_{il} \neq 0 \\ 0, & r_{il} = 0 \end{cases}$$

In an implementation, due to the loss of amplitude characteristics of the ℓ_0 norm, it is hard to minimize Equation 8.54 directly. Moreover, since the ℓ_1 norm is a convex envelope of the ℓ_0 norm, both of them have good approximate characteristics [68]. Therefore, the ℓ_1 norm approximately replacing ℓ_0 norm is often used to characterize the sparsity of impulse noise:

$$\Gamma(\boldsymbol{R}_l) = \| \boldsymbol{R}_l \|_{\ell_1} \tag{8.56}$$

By substituting Equation 8.56 into Equation 8.5, RFCM's objective function associated with impulse noise is rewritten as

$$J^{\mathrm{RFCM}}(\boldsymbol{U}, \boldsymbol{V}, \boldsymbol{R}) = \sum_{i=1}^{K} \sum_{j=1}^{c} u_{ij}^m \| \boldsymbol{x}_i - \boldsymbol{r}_i - \boldsymbol{v}_j \|^2 + \sum_{l=1}^{L} \beta_l \sum_{i=1}^{K} |r_{il}| \tag{8.57}$$

Obviously, Equation 8.57 is the same as DSFCM's objective function Equation 8.11. Therefore, DSFCM is a particular case of an RFCM framework. □

Remark 2. If only impulse noise is involved in an image, DSFCM is the same as RFCM. In addition, by Equations 8.54 and 8.56, we see that there are two choices of regularization terms suitable for impulse noise estimation, i.e., $\Gamma(\boldsymbol{R}_l) = \| \boldsymbol{R}_l \|_{\ell_0}$ and $\Gamma(\boldsymbol{R}_l) = \| \boldsymbol{R}_l \|_{\ell_1}$.

8.7 Experimental Studies

In this section, to show the performance, efficiency and robustness of WRFCM, we provide numerical experiments on synthetic, medical, and other real-world images. To highlight the superiority and improvement of WRFCM over conventional FCM, we also compare it with seven FCM variants, i.e., FCM_S1 [11], FCM_S2 [11], FLICM [14], KWFLICM [16], FRFCM [31], WFCM [23], and DSFCM_N [35]. They are the most representative ones in the field. At the last of this section, to further verify WRFCM's strong robustness, we compare WRFCM with two competing approaches unrelated to FCM, i.e., PFE [69] and AMR_SC [70]. For a fair comparison, we note that all experiments are implemented in Matlab on a laptop with Intel(R) Core(TM) i5-8250U CPU of (1.60 GHz) and 8.0 GB RAM.

8.7.0.1 Evaluation indicators

To quantitatively evaluate the performance of WRFCM, we adopt three objective evaluation indicators, i.e., segmentation accuracy (SA) [16], Matthews correlation coefficient (MCC) [71], and Sorensen-Dice similarity (SDS) [72,73]. Note that a single one cannot fully reflect true segmentation results. SA is defined as

$$\text{SA} = \sum_{j=1}^{c} |S_j \cap G_j| / K$$

where S_j and G_j are the j-th cluster in a segmented image and its ground truth, respectively. $|\cdot|$ denotes the cardinality of a set. MCC is computed as

$$\text{MCC} = \frac{T_P \cdot T_N - F_P \cdot F_N}{\sqrt{(T_P + F_P) \cdot (T_P + F_N) \cdot (T_N + F_P) \cdot (T_P + F_N)}}$$

where T_P, F_P, T_N, and F_N are the numbers of true positive, false positive, true negative, and false negative, respectively. SDS is formulated as

$$\text{SDS} = \frac{2T_P}{2T_P + F_P + F_N}$$

8.7.1 Dataset descriptions

Tested images except for synthetic ones come from four publicly available databases including a medical one and three real-world ones. The details are outlined as follows:

1) BrianWeb[1]: This is an online interface to a 3D MRI simulated brain database. The parameter settings are fixed to 3 modalities, 5 slice thicknesses, 6 levels of noise, and 3 levels of intensity non-uniformity. BrianWeb provides golden standard segmentation.

---

[1]http://www.bic.mni.mcgill.ca/brainweb/.

2) Berkeley Segmentation Data Set (BSDS)[2] [74]: This database contains 200 training, 100 validation and 200 testing images. Golden standard segmentation is annotated by different subjects for each image of size 321×481 or 481×321.

3) Microsoft Research Cambridge Object Recognition Image Database (MSRC)[3]: This database contains 591 images and 23 object classes. Golden standard segmentation is provided.

4) NASA Earth Observation Database (NEO)[4]: This database continually provides information collected by NASA satellites about Earth's ocean, atmosphere, and land surfaces. Due to bit errors appearing in satellite measurements, sampled images of size 1440×720 contain unknown noise. Thus, their ground truths are unknown.

5) PASCAL Visual Object Classes 2012 (VOC2012)[5]: This dataset contains 20 object classes. The training/validation data has 11,530 images containing 27,450 ROI annotated objects and 6,929 segmentations. Each image is of size 500×375 or 375×500.

8.7.2 Parameter settings

Prior to numerical simulations, we report the parameter settings of WRFCM and comparative algorithms. Since AMR_SC and PFE are not related to FCM, we follow their parameter settings introduced in their original articles [69, 70]. In the following, we focus on all FCM-related algorithms. Since spatial information is used in all algorithms, a filtering window of size $\omega = 3 \times 3$ is selected for FCM_S1, FCM_S2, KWFLICM, FRFCM, and WFCM and a local window of size $|\mathcal{N}_i| = 3 \times 3$ is chosen for FLICM, KWFLICM, DSFCM_N, and WRFCM. We set $m = 2$ and $\varepsilon = 1 \times 10^{-6}$ across all algorithms. The setting of c is presented in each experiment.

Except m, ε, and c, FLICM and KWFLICM are free of all parameters. However, the remaining algorithms involve different parameters. In FCM_S1 and FCM_S2, α is set to 3.8, which controls the impact of spatial information on FCM by following [11]. In FRFCM, an observed image is taken as a mask image. A marker image is produced by a 3×3 structuring element. WFCM requires one parameter $\mu \in [0.55, 0.65]$ only, which constrains the neighbor term. For DSFCM_N, λ is set based on the standard deviation of each channel of image data.

As to WRFCM, it requires two parameters, i.e., ξ in Equation 8.16 and β in Equation 8.19. By analyzing mixed noise distributions, ξ is experimentally

[2]https://www2.eecs.berkeley.edu/Research/Projects/CS/vision/grouping/resources.html.
[3]http://research.microsoft.com/vision/cambridge/recognition/.
[4]http://neo.sci.gsfc.nasa.gov/.
[5]http://host.robots.ox.ac.uk/pascal/VOC/voc2012/index.html.

set to 0.0008. The weighting results are portrayed in Figure 8.5. Since the standard deviation of image data is related to noise levels to some extent [35], we can set $\boldsymbol{\beta}$ in virtue of the standard deviation of each channel. Based on massive experiments, $\boldsymbol{\beta} = \{\beta_l : l = 1, 2, \cdots, L\}$ is recommended to be chosen as follows:

$$\beta_l = \frac{\phi \cdot \delta_l}{100} \quad \phi \in [5, 10]$$

where δ_l is the standard deviation of the l-th channel of $\boldsymbol{X}$. If ϕ is set, β_l is computed. Therefore, $\boldsymbol{\beta}$ is equivalently replaced by ϕ. In Figure 8.10, we cover an example to show the setting of ϕ associated with Figure 8.11.

Example 8.7.1. Consider two observed images, as shown in Figure 8.11(a1) and (a2). WRFCM is employed to segment them. As Figure 8.10(a) indicates, when coping with the first image, the SA value reaches its maximum gradually as the value of ϕ increases. Afterward, it decreases rapidly and tends to be stable. As shown in Figure 8.10(b), for the second image, after the SA value reaches its maximum, it has no apparent changes, implying that the segmentation performance is rather stable. In conclusion, for image segmentation, WRFCM can produce better and better performance as parameter ϕ increases from a small value.

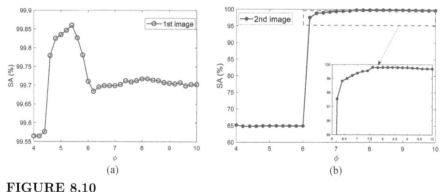

FIGURE 8.10
SA values versus ϕ.

8.7.3 Comparison with FCM-related methods

8.7.3.1 Results on synthetic images

In the first experiment, we representatively choose two synthetic images of size 256×256. A mixture of Poisson, Gaussian, and impulse noises is considered for all cases. To be specific, Poisson noise is first added. Then we add Gaussian noise with $\sigma = 30$. Finally, the random-valued impulse noise with $\zeta = 20\%$ is added since it is more difficult to detect than salt and pepper impulse noise. For two images, we set $c = 4$. The segmentation results are given in Figure 8.11 and Table 8.2. The best values are in bold.

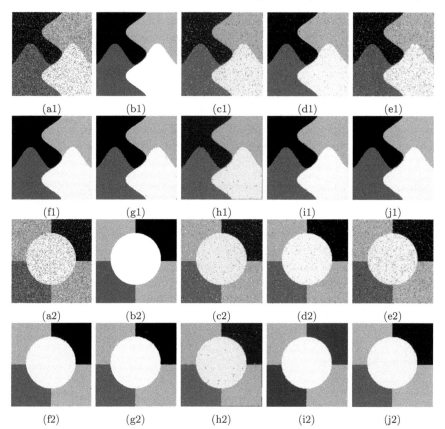

(a1) (b1) (c1) (d1) (e1)

(f1) (g1) (h1) (i1) (j1)

(a2) (b2) (c2) (d2) (e2)

(f2) (g2) (h2) (i2) (j2)

FIGURE 8.11
Visual results for segmenting synthetic images ($\phi_1 = 5.58$ and $\phi_2 = 7.45$). From (a) to (j): noisy images, ground truth, and results of FCM_S1, FCM_S2, FLICM, KWFLICM, FRFCM, WFCM, DSFCM_N, and WRFCM.

As Figure 8.11 indicates, FCM_S1, FCM_S2, and FLICM achieve poor results in presence of such a high level of mixed noise. Compared with them, KWFLICM, FRFCM, and WFCM suppress the vast majority of mixed noise. Yet they cannot completely remove it. DSFCM_N visually outperforms other peers mentioned above. However, it generates several topology changes such as merging and splitting. By taking the second synthetic image as a case, we find that DSFCM_N produces some unclear contours and shadows. Superiorly to seven peers, WRFCM not only removes all the noise but also preserves more image features.

Table 8.2 shows the segmentation results of all algorithms quantitatively. It assembles the values of all three indictors. Clearly, WRFCM achieves better SA, SDS, and MCC results for all images than other peers. In particular, its SA value comes up to 99.859% for the first synthetic image. Among its seven

TABLE 8.2

Segmentation performance (%) on synthetic images.

Algorithm	First synthetic image			Second synthetic image		
	SA	SDS	MCC	SA	SDS	MCC
FCM_S1	92.902	98.187	96.362	92.625	98.414	95.528
FCM_S2	96.157	98.999	97.991	96.292	99.127	97.520
FLICM	85.081	90.145	95.082	85.667	95.894	88.576
KWFLICM	99.706	99.858	99.715	99.730	99.904	99.725
FRFCM	99.652	99.920	99.839	99.675	99.895	99.698
WFCM	97.827	99.325	98.652	98.079	99.363	98.197
DSFCM_N	98.954	99.545	99.086	99.226	99.757	99.303
WRFCM	**99.859**	**99.937**	**99.843**	**99.802**	**99.958**	**99.792**

TABLE 8.3

Segmentation performance (%) on medical images in BrianWeb.

Algorithm	First medical image			Second medical image		
	SA	SDS	MCC	SA	SDS	MCC
FCM_S1	75.756	97.852	96.225	75.026	98.109	96.656
FCM_S2	75.769	98.119	96.664	74.970	98.176	96.765
FLICM	74.998	98.070	96.568	74.185	98.122	96.660
KWFLICM	74.840	98.259	96.878	73.839	97.860	96.190
FRFCM	75.853	97.620	95.775	75.514	97.660	95.830
WFCM	75.507	97.124	94.957	74.471	97.213	95.045
DSFCM_N	76.400	92.325	86.262	75.288	91.574	85.095
WRFCM	**82.317**	**98.966**	**98.147**	**82.141**	**98.298**	**96.970**

peers, KWFLICM obtains generally better results. In the light of Figure 8.11 and Table 8.2, we conclude that WRFCM performs better than its peers.

8.7.3.2 Results on medical images

Next, we representatively segment two medical images from BrianWeb. They are represented as two slices in the axial plane with 70 and 80, which are generated by T1 modality with slice thickness of 1mm resolution, 9% noise and 20% intensity non-uniformity. Here, we set $c = 4$ for all cases. The comparisons between WRFCM and its peers are shown in Figure 8.12 and Table 8.3. The best values are in bold.

By viewing the marked red squares in Figure 8.12, we find that FCM_S1, FCM_S2, FLICM, KWFLICM, and DSFCM_N are vulnerable to noise and intensity non-uniformity. They give rise to the change of topological shapes to some extent. Unlike them, FRFCM and WFCM achieve sufficient noise removal. However, they produce overly smooth contours. Compared with its seven peers, WRFCM can not only suppress noise adequately but also acquire accurate contours. Moreover, it yields the visual result closer to ground truth than its peers. As Table 8.3 shows, WRFCM obtains optimal SA, SDS, and MCC results for all medical images. As a conclusion, it outperforms its peers visually and quantitatively.

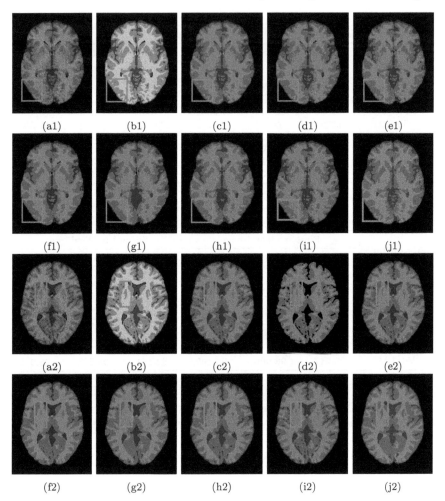

FIGURE 8.12
Visual results for segmenting medical images ($\phi = 5.35$). From (a) to (j): noisy images, ground truth, and results of FCM_S1, FCM_S2, FLICM, KWFLICM, FRFCM, WFCM, DSFCM_N, and WRFCM.

8.7.3.3 Results on real-world images

In order to demonstrate the practicality of WRFCM for other image segmentation, we typically choose two sets of real-world images in the last experiment. The first set contains five representative images from BSDS and MSRC. There usually exist some outliers, noise or intensity inhomogeneity in each image. For all tested images, we set $c = 2$. The results of all algorithms are shown in Figure 8.13 and Table 8.4.

Figure 8.13 visually shows the comparison between WRFCM and seven peers while Table 8.4 gives the quantitative comparison. Apparently,

FIGURE 8.13

Segmentation results on five real-world images in BSDS and MSRC ($\phi_1 = 6.05$, $\phi_2 = 10.00$, $\phi_3 = 9.89$, $\phi_4 = 9.98$, and $\phi_5 = 9.50$). From top to bottom: observed images, ground truth, and results of FCM_S1, FCM_S2, FLICM, KWFLICM, FRFCM, WFCM, DSFCM_N, and WRFCM.

TABLE 8.4

Segmentation performance (%) on real-world Images in BSDS and MSRC.

Algorithm	Figure 8.130 column 1			Figure 8.13 column 2			Figure 8.13 column 3			Figure 8.13 column 4			Figure 8.13 column 5		
	SA	SDS	MCC	SA	SDS	MCC	SA	SDS	MCC	SA	SDS	MCC	SA	SDS	MCC
FCM_S1	86.384	89.687	69.705	50.997	66.045	2.724	67.289	72.570	32.232	80.688	88.159	49.369	78.717	47.696	48.874
FCM_S2	86.138	79.701	69.208	51.433	12.089	2.951	67.105	59.523	31.941	80.657	47.557	49.256	78.365	86.449	47.881
FLICM	86.476	89.771	69.882	55.292	70.055	2.403	89.233	91.167	78.117	80.771	47.826	49.729	80.617	54.490	54.029
KWFLICM	87.119	90.278	71.283	48.252	63.432	1.554	64.617	66.081	30.820	80.484	46.723	48.777	77.963	44.791	46.755
FRFCM	97.701	**98.235**	94.941	99.690	97.436	97.273	99.380	99.467	98.732	83.974	89.927	58.683	96.985	97.861	92.987
WFCM	98.442	97.755	96.563	99.688	**99.834**	97.268	99.295	99.160	98.555	84.480	62.664	60.043	96.445	93.943	91.719
DSFCM_N	93.116	90.279	84.987	50.688	11.093	0.638	92.101	90.791	83.922	50.858	60.181	0.506	95.412	92.319	89.179
WRFCM	**98.732**	98.162	**97.201**	**99.746**	97.906	**97.771**	**99.442**	**99.520**	**98.857**	**99.826**	**99.888**	**99.074**	**99.869**	**99.789**	**99.694**

WRFCM achieves better segmentation results than its peers. FCM_S1, FCM_S2, FLICM, KWFLICM, and DSFCM_N obtain unsatisfactory results on all tested images. Superiorly to them, FRFCM and WFCM preserve more contours and feature details. From a quantitative point of view, WRFCM acquires optimal SA, SDS, and MCC values much more than its peers. Note that it merely gets a slightly smaller SDS value than FRFCM and WFCM for the first and second images, respectively.

The second set contains images from NEO. Here, we select two typical images. Each of them represents an example for a specific scene. We produce the ground truth of each scene by randomly shooting it for 50 times within the time span 2000–2019. The visual results of all algorithms are shown in Figures 8.14 and 8.15. The corresponding SA, SDS, and MCC values are given in Table 8.5.

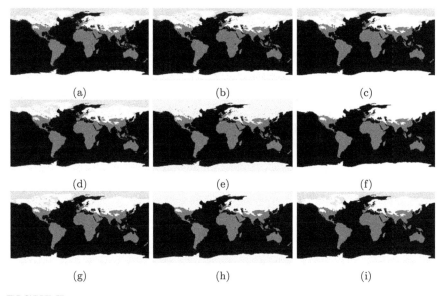

FIGURE 8.14
Segmentation results on the first real-world image in NEO ($\phi = 6.10$). From (a) to (i): observed image and results of FCM_S1, FCM_S2, FLICM, KWFLICM, FRFCM, WFCM, DSFCM_N, and WRFCM.

Figure 8.14 shows the segmentation results on sea ice and snow extent. The colors represent the land and ocean covered by snow and ice per week (here is February 7–14, 2015). We set $c = 4$. Figure 8.15 gives the segmentation results on chlorophyll concentration. The colors represent where and how much phytoplankton is growing over a span of days. We choose $c = 2$. As a whole, by seeing Figures 8.14 and 8.15, as well as Table 8.5, FCM_S1, FCM_S2, FLICM, KWFLICM, and WFCM are sensitive to unknown noise. FRFCM and DSFCM_N produce overly smooth results. Especially, they generate incorrect clusters when segmenting the first image in NEO. Superiorly to its seven

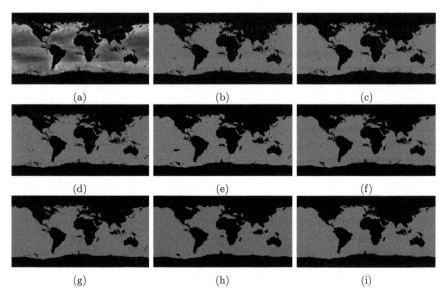

FIGURE 8.15

Segmentation results on the second real-world image in NEO ($\phi = 9.98$). From (a) to (i): observed image and results of FCM_S1, FCM_S2, FLICM, KWFLICM, FRFCM, WFCM, DSFCM_N, and WRFCM.

TABLE 8.5

Segmentation performance (%) on real-world images in NEO.

Algorithm	Figure 8.14			Figure 8.14		
	SA	SDS	MCC	SA	SDS	MCC
FCM_S1	90.065	97.060	95.106	80.214	92.590	90.329
FCM_S2	93.801	97.723	95.563	81.054	92.066	90.023
FLICM	90.234	97.056	95.781	81.582	92.352	90.236
KWFLICM	85.902	80.109	76.329	95.001	96.364	95.633
FRFCM	81.319	80.616	78.220	96.369	97.309	96.215
WFCM	95.882	98.854	97.293	97.342	97.430	97.178
DSFCM_N	80.131	81.618	79.597	96.639	97.936	96.436
WRFCM	**99.080**	**99.149**	**98.512**	**98.881**	**98.797**	**97.582**

peers, WRFCM cannot only suppress unknown noise well but also retain image contours well. In particular, it makes up the shortcoming that other peers forge several topology changes in the form of black patches when coping with the second image in NEO.

8.7.3.4   Performance improvement

Besides segmentation results reported for all algorithms, we also present the performance improvement of WRFCM over seven comparative algorithms in Table 8.6. Clearly, for all types of images, the average SA, SDS, and MCC

TABLE 8.6

Average performance improvements (%) of WRFCM over comparative algorithms.

Algorithm	Synthetic images			Medical images		
	SA	SDS	MCC	SA	SDS	MCC
FCM_S1	7.067	1.647	3.873	6.838	0.651	1.118
FCM_S2	3.606	0.884	2.062	6.859	0.484	0.844
FLICM	14.456	6.928	7.989	7.637	0.536	0.945
KWFLICM	0.113	0.067	0.098	7.889	0.572	1.024
FRFCM	0.167	0.040	0.049	6.545	0.992	1.756
WFCM	1.878	0.603	1.393	7.240	1.463	2.558
DSFCM_N	0.741	0.297	0.623	6.385	6.682	11.880

Algorithm	Real-world images in BSDS and MSRC			Real-world images in NEO		
	SA	SDS	MCC	SA	SDS	MCC
FCM_S1	26.708	26.221	57.938	13.841	4.148	5.329
FCM_S2	26.783	41.989	58.271	11.553	4.078	5.254
FLICM	21.045	28.391	47.687	13.073	4.269	5.038
KWFLICM	27.836	36.791	58.681	8.529	10.736	12.066
FRFCM	3.977	2.468	9.996	10.137	10.010	10.829
WFCM	3.853	8.381	9.690	2.369	0.831	0.811
DSFCM_N	23.088	30.120	46.673	10.596	9.196	10.030

improvements of WRFCM over other peers are within the value span 0.113%–27.836%, 0.040%–41.989%, and 0.049%–58.681%, respectively.

8.7.3.5 Overhead analysis

In the previous subsections, the segmentation performance of WRFCM is presented. Next, we provide the comparison of computing overheads between WRFCM and seven comparative algorithms in order to show its practicality. We note that K is the number of image pixels, c is the number of prototypes, t is the iteration count, ω represents the size of a filtering window, $|\mathcal{N}_i|$ denotes the size of a local window centralized at pixel j, and μ is the number of pixel levels in an image. Generally, $\mu \ll K$. Generally speaking, since FCM-related algorithms are unsupervised and their solutions are easily updated iteratively, they have low storage complexities of $O(K)$. In addition, we summarize the computational complexity of all algorithms in Table 8.7.

As Table 8.7 shows, FRFCM has lower computational complexity than its peers due to $\mu \ll K$. Except WFCM, the computational complexity of other algorithms is basically $O(K)$, the computational complexity of WRFCM is not high. To compare the practicability between WRFCM and its peers, we present the execution times of all algorithms for segmenting synthetic, medical, and real-world images in Table 8.8.

As Table 8.8 shows, for gray and color image segmentation, the computational efficiencies of FLICM and KWFLICM are far lower than those of others. In contrast, since gray level histograms are considered, FRFCM takes the least

TABLE 8.7
Computational complexity of all algorithms.

Algorithm	Computational complexity
FCM_S1	$O(K \times \omega + K \times c \times t)$
FCM_S2	$O(K \times \omega + K \times c \times t)$
FLICM	$O(K \times c \times t \times |\mathcal{N}_i|)$
KWFLICM	$O(K \times (\sqrt{\omega}+1)^2 + K \times c \times t \times |\mathcal{N}_i|)$
FRFCM	$O(K \times \omega + \mu \times c \times t)$
WFCM	$O(K \times \omega + 9 \times K \times \log K + K \times c \times t)$
DSFCM_N	$O(K \times c \times t \times |\mathcal{N}_i|)$
WRFCM	$O(K \times c \times t \times |\mathcal{N}_i|)$

TABLE 8.8
Comparison of execution time (in seconds) of all algorithms.

Algorithm	Figure 8.11 (Average)	Figure 8.12 (Average)	Figure 8.13 (Average)	Figure 8.14	Figure 8.15
FCM_S1	4.284	3.971	5.075	5.535	4.644
FCM_S2	3.674	3.362	5.346	5.288	4.182
FLICM	50.879	96.709	278.694	224.195	40.727
KWFLICM	67.838	128.946	371.592	298.926	54.303
FRFCM	0.263	0.247	1.162	5.815	1.786
WFCM	3.772	2.685	5.419	6.644	7.104
DSFCM_N	7.758	5.786	8.987	36.648	18.897
WRFCM	6.455	2.819	4.893	4.977	2.761

execution time among all algorithms. Due to the computation of a neighbor term in advance, FCM_S1 and FCM_S2 are more time-saving than most of other algorithms. Even though WFCM and DSFCM_N need more computing overheads than FRFCM, they are still very efficient. For color image segmentation, the execution time of DSFCM_N increases dramatically. Compared with most of seven comparative algorithms, WRFCM shows higher computational efficiency. In most cases, it only runs slower than FRFCM. However, the shortcoming can be offset by its better segmentation performance. In a quantitative study, for each image, WRFCM takes 0.011 and 2.642 seconds longer than FCM_S2 and FRFCM, respectively. However, it saves 0.321, 133.860, 179.940, 0.744, and 11.234 seconds over FCM_S1, FLICM, KWFLICM, FR-FCM, WFCM, and DSFCM_N, respectively.

8.7.4   Comparison with non-FCM methods

In this subsection, we compare WRFCM with two non-FCM methods, i.e., PFE [69] and AMR_SC [70]. We list the comparison results on VOC2012 in Figure 8.16 and Table 8.9. The computing times of the WRFCM, AMR_SC, and PFE are summarized in Table 8.10. The results indicate that WRFCM achieves better effectiveness and efficiency than AMR_SC and PFE.

FIGURE 8.16
Segmentation results on four real-world images in VOC2012 ($\phi_1 = 5.65$, $\phi_2 = 6.75$, $\phi_3 = 5.50$, $\phi_4 = 8.35$). From top to bottom: observed images, ground truth, and results of AMR_SC, PFE, and WRFCM.

TABLE 8.9
Segmentation performance (%) on real-world images in VOC2012.

Algorithm	Figure 8.16 column 1			Figure 8.16 column 2		
	SA	SDS	MCC	SA	SDS	MCC
AMR_SC	90.561	90.259	84.867	89.216	90.060	92.849
PFE	95.992	95.211	94.826	96.077	97.748	95.513
WRFCM	98.946	98.651	97.353	97.167	98.848	96.072

Algorithm	Figure 8.16 column 3			Figure 8.16 column 4		
	SA	SDS	MCC	SA	SDS	MCC
AMR_SC	85.095	88.615	80.878	91.148	92.414	90.156
PFE	90.358	94.516	90.923	93.273	94.950	93.743
WRFCM	92.756	96.132	93.330	95.436	96.869	97.012

TABLE 8.10
Comparison of execution time (in seconds) between WRFCM and two non-FCM methods.

Algorithm	Figure 13 column 1	Figure 13 column 2	Figure 13 column 3	Figure 13 column 4
AMR_SC	6.639	6.819	6.901	7.223
PFE	206.025	234.796	217.801	238.473
WRFCM	5.509	6.097	5.425	5.828

8.8 Conclusions and Future Work

For the first time, a residual-driven FCM (RFCM) framework is proposed for image segmentation, which advances FCM research. It realizes favorable noise estimation in virtue of a residual-related regularization term coming with an analysis of noise distribution. On the basis of the framework, RFCM with weighted ℓ_2-norm regularization (WRFCM) is presented for coping with image segmentation with mixed or unknown noises. Spatial information is also considered in WRFCM for making residual estimation more reliable. A two-step iterative algorithm is presented to implement WRFCM. In addition, a comparative study of two NFCM methods, namely DSFCM and RFCM, is completed. Two issues are carefully investigated: 1) RFCM realizes more accurate noise estimations DSFCM when different types of noise are encountered; and 2) DSFCM is a particular case of RFCM when only impulse noise is involved. Experiments reported for five benchmark databases demonstrate that it outperforms existing FCM variants and non-FCM methods. Moreover, differing from popular residual-learning methods, it is unsupervised and exhibits a high speed of clustering.

There are some open issues worth pursuing as future work. First, DSFCM's deficiencies can be overcome by using a transform method such as Fourier transform, orthonormal wavelets [36], translation-invariant wavelets [37], and framelets [38–41]. In some transformed domain, noise/outliers are sparse and generally located in high-frequency channels. Based on the transform, DSFCM can be applied. Second, as a competing approach for image segmentation, can WRFCM be extended to complex research fields such as remote sensing [75], ecological systems [76], transportation networks [77,78] and other applications [79–82]? How can the number of clusters be selected automatically? Answering them calls for more research effort.

Acknowledgments

This work was supported in part by the China National Postdoctoral Program for Innovative Talents under Grant No. BX2021249, in part by the Fellowship of China Postdoctoral Science Foundation under Grant No. 2021M702678, in part by the Guangdong Basic and Applied Basic Research Foundation under Grant No. 2021A1515110019, in part by the National Natural Science Foundation of China under Grant Nos. 61873342, 61672400, 62076189, 62206220, in part by the Recruitment Program of Global Experts, and in part by the Science and Technology Development Fund, MSAR, under Grant No. 0012/2019/A1.

Authors' biographies

Cong Wang received B.S. degree in automation and M.S. degree in mathematics from Hohai University, Nanjing, China, in 2014 and 2017, respectively. He received Ph.D. degree in mechatronic engineering from Xidian University, Xi'an, China in 2021. He is now an assistant professor in School of Artificial Intelligence, OPtics and ElectroNics (iOPEN), Northwestern Polytechnical University, Xi'an, China. He was a Visiting Ph.D. Student at the Department of Electrical and Computer Engineering, University of Alberta, Edmonton, AB, Canada, and the Department of Electrical and Computer Engineering, National University of Singapore, Singapore. He was also a Research Assistant at the School of Computer Science and Engineering, Nanyang Technological University, Singapore. His current research interests include information computation, G-image processing, fuzzy theory, as well as wavelet analysis and its applications. He has over 30+ papers and host/participate in 10+ research projects (over 2.5 million RMB). He also has a book and a chapter to be published. He is funded by the China National Postdoctoral Program for Innovative Talents and the Excellent Chinese and Foreign Youth Exchange Program of the China Association for Science and Technology. He serves as a Frequent Reviewer of 40+ international journals, including a number of the IEEE TRANSACTIONS and many international conferences. He is a member of Institute of Electrical and Electronics Engineers (IEEE), Chinese Institute of Electronics (CIE), China Computer Federation (CCF), China Society of Image and Graphics (CSIG) and China Society for Industrial and Applied Mathematics (CSIAM).

MengChu Zhou received his B.S. degree in Control Engineering from Nanjing University of Science and Technology, Nanjing, China in 1983, M.S. degree in Automatic Control from Beijing Institute of Technology, Beijing, China in 1986, and Ph.D. degree in Computer and Systems Engineering from Rensselaer Polytechnic Institute, Troy, NY in 1990. He joined New Jersey Institute of Technology (NJIT), Newark, NJ in 1990, and is now a Distinguished Professor of Electrical and Computer Engineering. His research interests are in Petri nets, intelligent automation, Internet of Things, big data, web services, and intelligent transportation.

His recently co-authored/edited books include Supervisory Control and
Scheduling of Resource Allocation Systems: Reachability Graph Perspective, IEEE Press/Wiley, Hoboken, Hoboken, New Jersey J, 2020 (with B.
Huang) and Contemporary Issues in Systems Science and Engineering,
IEEE/Wiley, Hoboken, New Jersey, 2015 (with H.-X. Li and M. Weijnen).
He has over 1000 publications including 14 books, 600+ journal papers
(500+ in IEEE TRANSACTIONS), 31 patents and 30 book-chapters. He
is the founding Editor of IEEE Press Book Series on Systems Science
and Engineering and Editor-in-Chief of IEEE/CAA Journal of Automatica Sinica. He is a recipient of Humboldt Research Award for US Senior
Scientists from Alexander von Humboldt Foundation, Franklin V. Taylor Memorial Award and the Norbert Wiener Award from IEEE Systems,
Man and Cybernetics Society. He is founding Chair/Co-chair of Enterprise
Information Systems Technical Committee (TC), Environmental Sensing,
Networking, and Decision-making TC and AI-based Smart Manufacturing Systems TC of IEEE SMC Society. He has been among most highly
cited scholars for years and ranked top one in the field of engineering
worldwide in 2012 by Web of Science/Thomson Reuters and now Clar-
ivate Analytics. He is a life member of Chinese Association for Science
and Technology-USA and served as its President in 1999. He is a Fellow
of International Federation of Automatic Control (IFAC), American Association for the Advancement of Science (AAAS), Chinese Association of
Automation (CAA) and National Academy of Inventors (NAI).

Witold Pedrycz received MS.c. degree in computer science and technology,
Ph.D. degree in computer engineering, and D.Sci. degree in systems science from the Silesian University of Technology, Gliwice, Poland, in 1977,
1980, and 1984, respectively. He is a Professor and the Canada Research
Chair in Computational Intelligence with the Department of Electrical and
Computer Engineering, University of Alberta, Edmonton, AB, Canada. He
is also with the Systems Research Institute of the Polish Academy of Sciences, Warsaw, Poland. He is a foreign member of the Polish Academy of
Sciences. He has authored 15 research monographs covering various aspects
of computational intelligence, data mining, and software engineering. His
current research interests include computational intelligence, fuzzy modeling, granular computing, knowledge discovery and data mining, fuzzy
control, pattern recognition, knowledge-based neural networks, relational
computing, and software engineering. He has published numerous papers
in the above areas. Dr. Pedrycz was a recipient of the IEEE Canada Computer Engineering Medal, the Cajastur Prize for Soft Computing from the
European Centre for Soft Computing, the Killam Prize, and the Fuzzy
Pioneer Award from the IEEE Computational Intelligence Society. He is
intensively involved in editorial activities. He is an Editor-in-Chief of Information Sciences, an Editor-in-Chief of WIREs Data Mining and Knowledge Discovery (Wiley) and the International Journal of Granular Computing (Springer). He currently serves as a member of a number of editorial

boards of other international journals. He is a fellow of the Royal Society of Canada.

ZhiWu Li received B.S. degree in mechanical engineering, the M.S. degree in automatic control, and Ph.D. degree in manufacturing engineering from Xidian University, Xi'an, China, in 1989, 1992, and 1995, respectively. He joined Xidian University in 1992. He is also currently with the Institute of Systems Engineering, Macau University of Science and Technology, Macau, China. He was a Visiting Professor with the University of Toronto, Toronto, ON, Canada, the Technion-Israel Institute of Technology, Haifa, Israel, the Martin-Luther University of Halle-Wittenburg, Halle, Germany, Conservatoire National des Arts et Métiers, Paris, France, and Meliksah Universitesi, Kayseri, Turkey. His current research interests include Petri net theory and application, supervisory control of discrete-event systems, workflow modeling and analysis, system reconfiguration, game theory, and data and process mining. Dr. Li was a recipient of an Alexander von Humboldt Research Grant, Alexander von Humboldt Foundation, Germany. He is listed in Marquis Who's Who in the World, 27th Edition, 2010. He serves as a Frequent Reviewer of 90+ international journals, including Automatica and a number of the IEEE TRANSACTIONS as well as many international conferences. He is the Founding Chair of Xi'an Chapter of IEEE Systems, Man, and Cybernetics Society. He is a member of Discrete-Event Systems Technical Committee of the IEEE Systems, Man, and Cybernetics Society and IFAC Technical Committee on Discrete-Event and Hybrid Systems, from 2011 to 2014.

Bibliography

[1] A. Baraldi and P. Blonda. A survey of fuzzy clustering algorithms for pattern recognition. I. *IEEE Trans. Syst. Man Cybern. Part B Cybern.*, 29(6):778–785, Dec. 1999.

[2] A. Baraldi and P. Blonda. A survey of fuzzy clustering algorithms for pattern recognition. II. *IEEE Trans. Syst. Man Cybern. Part B Cybern.*, 29(6):786–801, Dec. 1999.

[3] C. Subbalakshmi, G. Ramakrishna, and S. K. M. Rao. Evaluation of data mining strategies using fuzzy clustering in dynamic environment. In *Proc. Int. Conf. Adv. Comput., Netw., Informat.*, pages 529–536. Springer, 2016.

[4] X. Zhu, W. Pedrycz, and Z. Li. Granular encoders and decoders: A study in processing information granules. *IEEE Trans. Fuzzy Syst.*, 25(5):1115–1126, Oct. 2017.

[5] M. Yambal and H. Gupta. Image segmentation using fuzzy c means clustering: A survey. *Int. J. Adv. Res. Comput. Commun. Eng.*, 2(7):2927–2929, Jul. 2013.

[6] J. C. Dunn. A fuzzy relative of the ISODATA process and its use in detecting compact well-separated clusters. *J. Cybernet.*, 3(3):32–57, 1973.

[7] J. C. Bezdek. *Pattern Recognition with Fuzzy Objective Function Algorithms*. New York: Plenum Press, 1981.

[8] J. C. Bezdek, R. Ehrlich, and W. Full. FCM: The fuzzy c-means clustering algorithm. *Comput. Geosci.*, 10(2-3):191–203, 1984.

[9] C. Wang, W. Pedrycz, Z. Li, and M. Zhou. Residual-driven Fuzzy C-Means clustering for image segmentation. *IEEE/CAA J. Autom. Sinic*, 8(4):876–889, Apr. 2021.

[10] M. Ahmed, S. Yamany, N. Mohamed, A. Farag, and T. Moriarty. A modified fuzzy C-means algorithm for bias field estimation and segmentation of MRI data. *IEEE Trans. Med. Imag.*, 21(3):193–199, Aug. 2002.

[11] S. Chen and D. Zhang. Robust image segmentation using FCM with spatial constraints based on new kernel-induced distance measure. *IEEE Trans. Syst. Man Cybern. Part B Cybern.*, 34(4):1907–1916, Aug. 2004.

[12] L. Szilagyi, Z. Benyo, S. Szilagyi, and H. Adam. MR brain image segmentation using an enhanced fuzzy C-means algorithm. In *Proc. 25th Annu. Int. Conf. IEEE EMBS*, pages 724–726. IEEE, Sep. 2003.

[13] W. Cai, S. Chen, and D. Zhang. Fast and robust fuzzy c-means clustering algorithms incorporating local information for image segmentation. *Pattern Recognit.*, 40(3):825–838, Mar. 2007.

[14] S. Krinidis and V. Chatzis. A robust fuzzy local information c-means clustering algorithm. *IEEE Trans. Image Process.*, 19(5):1328–1337, Jan. 2010.

[15] T. Celik and H. K. Lee. Comments on "A robust fuzzy local information c-means clustering algorithm". *IEEE Trans. Image Process.*, 22(3):1258–1261, Mar. 2013.

[16] M. Gong, Y. Liang, J. Shi, W. Ma, and J. Ma. Fuzzy c-means clustering with local information and kernel metric for image segmentation. *IEEE Trans. Image Process.*, 22(2):573–584, Mar. 2013.

[17] K. P. Lin. A novel evolutionary kernel intuitionistic fuzzy C-means clustering algorithm. *IEEE Trans. Fuzzy Syst.*, 22(5):1074–1087, Aug. 2014.

[18] A. Elazab, C. Wang, F. Jia, J. Wu, G. Li, and Q. Hu. Segmentation of brain tissues from magnetic resonance images using adaptively regularized kernel-based fuzzy-means clustering. *Comput. Math. Method. M.*, 2015:1–12, Nov. 2015.

[19] F. Zhao, L. Jiao, and H. Liu. Kernel generalized fuzzy c-means clustering with spatial information for image segmentation. *Digit. Signal Process.*, 23(1):184–199, Jan. 2013.

[20] F. Guo, X. Wang, and J. Shen. Adaptive fuzzy c-means algorithm based on local noise detecting for image segmentation. *IET Image Process.*, 10(4):272–279, Apr. 2016.

[21] Z. Zhao, L. Cheng, and G. Cheng. Neighbourhood weighted fuzzy c-means clustering algorithm for image segmentation. *IET Image Process.*, 8(3):150–161, Mar. 2014.

[22] X. Zhu, W. Pedrycz, and Z. W. Li. Fuzzy clustering with nonlinearly transformed data. *Appl. Soft Comput.*, 61:364–376, Dec. 2017.

[23] C. Wang, W. Pedrycz, J. Yang, M. Zhou, and Z. Li. Wavelet frame-based fuzzy C-means clustering for segmenting images on graphs. *IEEE Trans. Cybern.*, 50(9):3938–3949, Sep. 2020.

[24] R. R. Gharieb, G. Gendy, A. Abdelfattah, and H. Selim. Adaptive local data and membership based KL divergence incorporating C-means algorithm for fuzzy image segmentation. *Appl. Soft Comput.*, 59:143–152, Oct. 2017.

[25] C. Wang, W. Pedrycz, Z. Li, and M. Zhou. Kullback-Leibler divergence-based Fuzzy C-Means clustering incorporating morphological reconstruction and wavelet frames for image segmentation. *IEEE Trans. Cybern.*, 52(8):7612–7623, Aug. 2022.

[26] J. Gu, L. Jiao, S. Yang, and F. Liu. Fuzzy double c-means clustering based on sparse self-representation. *IEEE Trans. Fuzzy Syst.*, 26(2):612–626, Apr. 2018.

[27] C. Wang, W. Pedrycz, M. Zhou, and Z. Li. Sparse regularization-based Fuzzy C-Means clustering incorporating morphological grayscale reconstruction and wavelet frames. *IEEE Trans. Fuzzy Syst.*, 29(7):1826–1840, Jul. 2021.

[28] L. Vincent. Morphological grayscale reconstruction in image analysis: applications and efficient algorithms. *IEEE Trans. Image Process.*, 2(2):176–201, Apr. 1993.

[29] L. Najman and M. Schmitt. Geodesic saliency of watershed contours and hierarchical segmentation. *IEEE Trans. Pattern Anal. Mach. Intell.*, 18(12):1163–1173, Dec. 1996.

[30] J. Chen, C. Su, W. Grimson, J. Liu, and D. Shine. Object segmentation of database images by dual multiscale morphological reconstructions and retrieval applications. *IEEE Trans. Image Process.*, 21(2):828–843, Feb. 2012.

[31] T. Lei, X. Jia, Y. Zhang, L. He, H. Meng, and A. K. Nandi. Significantly fast and robust fuzzy c-means clustering algorithm based on morphological reconstruction and membership filtering. *IEEE Trans. Fuzzy Syst.*, 26(5):3027–3041, Oct. 2018.

[32] T. Lei, P. Liu, X. Jia, X. Zhang, H. Meng, and A. K. Nandi. Automatic fuzzy clustering framework for image segmentation. *IEEE Trans. Fuzzy Syst.*, 28(9):2078–2092, Sept. 2020.

[33] C. Wang, W. Pedrycz, Z. Li, M. Zhou, and J. Zhao. Residual-sparse Fuzzy C-Means clustering incorporating morphological reconstruction and wavelet frame. *IEEE Trans. Fuzzy Syst.*, 29(12):3910–3924, Dec. 2020.

[34] X. Bai, Y. Zhang, H, Liu, and Z. Chen. Similarity measure-based possibilistic FCM with label information for brain MRI segmentation. *IEEE Trans. Cybern.*, 49(7):2618–2630, Jul. 2019.

[35] Y. Zhang, X. Bai, R. Fan, and Z. Wang. Deviation-sparse fuzzy c-means with neighbor information constraint. *IEEE Trans. Fuzzy Syst.*, 27(1):185–199, Jan. 2019.

[36] I. Daubechies. *Ten Lectures on Wavelets*. SIAM, 1992.

[37] R. R. Coifman and D. L. Donoho. Translation-invariant de-noising. In *Wavelets and Statistics*. A. Antoniadis and G. Oppenheim, Eds. New York: Springer-Verlag, 125–150, 1995.

[38] I. Daubechies, B. Han, A. Ron, and Z. Shen. Framelets: MRA-based constructions of wavelet frames. *Appl. Comput. Harmon. Anal.*, 14(1):1–46, Jan. 2003.

[39] C. Wang and J. Yang. Poisson noise removal of images on graphs using tight wavelet frames. *Visual Comput.*, 34(10):1357–1369, Oct. 2018.

[40] J. Yang and C. Wang. A wavelet frame approach for removal of mixed gaussian and impulse noise on surfaces. *Inverse Probl. Imaging*, 11(5):783–798, Oct. 2017.

[41] C. Wang, Z. Yan, W. Pedrycz, M. Zhou, and Z. Li. A weighted fidelity and regularization-based method for mixed or unknown noise removal from images on graphs. *IEEE Trans. Image Process.*, 29(1):5229–5243, Dec. 2020.

[42] A. Fakhry, T. Zeng, and S. Ji. Residual deconvolutional networks for brain electron microscopy image segmentation. *IEEE Trans. Med. Imaging*, 36(2):447–456, Feb. 2017.

[43] K. Zhang, W. Zuo, Y. Chen, D. Meng, and L. Zhang. Beyond a gaussian denoiser: Residual learning of deep CNN for image denoising. *IEEE Trans. Image Process.*, 26(7):3142–3155, Jul. 2017.

[44] F. Kokkinos and S. Lefkimmiatis. Iterative residual cnns for burst photography applications. In *Proc. IEEE Conf. Comput. Vis. Pattern Recognit. (CVPR)*, pages 5929–5938, Jun. 2019.

[45] D. Ren, W. Zuo, D. Zhang, L. Zhang, and M. H. Yang. Simultaneous fidelity and regularization learning for image restoration. *IEEE Trans. Pattern Anal. Mach. Intell.*, 43(1):284–299, Jan. 2021.

[46] Y. Zhang, X. Li, M. Lin, B. Chiu, and M. Zhao. Deep-recursive residual network for image semantic segmentation. *Neural Comput. Applic.*, 32(16):12935–12947, Aug. 2020.

[47] J. Jiang, L. Zhang, and J. Yang. Mixed noise removal by weighted encoding with sparse nonlocal regularization. *IEEE Trans. Image Process.*, 23(6):2651–2662, Jun. 2014.

[48] P. Zhou, C. Lu, J. Feng, Z. Lin and S. Yan. Tensor low-rank representation for data recovery and clustering. *IEEE Trans. Pattern Anal. Mach. Intell.*, 43(5):1718–1732, May 2021.

[49] L. I. Rudin, S. Osher, and E. Fatemi. Nonlinear total variation based noise removal algorithms. *Phys. D*, 60(1-4):259–268, 1992.

[50] A. Chambolle. An algorithm for total variation minimization and applications. *J. Math. Imaging Vision*, 20(1):89–97, Jan. 2004.

[51] C. Clason, B. Jin, and K. Kunisch. A duality-based splitting method for ℓ^1-TV image restoration with automatic regularization parameter choice. *SIAM J. Sci. Comput.*, 32(3):1484–1505, May 2010.

[52] J. Yang, Y. Zhang, and W. Yin. An efficient TVL1 algorithm for deblurring multichannel images corrupted by impulsive noise. *SIAM J. Sci. Comput.*, 31(4):2842–2865, Jul. 2009.

[53] C. Clason. L^∞ fitting for inverse problems with uniform noise. *Inverse Problems*, 28(10):104007, Oct. 2012.

[54] P. Weiss, G. Aubert, and L. Blanc-Féraud. *Some applications of l^∞-constraints in image processing.* PhD thesis, Nov. 2006. RR-6115, HAL Id: inria-00114051.

[55] T. Le, R. Chartrand, and T. J. Asaki. A variational approach to reconstructing images corrupted by poisson noise. *J. Math. Imag. Vis.*, 27(3):257–263, Apr. 2007.

[56] G. Steidl and T. Teuber. Removing multiplicative noise by douglas-rachford splitting methods. *J. Math. Imaging Vision*, 36(2):168–184, 2010.

[57] G. Aubert and J. F. Aujol. A variational approach to removing multiplicative noise. *SIAM J. Appl. Math.*, 68(4):925–946, Jan. 2008.

[58] H. Woo and S. Yun. Proximal linearized alternating direction method for multiplicative denoising. *SIAM J. Sci. Comput.*, 35(2):B336–B358, Mar. 2013.

[59] M. V. Afonso and J. M. R. Sanches. Blind inpainting using ℓ_0 and total variation regularization. *IEEE Trans. Image Process.*, 24(7):2239–2253, Jul. 2015.

[60] J. Seabra, J. Xavier, and J. Sanches. Convex ultrasound image reconstruction with log-euclidean priors. In *Proc. Eng. Med. Biol. Soc. (EMBS)*, pages 435–438. IEEE, Aug. 2008.

[61] P. J. Huber. Robust regression: Asymptotics, conjectures and Monte Carlo. *Ann. Stat.*, 1(5):799–821, 1973.

[62] P. J. Huber. Robust Statistics. *Wiley-Interscience*, 1981.

[63] I. Daubechies, M. Defrise, and C. De Mol. An iterative thresholding algorithm for linear inverse problems with a sparsity constraint. *Commun. Pure Appl. Math.*, 57:1413–1457, Nov. 2004.

[64] J. Sun and Z. Xu. Color image denoising via discriminatively learned iterative shrinkage. *IEEE Trans. Image Process.*, 24(11):4148–4159, Nov. 2015.

[65] V. W. Macrelo, S. H. Elias, and R. P. Daniel. Accelerating overre-laxed and monotone fast iterative shrinkage-thresholding algorithms with line search for sparse reconstructions. *IEEE Trans. Image Process.*, 26(7):3569–3578, Jul. 2017.

[66] G. Corliss. Which root does the bisection algorithm find? *SIAM Rev.*, 19(2):325–327, 1977.

[67] T. W. Anderson. An introduction to multivariate statistical analysis. *New York: Wiley*, 1984.

[68] G. Yuan and B. Ghanem. ℓ_0TV: A sparse optimization method for impulse noise image restoration. *IEEE Trans. Pattern Anal. Mach. Intell.*, 41(2):352–364, Feb. 2019.

[69] C. Fang, Z. Liao, and Y. Yu. Piecewise flat embedding for image segmentation. *IEEE Trans. Pattern Anal. Mach. Intell.*, 41(6):1470–1485, Jun. 2019.

[70] T. Lei, X. Jia, T. Liu, S. Liu, H. Meng, and A. K. Nandi. Adaptive morphological reconstruction for seeded image segmentation. *IEEE Trans. Image Process.*, 26(11):5510–5523, Nov. 2019.

[71] D. N. H. Thanh, D. Sergey, V. B. S. Prasath, and N. H. Hai. Blood vessels segmentation method for retinal fundus images based on adaptive principal curvature and image derivative operators. *Int. Arch. Photogramm. Remote Sens. Spatial Inf. Sci.*, XLII-2/W12:211–218. May 2019. doi: 10.5194/isprs-archives-XLII-2-W12-211-2019.

[72] D. N. H. Thanh, U. Erkan, V. B. S. Prasath, V. Kumar, and N. N. Hien. A skin lesion segmentation method for dermoscopic images based on adaptive thresholding with normalization of color models. In *Proc. IEEE 6th Int. Conf. Electr. Electron. Eng.*, pages 116–120, Apr. 2019.

[73] A. A. Taha and A. Hanbury. Metrics for evaluating 3D medical image segmentation: Analysis, selection, and tool. *BMC Med. Imaging*, 15(29):1–29, Aug. 2015.

[74] P. Arbelaez, M. Maire, C. Fowlkes and J. Malik. Contour detection and hierarchical image segmentation. *IEEE Trans. Pattern Anal. Mach. Intell.*, 33(5):898–916, May 2011.

[75] T. Xu, L. Jiao, and W. J. Emery. SAR image content retrieval based on fuzzy similarity and relevance feedback. *IEEE J. Sel. Topics Appl. Earth Observ. Remote Sens.*, 10(5):1824–1842, May 2017.

[76] C. Wang, J. Chen, Z. Li, E. Nasr, and A. M. El-Tamimi. An indicator system for evaluating the development of land-sea coordination systems: A case study of lianyungang port. *Ecol. Indic.*, 98:112–120, Mar. 2019.

[77] Y. Lv, Y. Chen, X. Zhang, Y. Duan, and N. Li. Social media based transportation research: The state of the work and the networking. *IEEE/CAA J. Autom. Sinica*, 4(1):19–26, Jan. 2017.

[78] H. Han, M. Zhou, and Y. Zhang. Can virtual samples solve small sample size problem of KISSME in pedestrian re-identification of smart transportation? *IEEE Trans. Intell. Transp. Syst.*, 21(9):3766–3776, Sep. 2020.

[79] W. Hu, Y. Huang, F. Zhang, and R. Li. Noise-tolerant paradigm for training face recognition cnns. In *Proc. IEEE/CVF Conf. Comput. Vis. Pattern Recognit. (CVPR)*, pages 11879–11888, Jun. 2019.

[80] Z. Cao, X. Xu, B. Hu, M. Zhou, and Q. Li. Real-time gesture recognition based on feature recalibration network with multi-scale information. *Neurocomputing*, 347:119–130, Jun. 2019.

[81] C. Wang, W. Pedrycz, Z. Li, M. Zhou, and S. S. Ge. G-image segmentation: Similarity-preserving fuzzy c-means with spatial information constraint in wavelet space. *IEEE Trans. Fuzzy Syst.*, 29(12):3887–3898, Dec. 2021.

[82] C. Wang, M. Zhou, W. Pedrycz, and Z. Li. Comparative study on noise-estimation-based Fuzzy C-Means for image segmentation. *IEEE Trans. Cybern.* to be published, doi:10.1109/TCYB.2022.3217897.

9

Sample Problems in Person Re-Identification

Hua Han

Shanghai University of Engineering Science

MengChu Zhou

New Jersey Institute of Technology

CONTENTS

9.1   Introduction 303
9.2   Person Re-ID under Small-Sample Learning 304
    9.2.1   Sample generation methods 305
        9.2.1.1   KISSME fundamentals 305
        9.2.1.2   Regularized discriminant analysis 306
        9.2.1.3   Stability analysis 307
    9.2.2   Pseudo-label learning method for generated samples: a
    semi-supervised learning method 314
        9.2.2.1   Image-to-image translation 316
        9.2.2.2   Semi-supervised pedestrian Re-ID 317
9.3   Conclusion 326

9.1   Introduction

At present, in the process of large-scale popularization and installation of video surveillance, the amount of video data that can be obtained and stored by surveillance video is constantly increasing. Based on the analysis from an ideal perspective, this kind of video mainly includes total information in the real world, which is of great value to our daily management and safety work. However, based on a realistic perspective,it is very difficult to manually analyze a large number of video data sets and obtain corresponding information from them. Therefore, most of the video data obtained and stored in surveillance systems have become unusable data stored in hard disks, thus leading to a scene of data explosion.

DOI: 10.1201/9781003053262-9

In addition, because there is a semantic gap between video information and the information used by people, i.e., low-level image features in computers are different from high-level semantic information grasped by humans. Specifically, for example, in the process of watching surveillance video, people can judge a series of information such as the crowd, pedestrians, and relationships among and emotions of people in video. A computer can only obtain a series of image features such as regional texture, image color blocks, and motion direction. In order to make full use of data in urban surveillance video, intelligent video analysis technology has become an effective way and inevitable choice.

The process of intelligent video analysis mainly includes: separating background and targets in a scene, identifying a real target, removing background interference (such as leaf shaking, water surface waves, and light changes), and then identifying specific targets in a camera scene, tracking and analyzing its behavior. This chapter focuses on a hot issue in intelligent video analysis: person re-identification (Re-ID).

Person Re-ID is a new technology that has emerged in the field of intelligent video analysis in recent years. It belongs to the category of image processing and analysis in complex video environments. It is the main task in many surveillance and security applications [1-3], and it has gained more and more attention in the field of computer vision [4-8]. Person Re-ID refers to identifying a target person in video sequences of existing possible sources and non-overlapping camera field of views.

The research on person Re-ID can be traced back to cross-camera multi-target tracking. In 2005, the literature [9] put forward a problem of how to match a target person in videos of other cameras when it was lost in video of a certain camera. In 2006, literature [10] first proposed the concept of person Re-ID, which was extracted from the cross-camera multi-target tracking problem and studied as an independent problem. Early research on person Re-ID usually used traditional methods, such as extracting hand-crafted features. Since 2014, deep learning has developed rapidly. Scholars have tried to apply deep-learning technology to the field of person Re-ID, and achieved better results than the traditional learning methods. However, whether it is hand-crafted feature-based methods or deep learning-based ones, to achieve the purpose of being adequately trained, they require a sufficient number of samples to participate in training. In many scenarios, it is very expensive, difficult, or even impossible to collect a large amount of labeled data. Therefore, how to use a small amount of labeled data to train a good model has become a very important topic in the development of machine learning, which is what we call small sample size problem (S^3).

9.2   Person Re-ID under Small-Sample Learning

Small-sample learning (SSL) can be traced back to the beginning of 2000 [11], also known as few-shot learning or one-shot learning, this concept is orthogonal to zero-shot learning (ZSL). Small-sample learning is of great significance and challenge in person Re-ID. Whether it has the ability to learn and generalize from S^3 data is an obvious dividing point between artificial intelligence and human intelligence. Humans can easily build the awareness of new things with only one or a few samples. Yet machine-learning algorithms usually require tens of thousands of supervised samples to ensure their generalization ability.

The theoretical and practical significance of studying small-sample learning mainly comes from the following three aspects:

(1) SSL does not rely on large-scale training samples, thus avoiding the high cost of specific data preparation;

(2) SSL can narrow the distance between human intelligence and artificial intelligence, which is the only way to develop general-purpose artificial intelligence;

(3) SSL can realize low-cost and fast model deployment for a new task that can only collect few samples.

Person Re-ID faces the problem of fast and low-cost learning in the case of small samples. In recent years, many research results have emerged regarding the problem of small-sample learning in person Re-ID. They include data augmentation techniques, metric learning-based methods, and meta-learning approaches. This chapter focuses on metric learning, data augmentation and pseudo-label generation method.

9.2.1 Sample generation methods

Before deep learning was introduced into the field of person Re-ID, research on Re-ID was mainly based on hand-crafted methods. Therefore, at this time, the research on S^3 is mostly based on hand-crafted methods. It has been mostly conducted from the perspective of metric learning, by solving inverse matrix instability problems of covariance matrices, or from the relationship between feature dimension and the number of training samples. Most of such research, e.g., [12][13] is to reduce the impact of insufficient samples by generating diverse, realistical new virtual samples. Before introducing these methods, we briefly introduce a baseline metric learning algorithm: Keep It Simple and Straightforward Metric (KISSME) [14].

9.2.1.1 KISSME fundamentals

KISSME [14] suggests that the samples after classification can fall into either same category (named intra-samples whose set is denoted as $\mathbf{\Omega}_0$) or different one (named inter-samples whose set is denoted as $\mathbf{\Omega}_1$). It is assumed that the difference vector of two samples with pairwise constraints conforms to the Gaussian distribution:

$$P(\mathbf{\Delta} \,|\mathbf{\Omega}_k) = (1/(2\pi^{d/2} \,|\mathbf{\Sigma}_k|^{1/2}))e^{(-1/2)\mathbf{\Delta}^T \mathbf{\Sigma}_0^{-1} \mathbf{\Delta}} \qquad (9.1)$$

where $k \in \{0, 1\}$, $\boldsymbol{\Delta} = \mathbf{x}_i - \mathbf{x}_j$ represents the difference between two feature vectors. Finally, the ratio of the probability $P(\boldsymbol{\Delta} | \boldsymbol{\Omega}_0)/P(\boldsymbol{\Delta} | \boldsymbol{\Omega}_1)$ is used to determine whether a sample pair is in the same class or not. For classification, if the ratio is larger than 1, $\mathbf{x}_i$ and $\mathbf{x}_j$ represent a same person, otherwise $\mathbf{x}_i$ and $\mathbf{x}_j$ represent two different persons.

KISSME can also be used to solve other multi-classification problems. By applying the Bayesian rule, we have the final simplified and approximated as $f(\boldsymbol{\Delta}) = \boldsymbol{\Delta}^T (\boldsymbol{\Sigma}_0^{-1} - \boldsymbol{\Sigma}_1^{-1}) \boldsymbol{\Delta}$, and the corresponding distance between $\mathbf{x}_i$ and $\mathbf{x}_j$ can be derived as:

$$d(\mathbf{x}_i, \mathbf{x}_j) = f(\boldsymbol{\Delta}) = \boldsymbol{\Delta}^T (\boldsymbol{\Sigma}_0^{-1} - \boldsymbol{\Sigma}_1^{-1}) \boldsymbol{\Delta} \tag{9.2}$$

Finally, the metric learning distance is equivalent to estimating:

$$M = \boldsymbol{\Sigma}_0^{-1} - \boldsymbol{\Sigma}_1^{-1} \tag{9.3}$$

Since KISSME requires no repeated iteration, it is efficient even if it faces a large number of samples. However, it requires additional effort in the process of calculating the kernel function when facing a few samples only [15]. With a few samples only, it easily suffers from singularity and inverse matrix instability problems of covariance matrices.

9.2.1.2 Regularized discriminant analysis

As for avoiding the singularity problem, we can reduce the dimension of feature vectors to meet the condition: $n > d$ where n is the number of samples and d is the dimension of feature vector. Yet, the dimension of feature vectors cannot be too small in order that the dimension-reduced feature vectors retain more than 95% of the original image information. Therefore, we need other technical means to further ensure that there is no singularity problem in covariance matrices. Regularized Discriminant Analysis (RDA) [16] is proved to be an effective way to avoid the singularity of covariance matrices.

First, diagonalize covariance matrix $\boldsymbol{\Sigma}_k$:

$$\boldsymbol{\Sigma}_k = \boldsymbol{\Phi}_k \boldsymbol{\Lambda}_k \boldsymbol{\Phi}_k^T \tag{9.4}$$

$$\boldsymbol{\Lambda}_k = diag[\lambda_{k1}, \lambda_{k2}, \lambda_{k3}, \cdots, \lambda_{kd}] \tag{9.5}$$

$$\boldsymbol{\Phi}_k = [\mathbf{v}_{k1}, \mathbf{v}_{k2}, \mathbf{v}_{k3}, \cdots, \mathbf{v}_{kd}] \tag{9.6}$$

where λ_{ki} ($i = 1, 2, 3, \cdots, d$) is an eigenvalue of $\boldsymbol{\Sigma}_k$, $k \in \{0, 1\}$. $\mathbf{v}_{ki}$ ($i = 1, 2, 3, \cdots, d$) is the ith eigenvector of $\boldsymbol{\Sigma}_k$:

$$\boldsymbol{\Sigma}_k^{-1} = \sum_{i=1}^{d} (\boldsymbol{\Phi}_k \boldsymbol{\Phi}_k^T)/\lambda_{ki} \tag{9.7}$$

Referring to RDA, the covariance matrix in Equation 9.4 can be interpolated by a unit matrix, and approximated as:

$$\tilde{\boldsymbol{\Sigma}}_k = (1 - \beta)\boldsymbol{\Sigma}_k + \beta\varepsilon_k \mathbf{I}_d = \boldsymbol{\Phi}_k[(1 - \beta)\boldsymbol{\Lambda}_k + \beta\varepsilon_k \mathbf{I}_d]\boldsymbol{\Phi}_k^T \tag{9.8}$$

where $\varepsilon_k = \frac{1}{d}Tr(\boldsymbol{\Sigma}_k)$, $Tr(\boldsymbol{\Sigma}_k)$ is the trace of $\boldsymbol{\Sigma}_k$. $\mathbf{I}_d$ is a $d \times d$ unit matrix. $0 < \beta < 1$ is a shrinkage parameter. β can be used to further adjust $\boldsymbol{\Sigma}_k$. This can make the covariance matrix $\tilde{\boldsymbol{\Sigma}}_k$ approach the unit matrix.

By using $\tilde{\boldsymbol{\Sigma}}_k$ to replace $\boldsymbol{\Sigma}_k$ of Equation 9.2, we have:

$$d(\mathbf{x}_i, \mathbf{x}_j) = \boldsymbol{\Delta}(\tilde{\boldsymbol{\Sigma}}_0^{-1} - \tilde{\boldsymbol{\Sigma}}_1^{-1})\boldsymbol{\Delta}^T \tag{9.9}$$

It can be rewritten as follows:

$$d(\mathbf{x}_i, \mathbf{x}_j) = \boldsymbol{\Delta}\tilde{\mathbf{M}}\boldsymbol{\Delta}^T \tag{9.10}$$

where $\tilde{\mathbf{M}} = \tilde{\boldsymbol{\Sigma}}_0^{-1} - \tilde{\boldsymbol{\Sigma}}_1^{-1}$. When sample $\mathbf{x}_i$ is given, the ranking result of candidate sample $\mathbf{x}_j$ can be calculated according to Equation 9.9. The smaller $d(\mathbf{x}_i, \mathbf{x}_j)$, the more similar candidate samples. At this point, the singularity problem can be solved to a large extent, but the inverse matrix instability problem still exist. Next we discuss how to solve it.

9.2.1.3 Stability analysis

From Equation 9.7 we know that $\boldsymbol{\Sigma}_k^{-1}$ is more influenced by small eigenvalues and their eigenvectors. The effect of too small eigenvalues is exaggerated. By plugging $\boldsymbol{\Lambda}_k = diag[\lambda_{k1}, \lambda_{k2}, \lambda_{k3}, \cdots, \lambda_{kd}]$ into Equation 9.8, we have the covariance matrix after RDA:

$$\tilde{\boldsymbol{\Sigma}}_k = \boldsymbol{\Phi}_k\{diag[(1-\beta)\lambda_{k1} + \beta\varepsilon_k, (1-\beta)\lambda_{k2} + \beta\varepsilon_k, \cdots, (1-\beta)\lambda_{kd} + \beta\varepsilon_k]\boldsymbol{\Phi}_k^T \tag{9.11}$$

The diagnoal matrix containing all eigenvalues of the covariance matrix is:

$$\tilde{\boldsymbol{\Lambda}}_k = diag[(1-\beta)\lambda_{k1} + \beta\varepsilon_k, (1-\beta)\lambda_{k2} + \beta\varepsilon_k, \cdots, (1-\beta)\lambda_{kd} + \beta\varepsilon_k] \tag{9.12}$$

Then $\tilde{\boldsymbol{\Sigma}}_k^{-1}$ can be rewritten as:

$$\tilde{\boldsymbol{\Sigma}}_k^{-1} = \sum_{i=1}^{d} \frac{\boldsymbol{\Phi}_k\boldsymbol{\Phi}_k^T}{(1-\beta)\lambda_{ki} + \beta\varepsilon_k} \tag{9.13}$$

The instability of $\tilde{\boldsymbol{\Sigma}}_k^{-1}$ is caused by the smallest eigenvalues and their eigenvectors. Let $\tilde{\mathbf{M}} = \sum_{i=1}^{d}(\boldsymbol{\Psi}\boldsymbol{\Psi}^T)/\lambda_i$. The stability of $\tilde{\mathbf{M}}$ is affected by its own small eigenvalues. Although RDA can overcome the singularity problem of covariance matrices, it cannot overcome the inverse matrix instability problem. We must thus identify another way to do so. One idea is to increase the number of training samples by making virtual samples based on the original ones. There are two methods to do so:

Method 1: Feature vector generation in subspace

As mentioned above, increasing the size of samples is a potential way to stabilize eigenvalues (i.e., avoiding too small eigenvalues). According to [17], the number of training samples should be greater than 5–10 times the number of feature space dimensions, so as to obtain higher pedestrian Re-ID accuracy. Generating virtual samples based on the original ones is one way to achieve sample amplification. The use of virtual samples can reduce the instability caused by the singular value in the process of calculating the corresponding covariance matrix, so as to improve the stability of the inverse matrix and reduce the over-fitting.

The corresponding standard deviation of intra-samples is as follows:

$$r_k = [(1/N_k) \sum_{i=1}^{N_k} (\mathbf{x}_{ki} - \mathbf{u}_k)^T (\mathbf{x}_{ki} - \mathbf{u}_k)]^{1/2} \tag{9.14}$$

where $k \in \{0, 1\}$, and 0 represents intra-samples meeting pairwise constraints, while 1 inter-samples. N_0 represents the number of similar feature vectors. N_1 represents the number of dissimilar vectors. $\mathbf{x}_{ki}$ represents the feature vector of class-k, the ith training sample. All kinds of samples can be seen as being distributed on a hypersphere with center $\mathbf{u}_k$, and radius r_k. The corresponding covariance matrix can be written as:

$$\boldsymbol{\Sigma}_k = (1/N_k) \sum_{i=1}^{N_k} (\mathbf{x}_{ki} - \mathbf{u}_k)(\mathbf{x}_{ki} - \mathbf{u}_k)^T \tag{9.15}$$

Let $\mathbf{y}_{ki} = \mathbf{x}_{ki} - \mathbf{u}_k$. Then Equations 9.14 and 9.15 can be converted into:

$$r_k = [(1/N_k) \sum_{i=1}^{N_k} \mathbf{y}_{ki}^T \mathbf{y}_{ki}]^{1/2} \tag{9.16}$$

$$\boldsymbol{\Sigma}_k = (1/N_k) \sum_{i=1}^{N_k} \mathbf{y}_{ki} \mathbf{y}_{ki}^T \tag{9.17}$$

Samples in Equations 9.16 and 9.17 are decentralized. $\mathbf{y}_{ki}$ is the training sample after decentralization, whose mean is 0, and standard deviation is r_k.

A traditional virtual sample is generally a simple geometric transformation of the original image, such as translation, rotation, and symmetry transformation. Because it is calculated in a high-dimensional space, the amount of computation is very large. A virtual sample in our work is generated in a feature space whose dimension is relatively small. It should satisfy the conditions that its mean is zero, and the standard deviation is less than or equal to r_k. That is, a virtual training sample $\mathbf{x}'_{ki}$ falls in the hypersphere whose center is at the origin, and radius r_k is the standard deviation of $\boldsymbol{\Sigma}_k$. In other words, $\mathbf{x}'_{ki}$ meets the following requirements:

$$\overline{X}' = (1/N'_k) \sum_{i=1}^{N'_k} \mathbf{x}'_{ki} = 0 \tag{9.18}$$

$$r'_{ki} = \sqrt{\frac{1}{N'_k} \sum_{i=1}^{N'_k} (\mathbf{x}'_{ki})^T \mathbf{x}'_{ki}} \le r_k \qquad (9.19)$$

where N'_k is the number of class-k virtual samples. $\bar{\mathbf{X}}'$ is the mean of the virtual samples and r'_{ki} is the square root of the inner product. There are many ways to generate random virtual samples that satisfy Equations 9.18 and 9.19. In this article, we use a set of basis vector $\mathbf{e}'_1 = (1, 0, 0 \cdots 0)$, $\mathbf{e}'_2 = (0, 1, 0, \cdots 0)$, and $\mathbf{e}'_d = (0, 0, 0, \cdots, 1)$ in d-dimensional subspaces, and then select another set of symmetric vectors $-\mathbf{e}'_1$, $-\mathbf{e}'_2$, ..., and $-\mathbf{e}'_d$. The mean of these two sets of vectors is 0, thus satisfying Equation 8.12. Because $\mathbf{e}'_d \mathbf{e}'^T_d = (-\mathbf{e}'_d)(-\mathbf{e}'_d)^T$, when computing covariance matrix, we can just take $\mathbf{e}'_1$, $\mathbf{e}'_2$, ..., and $\mathbf{e}'_d$. Let the i-th virtual sample of class-k be:

$$\mathbf{x}'_{ki} = R_d e_d \mathbf{y}_{ki} \qquad (9.20)$$

The following theorem shows that the standard deviation of $\mathbf{x}'_{ki}$, $k \in \{0, 1\}$ satisfies Equation 9.19.

Theorem 9.1. Let $\mathbf{x}'_{ki} = R_d e_d \mathbf{y}_{ki}$, $k \in \{0, 1\}$, $i = 1, 2, \cdots, d$, where d represents a d-dimensional space, R_d represents a random number in $[0, 1]$, and r_k represents the standard deviation of class k. Then the standard deviation r'_{ki} of $\mathbf{x}'_{ki}$ satisfies $r'_{ki} = R_d r_k \le r_k$.

Proof:

$$r'_{ki} = \sqrt{\frac{1}{N'_k} \sum_{i=1}^{N'_k} (\mathbf{x}'_{ki})^T \mathbf{x}'_{ki}}$$

$$= \sqrt{\frac{1}{N'_k} \sum_{i=1}^{N'_k} (R_d e_d \mathbf{x}_{ki})^T (R_d e_d \mathbf{x}_{ki})}$$

$$= R_d \sqrt{\frac{1}{N'_k} \sum_{i=1}^{N'_k} \mathbf{x}_{ki}^T e_d^T e_d \mathbf{x}_{ki}}$$

$$= R_d \sqrt{\frac{1}{N'_k} \sum_{i=1}^{N'_k} \mathbf{x}_{ki}^T \mathbf{x}_{ki}}$$

$$= R_d r_k \qquad (9.21)$$

Because $R_d \le 1$, $r'_{ki} \le r_k$. ◆

According to Equation 9.20, each pattern can generate d virtual samples, and a total of $d \cdot C$ virtual samples can be generated (C is the number of patterns). So let $\hat{\boldsymbol{\Sigma}}_k$ represent the new covariance matrix calculated by using the newly generated virtual samples and original samples. By using $\overset{\bullet}{\boldsymbol{\Sigma}}_k$ to replace $\boldsymbol{\Sigma}_k$ in Equation 9.8, we have:

$$\hat{\boldsymbol{\Sigma}}_k = (1 - \beta) \overset{\bullet}{\underset{k}{\boldsymbol{\Sigma}}} + \beta \varepsilon_k \mathbf{I}_d \qquad (9.22)$$

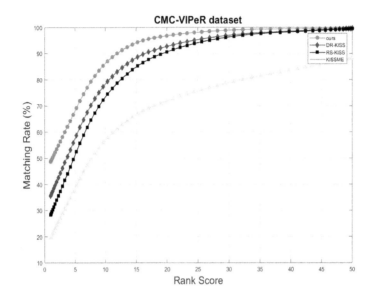

FIGURE 9.1
Comparison among three KISS-related algorithms.

We can rewrite Equation 9.9 and $\tilde{\mathbf{M}}$ as:

$$d(\mathbf{x}_i, \mathbf{x}_j) = \mathbf{\Delta}(\hat{\mathbf{\Sigma}}_0^{-1} - \hat{\mathbf{\Sigma}}_1^{-1})\mathbf{\Delta}^T$$

$$\tilde{\mathbf{M}} = \hat{\mathbf{\Sigma}}_0^{-1} - \hat{\mathbf{\Sigma}}_1^{-1} \tag{9.23}$$

This new method based on KISSME and virtual sample generation method is named KISS+. In order to highlight the performance improvement and optimization in the corresponding steps, we compare KISS+ with three KISS-related algorithms. They are original KISSME [14], RS-KISS [16], and DR-KISS(dual-regularized KISS) [18]. The comparison is based on VIPeR, and the experimental results are shown in Figure 9.1.

Figure 9.1 shows their CMC curves based on the VIPeR dataset, where the green curve with triangles represents KISSME, the black curve with squares represents RS-KISS, the blue with diamonds represents DR-KISS, and the red one with dots represents KISS+. From them we can visually see that the performance of KISS+ is superior to the others. Its matching rate at Rank=1 is 48.7%, which is much higher than 19.6% of KISSME, 28.3% of RS-KISS, and 35.6% of RS-KISS.

To answer the roles of stable eigenvalues played in KISS+ and KISSME, we compare the eigenvalues of their covariance matrices, and consider the top 20 smallest ones as sorted in descending order, as shown in Figure 9.2.

It can be seen from Figure 9.2 that the eigenvalues of the optimized co-variance matrix are significantly larger than those corresponding to KISSME

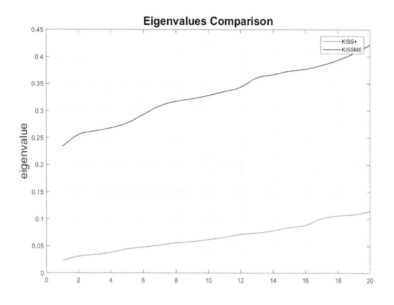

FIGURE 9.2
Eigenvalues (top 20 smallest ones) comparison between KISSME and KISS+.

because KISS+ uses not only original samples as KISSME but also virtual ones. The stability of the inverse matrix of the covariance one after optimization is greatly improved, which in turn provides favorable conditions for high-accuracy pedestrian Re-ID.

Method 2: New sample generated by genetic algorithm

When training samples increase, it is necessary to ensure that the newly added training samples are different from the original ones and yet retain the latter's basic characteristics. Genetic Algorithm (GA) [19] can use the original samples as initial population, and generate offspring samples to meet such requirements well.

GA [19] was proposed by Holland in the 1970s. It is a search and optimization algorithm developed for the simulation of biological genetics and long-term evolution and has gained many applications [20-24]. We can adopt it to generate virtual samples. We propose to use the feature vectors of original samples as its initial population. The probability of each sample feature being selected for mutation and crossover operations is made equal. The fitness value of the generated offspring sample is set as the maximum value of Mahalanobis distance between the new sample and original ones. Only if the value is less than the maximum distance among all the original samples, is the newly generated sample added to the sample set; otherwise, it is discarded because it is not close enough to the original samples. Suppose that the samples selected for crossover are $\mathbf{x}_a$ and $\mathbf{x}_b$, and for mutation is $\mathbf{x}_c$. Assume that the number of samples is n, and a, b, and c are positive integers satisfying that $0 < a, b, c < n$. $\mathbf{x}_a$, $\mathbf{x}_b$, and $\mathbf{x}_c$ are rewritten as $(x_{a1}, x_{a2}, x_{a3}, \ldots, x_{ad})$, $(x_{b1}, x_{b2}, x_{b3}, \ldots, x_{bd})$, and $(x_{c1}, x_{c2}, x_{c3}, \ldots, x_{cd})$, respectively, where d is the dimension of a feature vector. Then crossover and mutation operators are as follows:

Crossover:

$$\begin{aligned} \mathbf{x}_a' &= (x_{a1}, x_{a2}, x_{a3}, \ldots, x_{bi}, \ldots, x_{ad}) \\ \mathbf{x}_b' &= (x_{b1}, x_{b2}, x_{b3}, \ldots, x_{ai}, \ldots, x_{bd}) \end{aligned} \tag{9.24}$$

where $\mathbf{x}_a'$ and $\mathbf{x}_b'$ are the generated offspring samples, and $i \in \{1, 2, 3, \ldots, d\}$ is the randomly selected to operate crossover.

Mutation:

$$\mathbf{x}_c' = (x_{c1}, x_{c2}, x_{c3}, \ldots, x_{cj} + \delta, \ldots, x_{cd}) \tag{9.25}$$

where $\mathbf{x}_c'$ is the generated offspring sample, $j \in \{1, 2, 3, \ldots, d\}$ is the randomly selected to operate mutation, and δ is the coefficient of mutation.

Assume that $\mathbf{x}'$ is a newly generated sample. Our fitness function is:

$$F(\mathbf{x}') = \max_i \{\mathbf{d}(\mathbf{x}', \mathbf{x}_i)\} \tag{9.26}$$

where $\mathbf{d}(\mathbf{x}', \mathbf{x}_i)$ is the Mahalanobis distance between two samples $\mathbf{x}'$ and $\mathbf{x}_i$, $i \in \{1, 2, 3, \ldots, m\}$, m is the number of samples of a same class or with a same identity. If the maximum distance among all the original samples is denoted as d^M, and $d^M = \max_{i,j} \{\mathbf{d}(\mathbf{x}_i, \mathbf{x}_j)\}$ (i, j$\in$\{1, 2, 3, \ldots, m\}), then

$$\begin{cases} F(\mathbf{x}') \leq d^M, & \quad accepted \\ F(\mathbf{x}') > d^M, & \quad discarded \end{cases} \tag{9.27}$$

According to Equation 9.27, if the fitness value is no larger than d^M, the corresponding newly generated sample is added to the sample set; otherwise, discarded. Next, we use the updated sample set to train every covariance matrix. Assuming that the new covariance matrix is $\hat{\boldsymbol{\Sigma}}_k$. By using $\hat{\boldsymbol{\Sigma}}_k$ to replace $\tilde{\boldsymbol{\Sigma}}_k$ of Equation 9.9, we have

$$\mathbf{d}(\mathbf{x}_i, \mathbf{x}_j) = \boldsymbol{\Delta}(\hat{\boldsymbol{\Sigma}}_0^{-1} - \hat{\boldsymbol{\Sigma}}_1^{-1})\boldsymbol{\Delta}^T \tag{9.28}$$

where $\hat{\mathbf{M}} = \hat{\boldsymbol{\Sigma}}_0^{-1} - \hat{\boldsymbol{\Sigma}}_1^{-1}$. We can name the proposed method as G-KISS where G represents Genetic Algorithm.

Based on the characteristics of available databases, we chose i-LIDS to validate the proposed algorithm, because i-LIDS only contains 476 images of 119 people. The number of samples in the database is relatively modest, which leads to an S^3 problem (the number of samples n needs to satisfy d < n < 10d (d is the feature space dimension)). That is why we choose i-LIDS.

The high-dimensional feature extracted in experiments is Local Binary Patterns (LBP) [25], which is a simple yet very efficient texture operator. It compares each pixel with its nearby pixels and saves the result as a binary number. Because of its strong discriminating power and computational simplicity, LBP has been applied in different scenarios [26]. Its most important property is the robustness to changes in grayscale caused by illumination and other changes. Another important property is its computational simplicity, which allows it to analyze images in real time.

The experimental results are shown in Figure 9.3–9.4 and Table 9.1, where G0 represents KISSME that does not adopt any virtual samples from GA. From Figure 9.3, the effect of virtual samples on the improvement of KISSME is clear. As the number of generations increases, the number of generated virtual samples also increases, and in general, so does the matching rate. Yet this increase is not unlimited, as can be seen more clearly in Table I, when the generation count reaches about 400, CMC at Rank = 1 can achieve its optimal value of 51%. At this moment, there are 400 accepted virtual samples in the training set. When the generation count reaches 450 and 500, the matching rate has decreased. The reason for this may be that there are too many virtual samples in the set, and the new samples produce features that do not conform to the characteristics of the original samples. It should be noted that only one sample per generation in this experiment is involved in mutation and only two samples are involved in crossover. Therefore, if there are different number of samples involved in crossover and mutation, the number of needed generations to obtain similar results is different. For example, if two samples per generation are involved in mutation and four ones involved in crossover, the number of generations for obtaining a similar matching rate may be less than 200. Although G-KISS may be over-evolved, we can control the evolution in a way that keeps G-KISS at its high performance.

Figure 9.3 and Table 9.1 show the external performance of G-KISS in terms of matching rate improvement. We can further observe the contribution of G-KISS to eigenvalue stabilization by observing how eigenvalues of various covariance matrices vary. As shown in Figure 9.4, the bold dark blue curve in Figure 9.4 represents them at G0, i.e., KISSME without using virtual samples. The bold red curve represents them at G400, i.e., at GA generation 400. It can be seen from Figure 9.4 that the eigenvalues of various covariance matrices of G-KISS with virtual samples are generally greater than those of KISSME. The improvement of eigenvalue stabilization also brings about higher matching rate. Thus, virtual samples are valid for improving KISSME. However, this does not mean that eigenvalues could be as large as possible. Large eigenvalues would be overestimated, and as such, the performance can degrade. The results in Figure 9.4 are in agreement with the experimental results of Figure 9.3 and Table 9.1.

Experimental results show that the above two virtual sample generation methods are effective and can be used to improve KISSME's performance and overcome its S^3 problem.

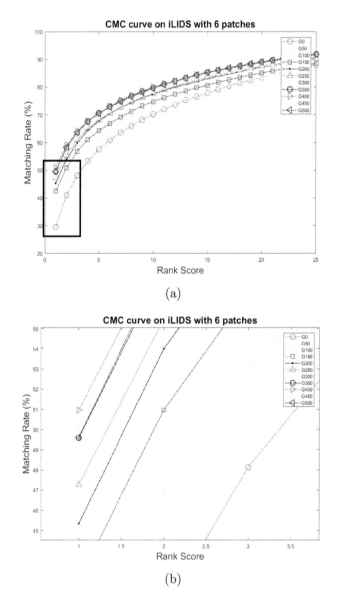

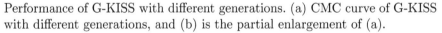

FIGURE 9.3
Performance of G-KISS with different generations. (a) CMC curve of G-KISS
with different generations, and (b) is the partial enlargement of (a).

9.2.2   Pseudo-label learning method for generated samples: a semi-supervised learning method

With the vigorous development of deep learning, especially the great success of
CNN in visual tasks, many SSL researchers have begun to shift their focus from

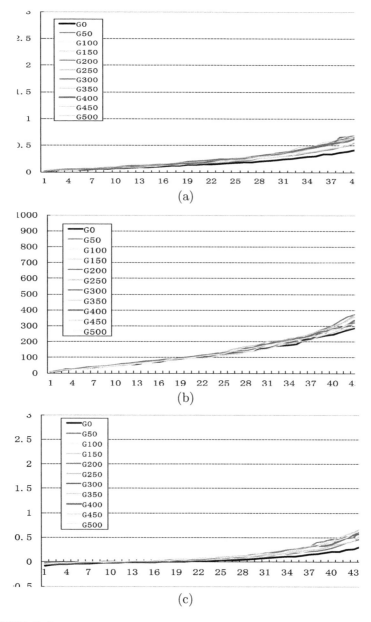

FIGURE 9.4

Eigenvalue variation of covariance matrices. (a) Eigenvalues ($*10^{-3}$) of inverse matrix of intra-covariance matrix, (b) Eigenvalues of inverse matrix of inter-covariance matrix, and (c) Eigenvalues of $\hat{M}$. The X-axis represents the sequence number of eigenvalues, and the Y-axis represents the eigenvalues.

TABLE 9.1
Matching rate at rank = 1, 5, 10, 20 and PUR scores of G-KISS with different generations and the number of accepted virtual samples.

Generation	p=59, 6 patches					
	Rank =1	Rank =5	Rank =10	Rank =20	PUR	Accepted virtual samples
0	29.6	57.5	70.2	83.4	27.0	0
50	35.5	59.5	71.6	83.6	29.8	72
100	37.5	61.3	73.2	85.4	31.7	93
150	42.5	64.3	74.7	85.0	34.8	132
200	45.3	67.8	77.4	87.2	38.2	176
250	47.3	68.1	77.6	87.5	39.3	250
300	48.2	68.8	78.3	87.4	40.2	276
350	49.6	70.6	79.7	89.0	42.0	303
400	51.0	70.7	79.9	88.8	42.8	400
450	50.1	70.0	79.0	88.0	41.7	431
500	49.6	70.4	79.5	89.1	42.1	469

non-deep models to deep ones. In 2015, G. Koch et al. [27] proposed a Siamese network to learn a class-irrelevant similarity metric on pairwise samples. It was the first to incorporate deep learning into the solution for an SSL problem. After that, subsequent SSL methods make full use of the advantages of deep neural networks in feature representation and end-to-end model optimization, and solve SSL problems from different angles, such as data augmentation. A typical method is Cycle-GAN [28]. New training samples generated by Cycle-GAN can be used to alleviate the problem of data imbalance in pedestrian Re-ID. Yet they are unlabeled. Therefore, a pseudo-label generation method [29] (semi-supervised learning method) was proposed. Before introducing this method, We first review Cycle-GAN and introduce a data generation process.

9.2.2.1 Image-to-image translation

For domains $\mathbf{X}$ and $\mathbf{Y}$, given training samples $\{\mathbf{x}_i\}_{i=1}^n \in \mathbf{X}$ and $\{\mathbf{y}_j\}_{j=1}^m \in \mathbf{Y}$, a column of matrices $\mathbf{X}$ and $\mathbf{Y}$ represents a sample. The goal of Cycle-GAN is to learn mapping functions between these two domains. It contains two mappings (G, F) and their adversarial discriminators $(D_{\mathbf{X}}, D_{\mathbf{Y}})$. G : $\mathbf{X} {\rightarrow} \mathbf{Y}$ tries to generate images $G(x)$ that look similar to images from domain $\mathbf{Y}$, while $D_{\mathbf{Y}}$ aims to distinguish whether images $G(x)$ are translated from domain $\mathbf{Y}$. $F{:}\mathbf{Y}{\rightarrow}\mathbf{X}$ is similar as G, tries to generate images $F(y)$ that look similar to images from domain $\mathbf{X}$, and its adversarial discriminators $D_{\mathbf{X}}$ is proposed to distinguish whether images $F(y)$ are translated from domain $\mathbf{X}$. The adversarial loss of generator G and its adversarial discriminator $D_{\mathbf{Y}}$ are as follows:

$$\Gamma_{GAN}(G, D_{\mathbf{Y}}, \mathbf{X}, \mathbf{Y}) = E_{y \sim p_{data}(y)}[\log D_{\mathbf{Y}}(y)]$$
$$+ E_{x \sim p_{data}(x)}[\log(1 - D_{\mathbf{Y}}(G(x)))] \quad (9.29)$$

For generator F and its adversarial discriminator $D_{\mathbf{X}}$, the adversarial loss is:

$$\Gamma_{GAN}(F, D_{\mathbf{X}}, \mathbf{Y}, \mathbf{X}) = E_{x \sim p_{data}(x)}[\log D_{\mathbf{X}}(x)]$$
$$+ E_{y \sim p_{data}(y)}[\log(1 - D_{\mathbf{Y}}(F(y)))] \quad (9.30)$$

To recover the original image after a cycle translation, Cycle-GAN introduces the *cycle consistent loss*:

$$\Gamma_{cyc}(G, F) = E_{x \sim p_{data}(x)}[|| F(G(x)) - x ||_1] + E_{y \sim p_{data}(y)}[|| G(F(y)) - y ||_1] \quad (9.31)$$

The full loss function is:

$$\Gamma(G, F, D_{\mathbf{X}}, D_{\mathbf{Y}}) = \Gamma_{GAN}(G, D_{\mathbf{Y}}, \mathbf{X}, \mathbf{Y}) + \Gamma_{GAN}(F, D_{\mathbf{X}}, \mathbf{Y}, \mathbf{X}) + \lambda \Gamma_{cyc}(G, F) \quad (9.32)$$

where $\Gamma_{GAN}(G, D_{\mathbf{Y}}, \mathbf{X}, \mathbf{Y})$ and $\Gamma_{GAN}(F, D_{\mathbf{X}}, \mathbf{Y}, \mathbf{X})$ are the loss functions for the mapping function G and F, and their adversarial discriminators $D_{\mathbf{Y}}$ and $D_{\mathbf{X}}$; while $E_{x \sim p_{data}(x)}$ and $E_{y \sim p_{data}(y)}$ are $L1$ loss functions. $\Gamma_{cyc}(G, F)$ is the cycle consistency loss function that is used to ensure that each image can be reconstructed after a cycle mapping. λ is used to indicate the relative importance of Γ_{GAN} and Γ_{cyc}.

Given a pedestrian Re-ID dataset whose images are collected from K different cameras in a camera network, the work of Cycle-GAN is to learn image-to-image translation for each camera pair. A different camera is treated as a different domain in the process of generating new images. With the learned models, it can generate K-1 new images for a training image. It has added an *identity mapping loss* [28] in the Cycle-GAN loss function Equation 9.32 which aims to force the generator to be near an identity mapping when using real images of the target domain as input to the generator:

$$\Gamma_{identity}(G, F) = E_{y \sim p_{data}(y)}[|| G(y) - y ||_1] + E_{x \sim p_{data}(x)}[|| F(x) - x ||_1] \quad (9.33)$$

9.2.2.2 Semi-supervised pedestrian Re-ID

New training samples generated by Cycle-GAN can be used to alleviate the problem of data imbalance in pedestrian Re-ID. Yet they are unlabeled. We propose a semi-supervised learning framework to learn the pseudo pairwise relation between unlabeled generated samples and labeled ones. In this way, the original training images and generated ones can be connected and combined to form a new training set.

The dimension of the original feature space is usually large while a low dimensional space is preferred for classification. In this chapter, we aim to learn a discriminant subspace and at the same time learn a distance function in the low dimensional subspace to solve the pedestrian Re-ID problem. We utilize self-paced learning to select reliable unlabeled samples for pedestrian Re-ID with semi-supervised learning. The work [30] uses a self-paced learning method to learn the pairwise relationship among unlabeled samples for image matching and achieves great results.

Given a training set $\mathbf{X}_l = (\mathbf{x}_1, \mathbf{x}_2, ..., \mathbf{x}_n) \in \mathbf{R}^{d \times n}$ containing n labeled training samples in a d-dimension space, images of the same person are captured from different camera in a camera network. $\mathbf{Z} = \{\mathbf{z}_i\} \in \mathbf{R}^n$ is the label set of labeled training samples. Each column of $\mathbf{Z}$ represents a labeled training sample. We define that image pair (x_i, x_j) is a *positive* one if $z_i = z_j$, which indicates that x_i and x_j belong to a same person; and otherwise a *negative* one if $z_i \neq z_j$, which indicates that x_i and x_j belong to different persons. There is a problem, i.e., the number of positive samples is very small, while the number of negative samples is large. To overcome this data imbalance problem, we employ Cycle-GAN to generate m training samples that are unlabeled, i.e., $\mathbf{X}_u = (x_{n+1}, x_{n+2}, ..., x_{n+m}) \in \mathbf{R}^{d \times m}$. The original training set and generated sample set can form a new training set $\mathbf{X} = (x_1, x_2, ..., x_n, x_{n+1}, ..., x_{n+m}) \in \mathbf{R}^{d \times (n+m)}$.

The task is to explore a transformation U to map data into a low-dimensional space in which we can better separate images of one person from those of other persons. The mapping transforms $\mathbf{X}$ from an original input space into a low-dimensional space: $\mathbf{X} = \mathbf{U}^\mathrm{T}\mathbf{X}$. The squared distance function $d_{\mathbf{U}}^2(\mathbf{x}_i, \mathbf{x}_j)$ can be defined as follows:

$$d_{\mathbf{U}}^2(\mathbf{x}_i, \mathbf{x}_j) = ||\mathbf{U}^\mathrm{T}\mathbf{x}_i - \mathbf{U}^\mathrm{T}\mathbf{x}_j||^2 \qquad (9.34)$$

where $\mathbf{U} \in \mathbf{R}^{d \times d'} (d' << d)$ and d' is the dimension of the projected subspace. By learning distance in a low dimensional space, we make the distances between positive sample pairs less than the ones between negative sample pairs.

In order to utilize unlabeled samples, a graph approach is used to establish the relationship among nearby data nodes. We add an edge between two nodes if they are close. First, we use a small labeled data set to initialize the model to learn a discriminate subspace and get an initial projection matrix $\mathbf{U}'$. Then we represent labeled data in a low-dimensional space: $\mathbf{X}_l' = (\mathbf{U}')^T\mathbf{X}_l$. Second, with projection matrix $\mathbf{U}'$, we can project the unlabeled data into a low-dimensional subspace, i.e., $\mathbf{X}_u' = (\mathbf{U}')^T\mathbf{X}_u$. Therefore, we can build the pseudo pairwise relations between unlabeled data and labeled one by constructing graph using $\mathbf{X}_u'$. We can encode the pseudo pairwise relationships into a pairwise constraint matrix $\mathbf{W} \in \mathbf{R}^{(n+m) \times (n+m)}$ for training samples. $\mathbf{W}_{ij} = 1$ if x_i and x_j belong to the same person and $\mathbf{W}_{ij} = 0$ otherwise.

Laplacian Eigenmaps are used to solve the projection matrix $\mathbf{U}'$ with labeled data. The distance is expected to be small among training samples belonging to the same person. Under this expectation, we can formulate the

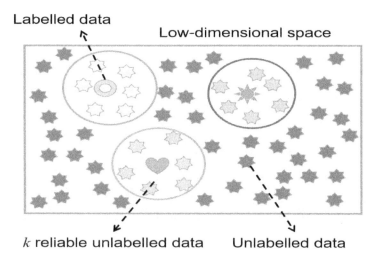

FIGURE 9.5
Overview of a graph approach for pseudo pairwise relationship estimation among unlabeled data and labeled ones.

problem as follows:

$$\mathbf{U}' = \arg \min_{U} \mathbf{\Phi}(\mathbf{X}_l, \mathbf{U}, \mathbf{W}_l)$$

$$= \frac{1}{2} \sum_{i=1}^{n} \sum_{j=1}^{n} \mathbf{W}_{ij}^l || \mathbf{U}^T \mathbf{x}_i - \mathbf{U}^T \mathbf{x}_j ||^2 = \mathrm{tr}(\mathbf{U}^T \mathbf{X}_l L^l \mathbf{X}_l^T \mathbf{U})$$

$$\text{s.t. } \mathrm{tr}(\mathbf{U}^T \mathbf{X}_l \mathbf{L}^l \mathbf{X}_l^T \mathbf{U}) = 1 \qquad (9.35)$$

where $\mathbf{W}_l$ is a pairwise constraint matrix. W_{ij}^l is an element of $\mathbf{W}_l \in \mathbf{R}^{n \times n}$ indicating whether samples x_i and x_j belong to a same person. If so, $W_{ij}^l = 1$, and otherwise 0. $tr(.)$ is a trace operator and $\mathbf{L}^l = \mathbf{D}^l - \mathbf{W}^l$ is a graph Laplacian matrix. $\mathbf{D}^l$ is a diagonal matrix whose diagonal elements equal to the sums of elements in the corresponding row of $\mathbf{W}_l$. The problem in Equation 9.35 can be solved with generalized eigen-decomposition. Therefore, we can learn a discriminant subspace with a projection matrix $\mathbf{U}'$ that is constructed by the resulting eigenvectors related to the r smallest eigenvalues. We can also get the expression of unlabeled data in this low-dimensional space with the projection matrix as shown in Figure 9.5.

In the pseudo pairwise relationship estimation step, we select k reliable candidates from unlabeled data to label data according to the distance in a low-dimensional space. Nodes with different colors in the low-dimensional space box denote samples with different identity. Therefore, we can build the pseudo pairwise relations between unlabeled data and labeled one. Then the pseudo pairwise relations can be encoded into a pairwise constraint matrix

$\mathbf{W}^u \in \mathbf{R}^{n \times m}$, i.e.,

$$\mathbf{W}_{ij}^U = \begin{cases} 1, & if \ \mathbf{x}_j \in N(\mathbf{x}_i) \\ 0 & otherwise \end{cases} \tag{9.36}$$

where x_i is a labeled sample, x_j is an unlabeled sample, and $N(x_i)$ denotes the reliable neighbor list of x_i. After obtaining pairwise constraint matrix $\mathbf{W}^u$, the generated samples can be combined with labeled data to form a new training set and then used to learn distance metric in a supervised way. Now we can formulate the objective function by utilizing the pseudo pairwise relations for labeled data and unlabeled one, i.e.,

$$\min_{\mathbf{U}} F = \frac{1}{2} \sum_{i=1}^{n+m} \sum_{j=1}^{n+m} \mathbf{W}_{ij} || \mathbf{U}^T \mathbf{x}_i - \mathbf{U}^T \mathbf{x}_j ||^2 = tr(\mathbf{U}^T \mathbf{X} \mathbf{L} \mathbf{X}^T \mathbf{U}) \tag{9.37}$$

$$s.t. \ tr(\mathbf{U}^T X L X^T \mathbf{U}) = 1$$

where $\mathbf{W}_{ij}$ is an element of pairwise constraint matrix $\mathbf{W} = [\mathbf{W}^l, \mathbf{W}^u]$. $\mathbf{L} = \mathbf{D} - \mathbf{W}$ is the graph Laplacian matrix. $\mathbf{D}$ is a diagonal matrix whose diagonal elements equal to the sum of the elements in the corresponding row of $\mathbf{W}$.

The problem of pedestrian Re-ID can be solved by solving the minimization problem in Equation 9.37 to obtain a projection matrix by generalized eigen-decomposition. We realize the proposed method in Algorithm 9.1.

Algorithm 3 Semi-supervised leaning-based pedestrian Re-ID

Input: Real labeled training set: $\mathbf{X}_l$;
                 Generated unlabeled training set $\mathbf{X}_u$;
                 The pairwise constraint matrix of labeled data: $\mathbf{W}_l$;
                 The maximal number of iterations: τ.
Output: The projection matrix $\mathbf{U}$.
   Initialization: Set $t=1$; Training set $\mathbf{X} = \mathbf{X}_l \cup \mathbf{X}u$; $\mathbf{U}'$ is initialized by learning a low dimensional subspace using a small amount of labeled training data.
   for $t = 1$ to τ **do**
      1. Projection of $\mathbf{X}u$: $\mathbf{X}^t{}_u' = (\mathbf{U}')^T \mathbf{X}_u$;
      2. Graph approach: build the pseudo pairwise relationships between labeled data and unlabeled one;
      3. Encode $\mathbf{W}_u$: encode the pseudo pairwise relationships into pairwise constraints matrix;
      4. $\mathbf{W}_l$: encode pairwise relationships into pairwise constraints matrix among labeled samples;
      5. $\mathbf{W} = [\mathbf{W}_l, \mathbf{W}_u]$;
      6. Update the projection matrix $\mathbf{U}$ by Eq (9.9);
      if $t > \tau$ or the stop projection is met **then**
      break;
   end
Output: The projection matrix $\mathbf{U}$.

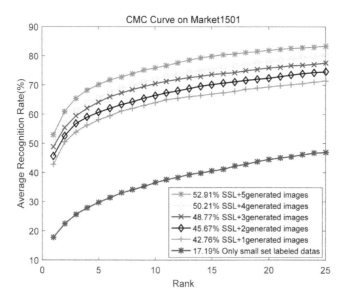

FIGURE 9.6
Performance comparison of the proposed approach with baseline on Market-1501 with different settings of the number of generate images. Rank-1 recognition rates are shown in the legends.

The pseudo pairwise relations between unlabeled samples and labeled ones play a crucial role in the proposed method. However, due to the change of viewpoint, it inevitably includes some mismatched pairwise relations. To solve this problem, we learn new pseudo pairwise relations via an iterative method and re-learn discriminant projection with these updated pseudo pairwise relations. This procedure is iterated until the pseudo pairwise relations remain unchanged.

The experimental results of the proposed method on Market1501 and DukeMTMC-ReID can be found in Figure 9.6 and Table 9.2.

Performance on Market1501:

In the experiment, by choosing 2 images for each person randomly we use only 11.61% labeled training samples. In order to reduce the cost of labeling, we employ Cycle-GAN to generate camera-style- transferred images for each person. Because Market1501 images are captured by 6 cameras, we can only generate 5 images at most. The selected labeled samples and newly generated unlabeled ones are combined to form a new training set. To better evaluate the performance, we conduct multiple experiments on Market1501 by setting the number of the newly generated images from 1 to 5, respectively.

The performance of the proposed approach and comparison with the state-of-the-art methods are shown in Table 9.2 and Figure 9.6. We can easily see

TABLE 9.2

Performance (CMC: rank = 1, 5, 10, 20 precision) of proposed method, and comparison with some state-of-the-art semi-supervised methods on Market 1501 and Duke MTMC-ReID.

| Method | Market-1501 | | | | | Duke MTMC-ReID | | | | |
	rank-1	rank-5	rank-10	rank-20	mAP	rank-1	rank-5	rank-10	rank-20	mAP
SSL+7 generated images	-	-	-	-	-	**37.70**	53.05	61.04	67.06	17.61
SSL+6 generated images	-	-	-	-	-	34.98	50.78	58.99	64.24	14.35
SSL+5 generated images	**52.91**	70.13	75.89	81.95	19.78	33.16	49.28	56.95	63.37	13.24
SSL+4 generated images	50.21	65.68	72.12	78.09	17.01	32.11	48.31	55.31	61.06	13.81
SSL+3 generated images	48.77	64.15	70.45	75.85	15.67	28.42	45.70	52.25	58.58	12.33
SSL+2 generated images	45.67	60.65	65.49	72.01	14.62	26.69	44.86	50.79	56.67	10.28
SSL+1 generated image	42.76	58.05	63.96	69.33	10.35	23.07	40.22	47.62	54.49	7.62
Baseline	**17.79**	**29.75**	**36.58**	**44.48**	**3.80**	**16.30**	**29.09**	**35.75**	**43.14**	**2.36**

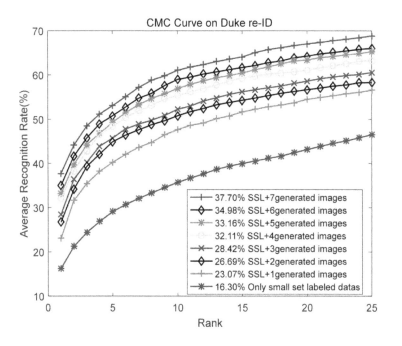

FIGURE 9.7
Performance comparison of the proposed approach with baseline on Duke MTMC-ReID with different settings of the number of generate images. Rank-1 recognition rates are shown in the legends.

that the proposed method performs well even when only very few labeled samples are used in a training process. But no noticeable improvement can be found in the case of a baseline that adds no newly generated images of the proposed SSL. It can also be seen in the case of 0 generated images in Table 9.2. The rank-1 accuracy of the baseline is only 17.79% because there are too few training samples. When the number of added generated images is 5, the performance gain is +35.12% over the baseline in terms of rank-1 accuracy. This indicates that the model can learn more information of samples from generated images. When the number of generated images added is 4, the performance gain is +32.42% over the baseline in rank-1 accuracy.

Performance on Duke MTMC-ReID:

In this experiment, by selecting 2 images for each person randomly, we use only 8.49% labeled training samples. Duke samples are captured by 8 cameras. We then generate 7 transferred images for each person. The newly generated unlabeled images and selected labeled ones from a training set are used to form a new training set. We conduct multiple experiments by setting the number of the added generated images from 1 to 7, respectively. The results are shown in Table 9.2 and Figure 9.7.

TABLE 9.3
Re-ID accuracy of SSL with different neighbors select when 5 generated samples added on Market-1501.

# neighbors(k)	1	2	3	4	5	6
rank-1	42.16	50.74	50.38	50.26	52.91	47.33
mAP	14.63	19.22	19.34	19.49	19.78	16.91

TABLE 9.4
Re-ID accuracy of SSL with different neighbors select when 7 generated samples added on Duke MTMC-reID.

# neighbors(k)	2	3	4	5	6	7
rank-1	30.15	32.16	33.13	35.24	37.48	37.7
mAP	10.5	12.1	12.8	14.4	15.3	17.6

Figure 9.7 shows the CMC performance with a different number of generated images. It can be seen that when using only a small amount of labeled data instead of unlabeled data for training, the rank-1 accuracy is only 16.30%. The performance gain is +21.4% over the baseline in rank-1 accuracy when 7 generated unlabeled samples are added since the model can learn more information of each person by generated images. For the case of adding 6 generated samples, the performance gain is +18.68% over the baseline in rank-1 accuracy and +16.86% for the case of adding 4 generated images.

Evaluation of parameters

We evaluate the impact of k in the graph approach on Market-1501 and Duke MTMC-ReID. It means the k reliable neighbor samples. Figure 9.8 and Tables 9.3 and 9.4 show the results of different k values and different number of generated images. In Figure 9.8, we can see that when k approaches the number of newly generated samples, we approach the best performance as expected. However, the performance is significantly reduced when k is larger than that because some negative samples are treated as positive ones and much noise is thus introduced. For example, when 5 generated samples for each labeled sample are added to Market-1501, Re-ID rate is 52.91% when k = 5. It reduces to 47.33% when k = 6 that exceeds the number of generated samples by just one.

Experimental results demonstrate that the value of k is critically important when learning the pseudo pairwise relations between unlabeled generated samples and labeled ones. The number of reliable neighbors (k) should be selected as the number of newly generated samples in order for SSL to gain the best performance. The performance drops sharply when it exceeds the number of positive samples added. This is because some negative samples would be treated as positive ones, thus introducing a great deal of noise.

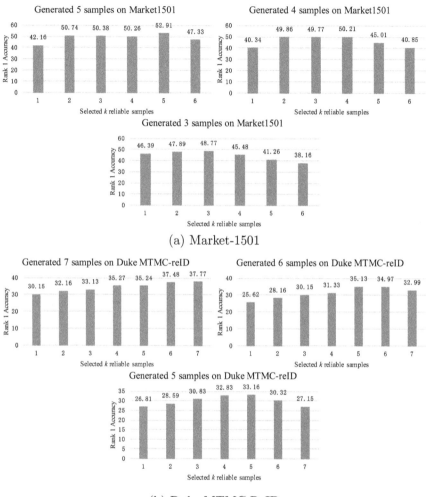

(a) Market-1501

(b) Duke MTMC-ReID

FIGURE 9.8

Re-ID matching rates (%) comparison with different k and generate samples on Market-1501 and Duke MTMC-ReID. (a) and (b) shown the rank-1 accuracy obtained on Market-1501 and Duke MTMC-ReID with different situations, respectively. Three systems are shown, which have 3, 4, and 5 generated samples for Market-1501 and 5, 6, and 7 generated samples for Duke MTMC-ReID, respectively.

9.3   Conclusion

This chapter focuses on a key topic in intelligent video analysis: person re-identification. First, it introduces the background of a person Re-ID problem in intelligent video analysis, and then overviews the Re-ID problem, as well as a difficult problem in the current Re-ID: small sample size problem. Next, it points out a conventional idea to solve S^3: data augmentation, and introduces two virtual sample generation methods based on KISSME. In recent years, deep learning has played a huge role in solving S^3 through data augmentation. This chapter thus introduces a method of learning pseudo-labels of new samples generated by deep learning.

In recent years, the research of person Re-ID has made much progress. The solution of many difficult problems makes it possible to realize practical applications. Person Re-ID technology can be used in the following important fields:

(1) Public safety field [31]: On the one hand, Re-ID can help quickly screen suspicious persons, establish a rapid response mechanism, and fight crimes with precision. On the other hand, in crowded public areas, such as airports and train stations, person Re-ID technology can be used to quickly find lost children and elderly people. It plays an incalculable role in creating a safe environment.

(2) New retail field [32][33]: Person Re-ID technology can be used to obtain customer's behavior trajectory, obtain customer's digital information, help businesses mine more commercial value, and provide customers with customized services. In addition to the application in offline retail solutions, Re-ID can be used to connect online and offline retail scenarios and provide a "one-stop" consumer service experience.

(3) Intelligent transportation field [34]-[36]: The use of person Re-ID technology can realize the connection between people and even vehicles. Help intelligent transportation systems to complete automatic closed loop of people, vehicles and roads together.

(4) Smart city [37][38]: Person Re-ID is an important technical link in a smart city. Re-ID technology cannot only be used to realize the statistics of people flow information, but also include the restoration of the flow trajectory of the whole scene and the comparison of people, which is convenient for real-time management and deployment of various terminal resources, saving many manpower and material resources.

Bibliography

[1] Y. Li, Z. Wu, S. Karnam, et al. Real-world re-identification in an airport camera network, *International Conference on Distributed Smart Cameras*. Venice, Italy, Nov. 4–7, 2014.

[2] S. Gong, M. Cristani, S. Yan, et al. *Person re-identification*. London, UK: Springer, 2014.

[3] O. Camps, M. Gou, T. Hebble, et al. From the lab to the real world: Re-identification in an airport camera network. *IEEE Transactions on Circuits and Systems for Video Technology*, no. 9, pp. 540–553, 2016.

[4] D. Gray and H. Tao, Viewpoint invariant pedestrian recognition with an ensemble of localized features. *European Conference on Computer Vision (ECCV)*. Marseill, France, pp: 262–275, 2008.

[5] B. Prosser, W. Zheng, S. Gong, et al. Person re-identification by support vector ranking. *The British Machine Vision Conference*. Aberystwyth, British, pp: 1–21, 2010.

[6] A. Mignon and F. Jurie, PCCA: a new approach for distance learning from sparse pairwise constraints. *IEEE Conference on Computer Vision and Pattern Recognition (CVPR)*, pp: 2666–2672, 2012.

[7] R. Zhao, W. Ouyang, and X. Wang, Unsupervised salience learning for person re-identification. *IEEE Conference on Computer Vision and Pattern Recognition (CVPR)*. Oregon, USA, pp: 3586–3593, 2013.

[8] W. Zheng, X. Li, and T. Xiang, Partial person re-identification. *IEEE International Conference on Computer Vision (CVPR)*. Santiago, Chile, pp: 4678–4686, 2015.

[9] W. Zajdel, Z. Zivkovic, and B. Krose, Keeping track of humans: have I seen this person before? *IEEE International Conference on Robotics and Automation*, Barcelona, Spain, pp: 2081–2086, 2005.

[10] N. Gheissari, T. Sebastian, and R. Hartley, Person reidentification using spatiotemporal appearance. *IEEE Conference on Computer Vision and Pattern Recognition (CVPR)*, New York, USA, pp: 1–8, 2006.

[11] E. G. Miller, N. E. Matsakis, and P. A. Viola, Learning from one example through shared densities on transforms, *CVPR*, pp. 464–471, 2000.

[12] H. Han, M. Zhou, X. Shang, Wei Cao, and Abdullah Abusorrah, KISS+ for rapid and accurate pedestrian re-identification, *IEEE Transactions on Intelligent Transportation Systems*, vol. 22, no. 1, pp: 394–403, 2021.

[13] H. Han, M. Zhou, and Y. Zhang, Can virtual samples solve small sample size problem of KISSME in pedestrian re-identification of smart transportation?, *IEEE Transactions on Intelligent Transportation Systems*, vol. 21, no. 9, pp. 3766–3776, 2020.

[14] M. Kostinger, M Hirzer, P. Wohlhart, P. M. Roth, and H. Bischof, Large scale metric learning from equivalence constraints, *CVPR*, Providence, RI, USA, pp. 2288–2295, 2012.

[15] L.-F. Chen, H.-Y. M. Liao, M.-T. Ko, J.-C. Lin, and G.-J. Yu, A new LDA-based face recognition system which can solve the small sample size problem, *Pattern Recogn.*, vol. 33, no. 10, pp. 1713–1726, 2000.

[16] D. Tao, L. Jin, Y. Wang, Y. Yuan, and X. Li, Person re-identification by regularized smoothing KISS metric learning, *IEEE Transactions on Circuits and Systems for Video Technology*, vol. 23, no. 10, pp. 1675–1685, 2013.

[17] A. K. Jain, and B. Chandrasekaran, 39 Dimensionality and sample size considerations in pattern recognition practice, *Handbook of Statistics*, vol. 2, no. 39, pp. 835–855, 1982.

[18] D. Tao, Y. Guo, M. Song, Y. Li, Z. Yu, and Y. Tang, Person re-identification by dual-regularized KISS metric learning, *IEEE Transactions on Image Processing*, vol. 25, no. 6, pp. 2726–2738, 2016.

[19] J. H. Holland, *Adaptation in Natural and Artificial Systems: An Introductory Analysis with Applications to Biology, Control and Artificial Intelligence*, Second edition, MIT Press, 1992.

[20] K. Xing, L. Han, M. C. Zhou, and F. Wang, Deadlock-free genetic scheduling algorithm for automated manufacturing systems based on deadlock control policy, *IEEE Transactions on Systems, Man & Cybernetics: Part B*, vol. 42, no. 3. pp. 603–615, 2012.

[21] Z. Ding, J. Liu, Y. Sun, C. Jiang, and M. Zhou, A transaction and QoS-aware service selection approach based on genetic algorithm, *IEEE Transactions on Systems, Man & Cybernetics: Systems*, vol. 45, no. 7. pp. 1035–1046, 2015.

[22] T. Mareda, L. Gaudard, and F. Romerio, A parametric genetic algorithm approach to assess complementary options of large scale windsolar coupling, *IEEE/CAA Journal of Automatica Sinica*, vol. 4, no. 2. pp. 260–272, 2017.

[23] X. Zuo, C. Chen, W. Tan, and M. C. Zhou, Vehicle scheduling of an urban bus line via an improved multiobjective genetic algorithm, *IEEE Transactions on Intelligent Transportation Systems*, vol. 16, no. 2, pp. 1030–1041, 2015.

[24] H. Han, Y. Ding, K. Hao, and X. Liang, An evolutionary particle filter with immune genetic algorithm for intelligent video target tracking, *Comput. Math. Appl.*, vol. 62, no.7, pp: 2685–2695, 2011.

[25] T. Ojala, M. Pietikäinen, and D. Harwood, A comparative study of texture measures with classification based on feature distributions, *Pattern Recogn.*, vol. 29, no.1, pp: 51–59, 1996.

[26] http://www.scholarpedia.org/article/Local_Binary_Patterns

[27] G. Koch, R. Zemel, and R. Salakhutdinov, Siamese neural networks for one-shot image recognition, *International Conference on Machine Learn (ICML)*, vol. 2, 2015.

[28] J.-Y. Zhu, T. Park, P. Isola, and A. A. Efros, Unpaired image-to-image translation using cycle-consistent adversarial networks, *IEEE International Conference on Computer Vision (ICCV)*, Venice, Italy, Oct. 22–29, pp: 2242–2251, 2017.

[29] H. Han, W. Ma, M. Zhou, Q. Guo, and Abusorrah Abdullah, A novel semi-supervised learning approach to pedestrian reidentification, *IEEE Internet of Things Journal*, vol. 8, no. 4, pp: 3042–3052, 2021.

[30] X. Yang, M. Wang, R. Hong, Q. Tian, and Y. Rui, Enhancing person re-identification in a self-trained subspace, *ACM Transactions on Multimedia Computing, Communications, and Applications*, vol. 13, no. 3, 27, 2017.

[31] M. Zhou, H. Dong, P. A. Ioannou, Y. Zhao, and F.-Y. Wang, Guided crowd evacuation: approaches and challenges, *IEEE/CAA Journal of Automatica Sinica*, vol. 6, no. 5, pp. 1081–1094, September 2019.

[32] I. Hwang and Y. J. Jang, Process Mining to Discover Shoppers' Pathways at a Fashion Retail Store Using a WiFi-Base Indoor Positioning System, *IEEE Transactions on Automation Science and Engineering*, vol. 14, no. 4, pp. 1786–1792, Oct. 2017.

[33] C. Liu, Q. Zeng, and M. Zhou, Comments and Corrections to Process mining to discover shoppers' pathways at a fashion retail store using a WiFi-base indoor positioning system [Oct 17, 1786–1792], *IEEE Transactions on Automation Science and Engineering*, vol. 17, no. 1, pp. 548–548, Jan. 2020.

[34] F. Zhang, M. Zhou, L. Qi, Y. Du, and H. Sun, A game theoretic approach for distributed and coordinated channel access control in cooperative vehicle safety systems, *IEEE Transactions on Intelligent Transportation Systems*, vol. 21, no. 6, pp. 2297–2309, June 2020.

[35] X. Wang et al. Privacy-preserving content dissemination for vehicular social networks: challenges and solutions, *IEEE Communications Surveys & Tutorials*, vol. 21, no. 2, pp. 1314–1345, Secondquarter 2019.

[36] Y.-S. Huang, Y.-S. Weng and M. Zhou, Critical scenarios and their identification in parallel railroad level crossing traffic control systems, *IEEE Transactions on Intelligent Transportation Systems*, vol. 11, no. 4, pp. 968–977, Dec. 2010.

[37] Z. Liu, N. Wu, Y. Qiao, and Z. Li, Performance evaluation of public bus transportation by using DEA models and shannon's entropy: an example from a company in a large city of China, *IEEE/CAA Journal of Automatica Sinica*, vol. 8, no. 4, pp. 779–795, April 2021.

[38] M. Ghahramani, M. Zhou, and G. Wang, Urban sensing based on mobile phone data: approaches, applications, and challenges, *IEEE/CAA Journal of Automatica Sinica*, vol. 7, no. 3, pp. 627–637, May 2020.

Index

3D body keypoints, 233

A
Acoustic Features, 7
Active Contour, 131
    snakes, 131
Acyclic Oriented Graph Matching,
    156
Adaptive Bidirectional Detection,
    104
Advanced Traffic Management
    Systems, 86
    ATMS, 86
Age groups, 5
Applications, 1, 25, 58, 85, 221
    activity recognition
    affective computing, 221
    automated behavior
        understanding, 221
    automatic license plate, 58
    automatic number plate
        recognition, 86
    classroom behavioral
        understanding, 223
    crowd behavior, 58
    crowd monitoring, 58
    demographic, 1
    face recognition, 58
    facial expressions, 222
    gender recognition, 4, 12
    human behavior understanding,
        221
    incident detection, 105
    Industry 4.0, 59
    intelligent traffic, 85
    intelligent transportation
        system, 86
    Internet of Things, 58
    market analysis, 2
    object detection, 7, 68
    object recogntion, 59
    object segmentation, 68
    object tracking, 100
    person recognition, 59
    route optimization, 58
    security
    sentiment analysis, 5, 15
    smart surveillance, 58
    surveillance, 25, 58
    vehicle tracking, 100
    video analytics, 85, 87
Area Under Curve, 233
Artificial intelligence, 87
Augmented Interaction Multiple
    Models Particle Filter, 147
Automatic fuzzy clustering
    framework, 257
Automatic Incident Detection, 105
    video-based, 106
Automatic number plate recognition,
    86

B
Background, 89
    Codebook, 89
    fuzzy, 89
    neural networks, 89
    non-parametric, 89
    PBAS, 89
    SACON, 89
    SOBS, 89
    statistical, 89
    subtraction, 89
    texture, 89
    Vibe, 89
Bayesian, 61

Behavioral characteristics, 222
Bimodal, 3
Body postures, 227

C
Cell, 129
    matching, 153
    segmentation, 129
    tracking, 129, 146
Cell segmentation, 130
    clustering, 134
    deep learning, 130
    feature-based, 133
    machine learning, 130
    model-based, 130
    morphological, 133
Cell Segmentation, Tracking and
        Quantification, 135
    CSTQ, 135
CellTrack, 151
Cell tracking, 146
    deep learning, 151
    detection, 147
    model evolution, 149
    probablisitic, 150
Cell Tracking Challenge, 137
Cluster changing time, 236
Clustering, 96, 133, 251
    C-means, 133
    DBSCAN, 135
    FCM, 253
    Fuzzy C-Means, 256, 257
    K-means, 134
    residual-driven, 258
Cognitive based, 66
    deep learning, 68
    fuzzy logic, 69
    machine learning, 67
    psychological models, 69
    visual attention and saliency, 69
Color-space, 94
    HSV, 94
    Hue-Saturation-Value, 94
    RGB, 105
Combined Local/Global Optical
        Flow Method, 152

Complexity, 14, 47
Computational Cost
Computer vision, 24, 87
Convergence, 269
Convolutional neural network, 255
Cost sensitive, 9
Covariance matrix, 306
Cross Modality, 28
Crowd, 65
    behavior, 65
    dynamics, 65
    patterns, 65
CSTQ, 152
Curse of Dimensionality, 7
Cycle-GAN, 321

D
Data annotation, 238
Databases, 3
    MoBio, 3
    SADAM, 5
Data collection, 237
Dataset, 27, 201, 229, 279, 321
    Berkeley Segmentation Data
        Set, 280
    BrianWeb, 279
    CASIA-B, 27, 36
    Celeb-reID, 30, 36
    COCO, 32
    CUHK01, 28
    CUHK03, 28
    Duke MTMC-ReID, 323
    FER-2013, 229
    iLIDSVID, 28
    Market-1501, 28, 36, 37, 321
    MARS, 28
    Microsoft Research Cambridge
        Object Recognition Image
        Database, 280
    MNIST, 201
    Motion-ReID, 36
    NASA Earth Observation
        Database, 280
    OU-MVLP, 29
    PASCAL Visual Object Classes
        2012, 280

PRID2011, 28
Student Engagement Dataset, 231
VIPeR, 28
DeepCell, 151
Deep Convolutinal Neural Network, 98
Deep learning, 68, 138, 139, 223
 AlexNet, 27
 architecture, 27
 autoencode, 139
 Convolutional LSTM, 68
 convolutional neural network (CNN), 68, 139, 226
 DeepLab, 31
 ensemble, 68
 fully Convolutional Neural Network, 141
 Generative Adversarial Network, 68
 inception-V3, 27
 LongReID, 38
 LSTM, 68
 Mask-RCNN, 141
 MTCNN, 223
 multi-task cascaded convolutional neural network, 223
 object detection, 68
 object segmentation, 68
 OpenFace, 224
 OpenPose, 227
 recurrent neural network, 139
 ResNet-50, 26, 27
 squeeze-and-excitation network (SENET), 227
 stacked autoencoder, 68, 139
 VGG16, 237
Deep Neural Network (DNN), 233
Demographic analysis, 2
Deviation-sparse FCM, 253, 271
Dialect, 5
 Egyptian, 5
 Gulf, 5
 Levantine, 5
 Maghrebi, 5

Differential Interference Contrast, 145
Dimensionality reduction, 154, 234
 Independent Component Analysis, 154
 Principal Component Analysis, 154
Distribution of Magnitude of Optical Flow, 62
 DMOF, 62
Dual-regularized KISS, 310

E
Edge detection, 191
EduSense, 229
Efficiency, 111
Embedding, 7
Emotional dimension, 229
Emotions, 223
Entanglement, 180
Evaluation methods, 144
Evaluation metrics
 accuracy, 41
 CMC, 38
 Cumulative Matching Characteristic, 38
 mAP, 40
 Mean Average Precision, 40
Event Attention CatcH, 68
 EACH, 68

F
Faster R-CNN, 98
FCM, 253
 deviation-sparse, 253
 enhanced, 256
 fast and robust, 257
 fast genealized, 256
 noise-estimation-based, 253
 residual-driven, 253
Feature Engineering, 231
Feature extraction, 5, 63, 193
 appearance, 24, 25
 biometric, 24, 25
 cognitive based, 68
 deep Learning, 24

discrete cosine transform, 62
Distribution of Magnitude of
    Optical Flow, 69
embedding, 15
fluid dynamics, energy models,
    66
hand-crafted, 24, 25, 29
histogram, 24
histogram of gradient (HOG), 66
Histogram of Optical Flow, 66
histogram of tracklets (HOT),
    66
Kanade-Lucas-Tomasi, 69
LBP, 43
Local Binary Pattern, 71
low-level patterns, 66
map, 33
multimodal, 5
particle force and particle
    advection, 66
physics based, 63
prosodic, 15
social force and particle
    dynamics, 67
spatiotemporal, 62
spectral, 15
texture, 24
visual descriptors, 69
Feature vector generation, 307
First-order logic, 109
Flexibility, 110
Foreground segmentation, 105
Free-standing conversation group, 71
FCG, 71
Friends-formation, 71
Fusion, 8, 28, 71
decision level, 71
feature level, 8, 28
score level, 28

G
Gait, 25
analysis, 30
cycle, 26
GEI, 32
Gait Energy Image, 26

GEI, 26, 28, 32
Gate, 178
CNOT, 179
Fidelity, 206
Hadamard, 178
Identity, 179
latency, 206
Not, 179
Pauli, 179
Phase, 179
quantum, 178
Gaussian, 89
distribution, 89
GMM, 89
Mixture Model, 89
Gaussian Mixture Model, 62
GMM, 62
Gazing state, 226
GEI
Generalized eigen-decomposition, 319
Generative Adversarial Network, 109
Genetic Algorithm, 311
crossover, 312
fitness, 312
mutation, 312
Geometric and morphological
    transformations, 190
G-KISS, 313
Global Foreground Modeling, 92, 109

H
Hidden Markov Model, 61
HMM, 61
Hierarchical Dirichlet Processes, 61,
    62
HDP, 61, 62
Histogram of Flow Energy Image, 29
HOFEI, 29
Histogram of Oriented Gradient, 98
HOG, 98
HOG, 233

I
Image, 251
medical, 283
real-world, 284

segmentation, 251
synthetic, 281
Image compression, 192
Image denoising, 193
Image encryption, 192
Image filtering, 193
Image matching, 192
Images
    RGB, 25
    RGB-Depth, 25
    RGB-Infrared, 25
Image sequence, 157
Image-to-image translation, 316
Image watermarking, 193
Imbalanced Dataset, 9
Imperative languages, 203
Incident detection, 105
    stopped vechicle, 107
    traffic accident, 108
    traffic anomaly, 109
    traffic congestion, 107
    wrong-way vechicle, 107
Independent Component Analysis, 154
Intelligent transportation system, 86
    GPS, 86
    ITS, 86
    sensing devices, 86
Interaction Multiple Models Particle Filter, 148
I-vectors, 3

K
Kalman filter, 98
Kalman filtering, 68
Kanade-Lucas-Tomasi, 62
    KLT, 62
Kernel distance, 253
Kinect SDK, 233
KISS+, 310
KISSME, 305
Kullback-Leibler divergence, 68, 252

L
Language, 5
    Arabic, 5

dialect, 13
Latent Dirichlet Allocation, 62
    LDA, 62
Lattice Boltzmann Model, 66
    LBM, 66
Level set, 132
    edge based, 132
    region based, 132
LibSVM, 9
Logistic Regression, 236
Loss
    cross-entropy, 35
    triplet, 35
Low-level features, 96

M
Machine learning, 67
    AdaBoost, 67
    clustering, 67, 71
    decision tree, 71
    deep learning, 68
    Expectation-Maximization, 67
    fuzzy Logic, 69
    K-Nearest Neighbor, 67
    support vector machine, 67
Machine-Learning, 231
MATLAB, 279
Matthew's correlation coefficient, 279
Mean Squared Displacement, 154
Mel Frequency Cepstral Coffecients, 7
Metrics, 8
    accuracy, 10
    area under curve, 10
    AUC, 10
    F1 score, 10
    false negative, 9
    false positive, 9
    geometric mean, 10
    Mattew's correlation coefficient, 10
    precision, 10
    recall, 10
    receiver operating characteristic, 10
    ROC, 10

true negative, 9
true positive, 9
Modality, 7
  audio, 7
  image, 7
  textual, 7
  video, 7
  visual, 7
Motion Interaction Field, 108
Motion model estimation, 152
Motion patterns, 67
Multilayer Perceptron, 234
Multilingual, 2
Multimodal, 2, 59, 70
  analytics, 2
  demographic, 2
  fusion, 5, 8
  sentiment, 2
  video, 71
  wearable sensors, 71
Multimodal sensing, 239

N
Noise, 251, 263
  estimation, 251
  Gaussian, 263
  mixed, 263
Noise-estimation-based FCM, 253
Non-maximum suppression, 141
Normalization, 26, 33
  batch, 26, 33
  instance, 26, 33

O
OpenCV, 9
OpenQASM, 203
Optical flow, 105
  Lucas-Kanade, 105
Optical flow, 62
  Distribution of Magnitude of
    Optical Flow, 62
  Kanade-Lucas-Tomasi, 62
Optical Flow, 62
  Distribution of Magnitude of
    Optical Flow, 62
  Kanade-Lucas-Tomasi, 62

P
Pairwise t-test, 14
Pan-tilt-zoom, 103, 110
Partial Differential Equation, 131
  PDE, 131
Particle swarm optimization, 135
  PSO, 135
Pattern recognition, 87
PayAudioAnalysis, 9
Performance, 110
Person, 303
  Re-ID, 304
  re-identification, 303
Person Re-ID, 24, 29, 33
Person Re-identification, 24, 33
Physiological, 2
Preprocessing, 4, 30, 135
Principal Component Analysis, 154, 234
ProjectQ, 204
Pseudo-label learning, 314
P-value, 14
Python, 9

Q
Quadratic unconstrained binary
  optimizer, 204
Quantum, 176
  computation, 174, 176
  computer vision, 180
  computing tools, 202
  gates, 178
  hardware, 205
  image analysis, 181
  image processing, 175, 190
  image representation, 181
  machine-learning, 195
  mechanics, 177
  model evaluation, 194
  model selection, 194
  model training, 194
  parallelism, 180
  states, 183
  teleportation, 174
Quantum image representation, 181
  bitplane, 187

Block Image, 188
Colored, 182
Digital RGB Multi-Channel, 185
Double, 188
Double Quantum Color Images
    Representation, 189
generalized, 185
Generalized Novel Enhanced,
    186
Hue, Saturation and Lightness,
    190
Improved Flexible, 188
Improved Novel Enhanced, 184
Indexed Images, 189
Log-Polar Images, 183
MCQI, 184
Multi-Channel, 182
multiple images, 186
Multi-Wavelength Images, 186
NCQI, 185
Normal Arbitrary Quantum
    Superposition State, 183
Novel Enhanced, 182
Optimized for Color Images, 186
order-encoded, 187
Quantum States for M Colors,
    183
Quantum States for N
    Coordinates position, 183
simple, 184
Quantum machine-learning, 195
deep learning, 197
Generative Adversarial
    Networks, 200
PQK, 195
QGAN, 200
QKNN, 195
QNN, 197, 202
QSVM, 196
Qubits, 176
Query Image, 24

R
R-CNN, 98
RecogNet, 151
Region of interest, 102

Regularization, 35, 261
Regularized discriminant analysis,
    306
    RDA, 306
Re-identification
    long-term, 29, 30
    short-term
Reliability, 110
Residual-driven FCM, 253
Responsiveness, 111
RFCM, 255
RGBD, 229
Robustness, 269
Root-mean-square error, 234
    RMSE, 234

S
Saliency map, 70
Sample generation, 305
Segmentation, 192
Segmentation accuracy, 157, 279
Semi-supervised learning, 314
Semi-supervised pedestrian Re-ID,
    317
Sensing devices, 86
    acoustic, 86
    CCTV, 98, 100
    Inductive Loop Detectors, 86
    infrared, 86
    intrusive, 86
    micro-loop probes, 86
    microphones, 86
    non-intrusive, 86
    piezoelectric cables, 86
    radar, 86
    ultrasonic, 86
    video cameras, 86
Shadow removal, 94
Silhouette, 30
Similarity analysis, 192
Similarity Metric, 35
    cosine, 35
Simple and Straightforward Metric,
    305
    KISSME, 305
Skip gram, 7

Small-Sample Learning, 304
  SSL, 304
Social Group Analysis, 71
Social groups, 60, 71
  analysis, 71
  clustering, 71
Social Media, 1
Social platforms, 2
  Facebook, 2
  Flicker, 2
  Instagram, 2
  Meta, 2
  MySpace, 2
  TikTok, 2
  Twitter, 2
  Vimeo, 2
  YouTube, 2
Sorensen-Dice similarity, 279
Spatial information, 256
Spatiotemporal, 63
Spatio-temporal diffusion, 136
Speaker, 4
  dependent, 4
  independent, 4
Spectral Flux, 7
Speeded up robust features, 102
  SURF, 102
Stability, 307
Structural context descriptor, 66
  SCD, 66
Support Vector Machine (SVM), 4,
    98, 233
  SVM, 98

T
Tracking, 101
  blob-tracking, 101
  contour-based, 101
  feature-based, 102
  model-based, 102
  multiple-object, 101
  region-based, 101
  vision-based, 101

Tracking accuracy, 156
Traffic, 107
  accident, 108
  anomaly, 109
  challenges, 109
  congestion, 107
  flow, 87
  intelligent, 85, 109
Transcribed Textual Features, 7
Transportation Management Center,
    87

U
Unmanned aerial vehicles, 58
Urbanization, 58

V
Vehicle, 85
  classification, 86
  detection, 86, 88
  tracking, 86, 100
Versatility, 110
Visual, 221
  analytics, 221
  sensing, 222
Visual features, 7

W
Watershed transform, 136
Wearable devices, 240
Wearable sensors, 71
Wireless Body Area Networks, 240
WRFCM, 255

Y
YOLO, 98
You Only Look Once, 107
  YOLO, 98, 107

Z
Zero-shot learning, 305
  ZSL, 305